ELEMENTS OF
Control Systems Analysis

Classical and Modern Approaches

CHIH-FAN CHEN

Professor of Electrical Engineering
University of Houston
Houston, Texas

I. JOHN HAAS

Associate Professor of Electrical Engineering
Christian Brothers College
Memphis, Tennessee

PRENTICE-HALL, INC., ENGLEWOOD CLIFFS, N.J.

PRENTICE-HALL ELECTRICAL ENGINEERING SERIES

WILLIAM L. EVERITT, *editor*

INSTRUMENTATION AND CONTROL SERIES

William W. Seifert, *editor*

CHEN & HAAS   *Elements of Control Systems Analysis: Classical and Modern Approaches*

KUO   *Analysis and Synthesis of Sampled-Data Control Systems*

OGATA   *State Space Analysis of Control Systems*

© 1968 by PRENTICE-HALL, INC., Englewood Cliffs, N.J.

Library of Congress Catalog Card Number: 68-11216

Current printing (last digit): 10   9   8   7   6   5   4   3   2   1

Printed in the United States of America

Foreword

This text by Professors Chen and Haas is uniquely different from other books in the control systems area in two distinct aspects: first, it contains the integrated use of digital computation and, secondly, it combines complex variable and state variable analysis. This approach has been encouraged enthusiastically by colleagues of Professors Chen and Haas inasmuch as it truly represents the coming trend in systems analysis. Student response has also been most gratifying. Lectures based on the book material have stimulated development of the intuitive as well as inductive analyses by the student. This book should find wide appeal as a modern, up-to-date, introductory text in the control systems area.

C. V. KIRKPATRICK
Dean, Cullen College of Engineering
University of Houston

Preface

Control engineers are confronted with a variety of problems, which may be classified as follows:

1. THE IDENTIFICATION PROBLEM: to measure variables and convert data for analysis.

2. THE REPRESENTATION PROBLEM: to describe a system by an analytic form or mathematical model from which further investigations may be developed.

3. THE SOLUTION PROBLEM: based on the mathematical model above, to determine system response, quantitatively.

4. THE STABILITY PROBLEM: to evolve a general, qualitative analysis of the system.

5. THE DESIGN PROBLEM: with the materials and energy sources available, to modify an existing system or develop a new one, to serve specific purposes or tasks.

6. THE OPTIMIZATION PROBLEM: from several alternative designs, to choose the best. An optimization problem is a direct synthesis.

Control engineering is concerned with techniques that are used to solve these six problems in the most efficient manner possible.

There are two basic approaches toward solutions, which we refer to as the *conventional* and the *modern* approaches. The conventional approach is based on the complex function theory of mathematics and is an electrically oriented approach. Principal contributors to this approach in the United States have been Nyquist, Bode, Bush, Guilleman and Wiener. The modern approach is based on state variable theory, having a mechanical orientation. The principal contributors have been Europeans such as Poincare, Liapunov, Birkhoff, Markov and Pontryagin.

The conventional group concentrates on high-order differential equations in representation, poles-zeros in analysis, frequency measurements in identification, trial-and-error in design, and the indirect approach in optimization. Their main mathematical tool is the Laplace transformation, and their most-used computing machine is the analog computer.

The modern group concentrates on a set of first-order differential equations in representation, linear transformation in analysis, Liapunov theory in stability study, time domain analysis in identification, and the direct approach in optimization. Their main mathematical tool is matrix theory, and their most-used computing machine is the digital computer. Their thinking originates in and extends from classical mechanics, starting with the Lagrangian function or Hamilton's principle.

Why two groups? Because, as Truxall has said, for a difficult problem two approaches are always better than one.

The situation has presently come to this: one cannot stay with one approach as a matter of preference; to learn both has become a necessity. Yet it is difficult to find an elementary text using both approaches even though educators are attempting to unify the classical and the modern techniques in their teaching. Moreover, in industry researchers decry the gap that exists between theory and practice. To help fill this gap and to bring about the unification, there is a need for a text such as this one, which treats both approaches but which provides an emphasis on the latter.

Chapters 1 through 4 treat of conventional control engineering in the usual way. However, several points usually ignored in contemporary texts have been emphasized. For example, Lagrange's formula is explained in detail in Chapter 1, The Representation Problem; variation of parameters is demonstrated in Chapter 2, The Solution Problem; Routh's algorithm and some digital computer techniques are treated in Chapter 3, The Stability Problem; Chapter 4, The Identification Problem, includes the up-to-date techniques of Wiener-Lee, Bush, and Levy. These particular topics were selected either because they are fundamental to the state space approach or because they are particularly suitable for use with digital computation. In other words, the treatment of the conventional techniques was done with the modern viewpoint in mind.

Chapters 5 through 8 present the state variable approach to the same problems.

In Chapter 5 the fundamentals of matrix theory are covered; Euclidean and vector space are explained and the state space concept is defined.

Chapter 6, The Representation Problem, gives a detailed method for setting up state equations and a technique for converting Lagrange equations into Hamilton's equations.

In Chapter 7, The Solution Problem, there is a detailed explanation of the classical and Laplace transformation methods for evaluating transition matrices, signal flow matrices, and related partition theory, concluding with detailed instructions for the specific solution of state equations by the numerical method of Runge and Kutta.

Liapunov's second method dominates the treatment of the stability problem in Chapter 8, which presents the basic notation and definitions of terms used in stability studies and then shows how to generate Liapunov functions. A derivation of the Routh-Hurwitz criterion through the Liapunov method is then made. Certain important properties such as controllability and observability of linear systems are explained.

The book is restricted to analysis; design and optimization problems are not covered in this volume.

Since the scope of the book is broad and much of the material is new, it must be presented in an effective manner. Instead of a logically attractive presentation to appeal to the mathematician, we have attempted psychologically effective writing to appeal to the student.

With this approach as a principle, the general procedure in each unit has been first, to describe and illustrate the problem under consideration; then, when the reader is familiar with the goal which we are searching for, to offer the available methods of attack. The statement of the problem and its various solutions is followed by an explanation of the physical meaning and a geometric interpretation. Difficult but important topics are documented with detailed mathematical derivations. On the other hand, topics which are commonly treated in most other texts and which are therefore easy to find in references are replete with applications and aids to understanding. We have, to some extent, de-emphasized what is usually emphasized and enlarged upon what has been missing. In effect, this is an elementary text for the specialized field of control engineering.

Finally we wish to express our gratitude to Professors W. Seifert of M.I.T., B. L. Thomas of LaSalle Institute, E. L. Michaels and J. Paskusz of the University of Houston, L. H. Chu of Cheng Kung University, N. F. Tsang of the University of Arkansas and Drs. T. Mulcahey, W. Lipinski and C. Hsu of Argonne National Laboratory for their comments and suggestions. We also wish to acknowledge the invaluable help of some of our students, particularly L. S. Shieh, Richard Dzielak and Daniel Gregonia, and the extensive assistance provided by Miss Irene Tkaczyk of Argonne National Laboratory, without whose help it is doubtful that the text would have been completed. The final manuscript was typed by Mrs. Joan Averwater, Patricia Averwater, and Katherine Ruby of Christian Brothers College. To these friends, we owe our deep appreciation.

C.F.C., I.J.H.

Contents

PART ONE

Complex Variable Analysis

Chapter **1.** THE REPRESENTATION PROBLEM   3

    I. Cause-Effect Approach   3
    II. Energy Approach   13
    III. Block Diagrams   25

Chapter **2.** THE SOLUTION PROBLEM   44

    I. Classical Method   44
    II. Laplace Transformations   52
    III. The Transform Method   59
    IV. Analog Computation   86

Chapter **3.** THE STABILITY PROBLEM   99

    I. Basic Concepts   99
    II. The Direct Method of Routh-Hurwitz   106
    III. The Root Locus Method of Evans   117
    IV. The Root Locus Method of Routh   134
    V. The Frequency Method of Nyquist   143
    VI. Performance   155

Chapter **4.** THE IDENTIFICATION PROBLEM   170

    I. Analysis of a Periodic Signal: Fourier Series   170
    II. Analysis of an Aperiodic Signal: Fourier Integral   179
    III. Bode's Decomposition   185

    IV. Wiener-Lee's Method of Decomposition   194
    V. Bush's Decomposition   203
    VI. Levy's Curve Fitting Technique   223

PART TWO

State Variable Analysis

Chapter **5.** FUNDAMENTALS OF MATRICES   239

    I. Matrix Algebra   239
    II. Gauss' Reduction and Premultiplication   246
    III. Elementary Transformations   256
    IV. Quadratic Forms   262
    V. Vectors   270
    VI. State Variable Concept   277

Chapter **6.** THE REPRESENTATION PROBLEM   285

    I. Energy Method   286
    II. Topological Approach   293
    III. Transfer Function Decomposition—State Diagrams   298
    IV. General Linear Systems   306
    V. Nonlinear Feedback Systems   310

Chapter **7.** THE SOLUTION PROBLEM   319

    I. Laplace Transformation Method   320
    II. Classical Method   330
    III. Linear Transformation Method   336
    IV. Flow Graphs   356
    V. Numerical Solution   369

Chapter **8.** THE STABILITY PROBLEM   386

    I. Equilibrium Points   386
    II. The Direct Method of Liapunov   405
    III. Construction of Liapunov Functions for Linear Systems   418
    IV. Routh-Hurwitz Criterion and Liapunov Functions   427
    V. Controllability and Observability   434

    APPENDIX   447

    BIBLIOGRAPHY   455

    INDEX   465

PART ONE

Complex Variable Analysis

CHAPTER 1

The Representation Problem

Control engineers are concerned with choosing the right materials out of which to build proper dynamic systems to perform specific tasks. In order to do this, they must have a means of describing the systems. Although descriptions involving size, color, or chemical properties might be valuable in other fields of endeavor, they are of no practical help to the control engineer. He needs a description that is, above all, analytical, because his interest is in the dynamic behavior of the system.

How best to describe a system in an analytical way is called the representation problem.

In this chapter, a cause-effect approach to the building of a useful mathematical model is developed; Lagrange's equations are derived; finally, the concepts regarding transfer functions and block diagrams are explained.

I. Cause-Effect Approach

1. System Function

An experimental study of a system can be carried out by applying some known physical agent to the system. The reaction produced by this external agent will reveal some of the properties of the system.

As a simple example, suppose we wish to find the mass of a block of stone. An external force may be applied to the stone and the acceleration that results from this could be measured. This information will reveal the mass of the stone; a large mass would give small acceleration and a small mass would give large acceleration.

The applied agent (or cause) is called the excitation function, and the corresponding reaction (or effect), the response function. That which describes how the system transforms the excitation function into the response function is called the system function. They are related as shown in the simple block diagram in Fig. 1-1.

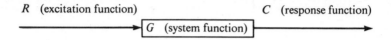

R (excitation function) C (response function)

G (system function)

Fig. 1-1. Excitation function × system function = response function.

Figure 1-2 provides a block diagram for the previous example. The mathematical equivalence of the block diagram can be written as follows:

$$f\frac{1}{M} = a \tag{1}$$

where M, f, and a are the mass, the force, and the acceleration, respectively.

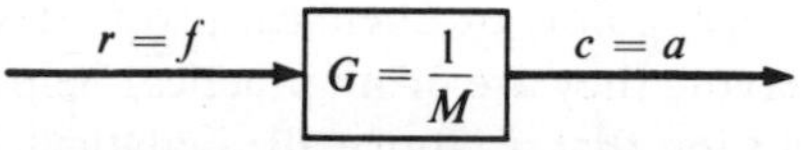

$$r = f \qquad G = \frac{1}{M} \qquad c = a$$

Fig. 1-2. A mechanical example.

Consider a second example: Find the resistance described in Fig. 1-3. A steady electromotive force may be connected across the terminals of the resistor, and the resulting current may be measured. If the current is large, the resistance is small, and vice versa. Figure 1-4 is a block diagram for the

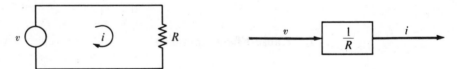

Fig. 1-3. A simple circuit.   **Fig. 1-4.** Block diagram for the electric circuit.

system, in which v, i, $1/R$ are the excitation function, the response function, and the system function, respectively. The mathematical relation is

$$v\frac{1}{R} = i \tag{2}$$

In both of the examples cited above, the following general expression holds:

excitation function × system function = response function

2. Experiments and Physical Laws

In the second example, the excitation function v and the response function i maintain a definite ratio to each other; this relationship is expressed by Eq. (2). In 1826 Georg Ohm made extensive measurements of the response function i, corresponding to the excitation function v, and established Eq. (2), which is known as Ohm's law. In a similar way Newton established the proportionality of force and acceleration, and it became one of his laws of motion. Thus Newton's and Ohm's laws give rise to the excitation-response relation (1) and (2).

Generally speaking, most physical laws are conclusions drawn from experimental data. These laws may be used in turn to investigate the physical quantities concerned.

Certain physical laws can be exhibited graphically. If Ohm's measurements on a certain electric circuit are plotted as in Fig. 1-5, the result would be a straight line.

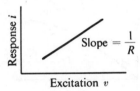

Fig. 1-5. Ohm's law.

About 1830, Joseph Henry performed extensive experiments on coils, through which he passed electric currents and produced electrokinetic momentums. He found the following:

$$i = \lambda \frac{1}{L} \tag{3}$$

The graphical representation of (3) is another straight line (Fig. 1-6).

Hooke experimented on springs, in which he observed that the elongation is proportional to the tension applied, resulting again in a straight line plot (Fig. 1-7).

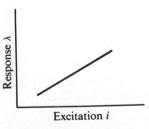

Fig. 1-6. Faraday's law.

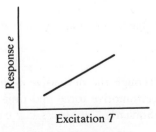

Fig. 1-7. Hooke's law.

In Fig. 1-5, 1-6, or 1-7 the excitation function is taken as the abscissa, and the response function as the ordinate. All the curves are straight lines, each representing some special properties of a system. These curves are therefore called the *characteristics*. The characteristic curves for the coil and the spring are Figs. 1-6 and 1-7, respectively.

All the above instances are deductions from experiments, and each one has extensive application. Thus they are called *laws*. Figure 1-5 shows Ohm's law; Fig. 1-6, Faraday's law; Fig. 1-7, Hooke's law.

3. Linear and Nonlinear Systems

As a matter of fact, all the foregoing laws have their limitations. Take the case of an incandescent lamp considered as a resistance. The complete plot of current vs. voltage is shown in Fig. 1-8(a), of which Fig. 1-5 is only a part. Ohm's law is applicable within the segment Oa_1, but not along a_1b_1.

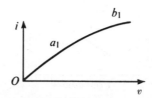

Fig. 1-8(a). A nonlinear system—incandescent lamp.

Since Ohm's law and Hooke's law are valid only within the linear region, they may be called laws applicable to linear systems.

Equation (3) may be rewritten as follows:

$$\lambda = Li \tag{3a}$$

The quantity λ, electrokinetic momentum, first observed by Henry, does not lend itself to convenient measurement. Instead of λ, its time rate of change is usually used; Eq. (3a) becomes

$$\frac{d\lambda}{dt} = L\frac{di}{dt} \tag{3b}$$

Define a new quantity $e = d\lambda/dt$. Then we have

$$e = L\frac{di}{dt} \tag{3c}$$

Although the derivative di/dt appears in Eq. (3c), the relationship between electromotive force and time rate of current change is still linear.

Similarly, Newton's law can be written as

$$f = Ma \tag{1a}$$

It can also be written as

$$f = M \frac{dv}{dt} \tag{1b}$$

or

$$f = M \frac{d^2 x}{dt^2} \tag{1c}$$

where v and x are velocity and displacement respectively. Both Eqs. (1b) and (1c) are but alternative forms of Eq. (1a).

It seems desirable here to define the scope of a linear relation. A system is linear if and only if

$$H(ax_1 + bx_2) = aHx_1 + bHx_2 \tag{4}$$

where a and b are any constants, and x_1 and x_2 are any excitation variables. H is called a linear operator.

The forgoing definition, with $a = b = 1$, becomes

$$H(x_1 + x_2) = Hx_1 + Hx_2 \tag{4a}$$

On the other hand, with $x_2 = 0$, Eq. (4) becomes

$$H(ax_1) = aHx_1 \tag{4b}$$

Equations (4a) and (4b) are called the property of *additivity* and the property of *homogeneity* respectively.

The common linear operations are: multiplication by a constant, differentiation, integration, and combinations of these.

In accordance with the previous definition, the laws of Newton, Henry, and Hooke give linear relations between the response and the excitation functions. On the other hand, if the response and the excitation functions exhibit a curved line when plotted, such as in a magnetic amplifier (Fig. 1-8(b)), or a diode (Fig. 1-8(c)), then the system is not linear, or *nonlinear*.

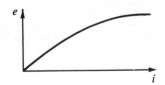

Fig. 1-8(b). A nonlinear system—magnetic amplifier.

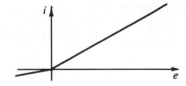

Fig. 1-8(c). A nonlinear system—diode.

4. Physical Systems and Differential Equations

In order to visualize the application of mathematical descriptions to physical systems, consider the series circuit of Fig. 1-9(a). The voltage equation reads

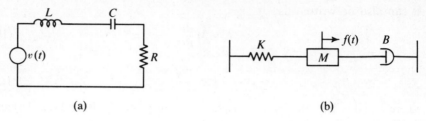

Fig. 1-9. (a) An electrical system; (b) A mechanical system.

$$L\frac{di}{dt} + Ri + \frac{1}{C}\int i\,dt = v(t) \tag{5}$$

$$v_l + v_r + \qquad v_c \quad = v(t) \tag{5a}$$

where v_l, v_r and v_c are the voltage drops across the inductor, the resistor, and the capacitor respectively. The above equation is established for the basic *RLC* circuit by the application of four laws.

$$v_l = L\frac{di}{dt} \tag{6}$$

is an alternative form of Henry's law;

$$v_r = Ri \tag{7}$$

is Ohm's law; and

$$v_c = \frac{1}{C}\int i\,dt \tag{8}$$

is the law governing capacitors. The equality of the various components to the total applied voltage is justified by Kirchhoff's second law.

Next, consider the mechanical system shown in Fig. 1-9. The equation of motion is

$$M\frac{d^2x}{dt^2} + B\frac{dx}{dt} + Kx = f(t) \tag{9}$$

$$f_m + f_b + f_k = f(t) \tag{10}$$

where $(-f_m)$, $(-f_b)$, and $(-f_k)$ are called the inertia force, the damping force, and the restoring force respectively. In setting up this differential equation, four laws are used. Here,

$$M\frac{d^2x}{dt^2} = f_m \tag{11}$$

is Newton's second law;

$$B\frac{dx}{dt} = f_b \tag{12}$$

is the law of viscous friction; and

$$Kx = f_k \qquad (13)$$

is Hooke's law. We may say that Eq. (10) is justified by Newton's third law: the equality of action and reaction. Note that Eqs. (5a) and (9) contain only multiplication by constants, differentiation, and integration, which are linear relations.

5. Analogs

In Eq. (5), if a new variable q, such that $i = dq/dt$, is introduced, then

$$L\frac{d^2q}{dt^2} + R\frac{dq}{dt} + \frac{q}{C} = v(t) \qquad (5b)$$

Compare the above equation with Eq. (9),

$$M\frac{d^2x}{dt^2} + B\frac{dx}{dt} + Kx = f(t) \qquad (9)$$

The two physically different systems satisfy equations which are similar to each other term by term. It can be concluded from the above that the two systems are not only mathematically similar, but also physically similar. Thus the electrical system described by Eq. (9) is the mechanical analog of system(5).

If the corresponding analogous system constants are equal and the excitation functions are the same, then the corresponding response functions will also be equal.

6. Physics and Engineering

Physicists and engineers both deal with problems of physical systems. Although no strict separation is necessary, it may be generally stated that physicists are concerned with the basic relations of physical quantities in comparatively simple systems, whereas the engineers are to apply these relationships to more complicated systems in order to attain certain practical results.

Consider, for example, electric circuits. Physicists are interested in the basic relationship of voltage to current in a resistor, an inductor, or a capacitor. Electrical engineers then make use of the general laws physicists established to solve problems involving more complicated networks.

With the more extended systems, such as multiple loops in a circuit, the problems become increasingly complicated. To deal with this type of complication, the engineers need to devise *methods* and *techniques* for efficient handling.

7. Analysis of Electric Circuits

Techniques developed for electrical networks which make it unnecessary to begin each problem with the basic physical laws are based upon several simple rules, or methods, among which the following two are well known:

(a) THE LOOP METHOD

Step 1. Choose the closed circuits, or loops.

Step 2. Indicate the direction of the circulating current around these loops, the loop currents.

Step 3. In accordance with the physical laws, set up the differential equation for each loop.

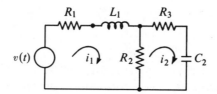

Fig. 1-10(a). The loop currents for analysis.

Applying these steps to the circuit of Figure 1-10(a) results in the following equations:

$$\left.\begin{aligned} R_1 i_1 + L_1 \frac{di_1}{dt} + R_2(i_1 - i_2) &= v \\ R_3 i_2 + \frac{1}{C_2} \int i_2 \, dt + R_2(i_2 - i_1) &= 0 \end{aligned}\right\} \tag{14}$$

In the above equations, v is the excitation function, and i_1 and i_2 are the response functions for loops 1 and 2, respectively. In terms of physical quantities, the problem is to find the response function resulting from the excitation function and the system function, whereas, from a mathematical point of view, the problem is to solve two simultaneous differential equations.

(b) THE NODE METHOD

Step 1. Choose the node pairs, such as AG, BG, etc.

Step 2. Indicate the relative potentials.

Step 3. With the known physical laws, set up the differential equations for each node pair.

Applying these steps to the circuit of Fig. 1-10(b), we obtain the following equations:

$$\frac{v_3 - v}{R_1} + \frac{1}{L_1} \int (v_3 - v_1) \, dt = 0$$

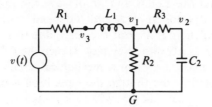

Fig. 1-10(b). The node voltages for analysis.

or

$$\frac{v_3}{R_1} + \frac{1}{L_1} \int (v_3 - v_1)\, dt = \frac{v}{R_1}$$

and

$$\frac{1}{L_1} \int (v_1 - v_3)\, dt + \frac{v_1}{R_2} + \frac{v_1 - v_2}{R_3} = 0 \qquad (15)$$

and

$$\frac{v_2 - v_1}{R_3} + C_1 \frac{dv_2}{dt} = 0$$

In Eq. (15), v/R_1 is the excitation function, and v_1, v_2, and v_3 are the corresponding response functions to be solved as functions of time.

8. Analysis of Mechanical Systems

In dealing with mechanical systems, the first step is to draw a diagram showing the actual locations and connections between the various elements, as in Fig. 1-11. By the laws of mechanics, the corresponding differential equations can be obtained. However, for an orderly reduction of more complicated systems an equivalent *mechanical circuit diagram* can be developed which will reduce the complexity before an attempt is made to write the equations.

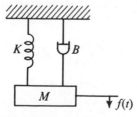

Fig. 1-11. An actual mechanical system.

As a first step, set up the reference points, as in Fig. 1-12. In this case there is only one point, which is indicated by the horizontal line x. Figure 1-11 shows that the upper ends of both K and B are fixed, and the lower ends are attached to M, so that these ends and M all have the same displacement x. Note that the forces developed by K and B depend on the relative displacements or velocities of their ends, but the force due to M is proportional to the rate of change of its absolute velocity. Hence the following general rule holds: One end of a mass shown in a mechanical circuit diagram is always attached to a fixed point, even though the mass itself is not fixed.

If the mechanical circuit diagram is constructed with displacements as references, then to apply electric circuit analogy, the velocities may be considered analogous to voltages, and the forces to currents. Since the mechanical laws are

$$f_b = Bv, \qquad f_k = K \int v \, dt, \quad \text{and} \quad f_m = M \frac{dv}{dt}$$

the following analogy holds:

$$R \sim \frac{1}{B}, \qquad L \sim \frac{1}{K}, \qquad C \sim M$$

With the above considerations, the mechanical system is accordingly transformed into a mechanical circuit diagram (Fig. 1-12) from which the corresponding equivalent electric circuit (Fig. 1-13) is derived. The system differential equation can then be written.

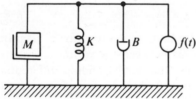

Fig. 1-12. The mechanical circuit diagram.

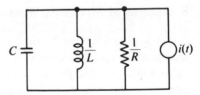

Fig. 1-13. The electrical analog.

$$M \frac{dv}{dt} + Bv + K \int v \, dt = f(t) \tag{16}$$

or

$$M \frac{d^2 x}{dt^2} + B \frac{dx}{dt} + Kx = f(t) \tag{9}$$

Since the current source is considered as analogous to an applied force, the differential equation corresponds to the electric circuit equation written on a nodal basis.

The introduction of a mechanical circuit diagram in this simple example does not present any particular advantage, as Eq. (9) can be easily derived from the laws of mechanics. But with a more complicated system, its advantages become evident.

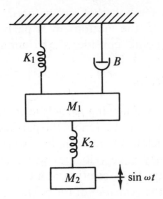

Fig. 1-14. A more complicated mechanical system.

Consider the system of Fig. 1-14. Designate the point of application of force as point 2, and the location of M_1 as point 1. By tracing downward, we establish the mechanical circuit diagram, and we can draw the corresponding equivalent electric circuit as Figs. 1-15 and 1-16. The node equations can then be written:

$$M_1 \frac{d^2 x_1}{dt^2} + K_1 x_1 + B \frac{dx_1}{dt} + K_2(x_1 - x_2) = 0 \qquad (17)$$

$$K_2(x_2 - x_1) + M_2 \frac{d^2 x_2}{dt^2} = \sin \omega t \qquad (18)$$

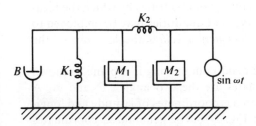

Fig. 1-15. The corresponding mechanical circuit of Fig. 1-14.

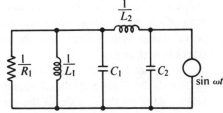

Fig. 1-16. The electrical analog of Fig. 1-14.

II. Energy Approach

Differential equation formulation may be derived from energy considerations instead of the cause and effect approach. Lagrange's work in classical mechanics is the most famous along these lines. Before we take up the derivation of Lagrange's equations, it would be well to consider the physical meaning of some of the symbols that will be used. Lagrange first analyzed the energy (and, therefore, the motion) of a particle; then he applied his findings to macroscopic bodies considered as a summation of small particles.

1. Coordinates and Velocities

To define the *position* of a moving particle in space, we may use Cartesian coordinates. The values x, y, and z denote the particle from a reference point. For a system of N particles, $3N$ coordinates are required.

Since the particles may have velocities which affect the positions after a time interval dt, a complete description of a dynamic system must include information concerning the *velocities* of the particles. When both the coordinate positions and the velocities at some instant are simultaneously specified, the performance of the system is uniquely defined.

Of course, the location of a moving particle may be described by other coordinate systems, such as spherical or cylindrical coordinates. The proper choice depends upon the nature of the problem.

2. Generalized Coordinates

Lagrange, considering the motion of any particle in space, used the so-called generalized coordinates:

$$q_i, \qquad i = 1, 2, \ldots, n$$

A set of generalized coordinates is any set of quantities by which the location of a particle can be specified. All types of quantities may be employed to serve this purpose, provided the set *uniquely specifies* the location of the particle or body.

As an example, consider a double pendulum moving in a plane as shown in Fig. 1-17. The generalized coordinates are the two angles, θ_1 and θ_2. A ship moving on the surface of the ocean has coordinates of latitude and longitude. In the electric circuit shown in Fig. 1-18, the generalized coordinates are the charges q_1 and q_2.

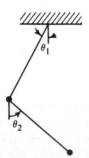

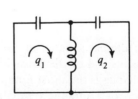

Fig. 1-17. A double pendulum.   Fig. 1-18. A two-loop circuit.

3. Degrees of Freedom

The number of independent quantities which must be specified in order that the position of a body can be uniquely defined is called the number of degrees of freedom.

Atwood's machine, shown in Fig. 1-19, has only one degree of freedom. For both the ship and the circuit, there were two degrees of freedom. A rigid body in space has six degrees of freedom: three translational and three rotational, whereas a rigid body rotating on an axis has only one degree of freedom. Precisely speaking, the number of degrees of freedom is the number of quantities which must be specified in order to determine the velocities of all the particles in the system for any motion which does not violate the constraints.

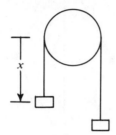

Fig. 1-19. Atwood's machine.

4. The Lagrangian

If the values of the coordinates and velocities at some instant are simultaneously specified, the performance of the system is uniquely defined. In other words, if q_i and $\dot{q}_i$ are given, the *state* of the system is completely determined.

Let a body occupy a position defined by q_1 at the instant t_1, and a position defined by q_2 at the instant t_2. Suppose the system changes from the first to the second state in such a way that the relationship

$$I = \int_{t_1}^{t_2} L(q, \dot{q}, t) \, dt \qquad (1)$$

takes the least possible value. Then the function L is called the *Lagrangian* of the system. The fact that dynamic systems will move in just this way (I of (1) takes the least possible value) is called *Hamilton's principle*.

Note that the Lagrangian of a system is a function of q and $\dot{q}$, but not a function of $\ddot{q}$, $\dddot{q}$, etc. This is another way of saying that the mechanical state of a system is completely determined when the *coordinates* and *velocities* are known.

Quantitatively, the Lagrangian of a system is determined by the energy functions T and V:

$$L = T(\dot{q}, q) - V(q) \qquad (2)$$

where T is the kinetic energy, and V the potential energy of the system.

Example. A speed governor is shown in Fig. 1-20. The mass M_2 moves on a vertical axis and the whole system rotates about this axis with constant angular velocity ω. The potential energy for this system is

$$V = -2m_1 gl \cos \theta - M_2 g 2l \cos \theta = 2(m_1 + M_2)gl \cos \theta$$

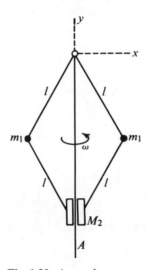

Fig. 1-20. A speed governor.

The kinetic energy T of the entire system is

$$T = T_1 + T_2$$

where

$$T_1 = \tfrac{1}{2}(2m_1)(l\dot{\theta})^2 + \tfrac{1}{2}(2m_1)\omega^2 l^2 \sin^2 \theta$$
$$T_2 = \tfrac{1}{2}M_2(2l\dot{\theta} \sin \theta)^2$$

The Lagrangian is

$$L = T - V$$
$$= m_1 l^2 \dot{\theta}^2 + m_1 \omega^2 l^2 \sin^2 \theta + 2M_2 l^2 \dot{\theta}^2 \sin^2 \theta + 2(m_1 + M_2)gl \cos \theta \qquad (3)$$

5. Lagrange's Equations

Rewriting Eq. (1) we have

$$I = \int_{t_1}^{t_2} L(q, \dot{q}, t)\, dt$$

Let q be the function for which I is a minimum. In other words, if q becomes $q + \delta q$, I will increase. The change in I when q is replaced by $q + \delta q$ is

$$\delta I = \int_{t_1}^{t_2} L(q + \delta q, \dot{q} + \delta \dot{q}) dt - \int_{t_1}^{t_2} L(q, \dot{q}) \, dt$$

The necessary condition for I to have a minimum is

$$\delta I = \delta \int_{t_1}^{t_2} L(q, \dot{q}) \, dt = 0 \qquad (4)$$

or

$$\int_{t_1}^{t_2} \left(\frac{\partial L}{\partial q} \delta q + \frac{\partial L}{\partial \dot{q}} \delta \dot{q} \right) dt = 0$$

which can be written as

$$\int_{t_1}^{t_2} \left(\frac{\partial L}{\partial q} \delta q + \frac{\partial L}{\partial \dot{q}} d \frac{\delta q}{dt} \right) dt = 0$$

Expanding gives

$$\int_{t_1}^{t_2} \frac{\partial L}{\partial q} \delta q \, dt + \int_{t_1}^{t_2} \frac{\partial L}{\partial \dot{q}} d \, \delta q = 0$$

Integrating the second term by parts:

$$\int_{t_1}^{t_2} \frac{\partial L}{\partial q} \delta q \, dt + \frac{\partial L}{\partial \dot{q}} \delta q \Big|_{t_1}^{t_2} - \int_{t_1}^{t_2} \delta q \frac{d}{dt} \frac{\partial L}{\partial \dot{q}} dt = 0$$

Rearranging, we have

$$\frac{\partial L}{\partial \dot{q}} \delta q \Big|_{t_1}^{t_2} + \int_{t_1}^{t_2} \left(\frac{\partial L}{\partial q} - \frac{d}{dt} \frac{\partial L}{\partial \dot{q}} \right) \delta q \, dt = 0$$

Since $\delta q(t_1) = \delta q(t_2) = 0$, the first term is equal to zero and the second term has to be zero. This can be true only if the integrand is zero. Therefore,

$$\frac{d}{dt} \left(\frac{\partial L}{\partial \dot{q}} \right) - \frac{\partial L}{\partial q} = 0 \qquad (5)$$

which is the Lagrange equation for a system with one degree of freedom.

Example. Atwood's machine is shown in Fig. 1-21. The kinetic energy of the system is

$$T = \tfrac{1}{2} M_1 \dot{x}^2 + \tfrac{1}{2} M_2 (-\dot{x})^2$$

The potential energy of the system is

$$V = -(M_1 \, gx - M_2 gx)$$

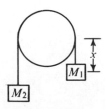

Fig. 1-21. Dynamics of Atwood's machine.

The Lagrangian is

$$L = T - V = \tfrac{1}{2} M_1 \dot{x}^2 + \tfrac{1}{2} M_2 \dot{x}^2 + M_1 g x - M_2 g x$$

Substituting the Lagrangian into (5) gives

$$\frac{d}{dt}(M_1 \dot{x} + M_2 \dot{x}) - (M_1 g - M_2 g) = 0$$

$$\ddot{x} = \frac{M_1 - M_2}{M_1 + M_2} g$$

6. Rayleigh Dissipation Function

We rewrite Eq. (5) as follows:

$$\frac{d}{dt}\left(\frac{\partial T}{\partial \dot{q}}\right) - \frac{\partial T}{\partial q} = -\frac{\partial V}{\partial q} + \frac{d}{dt}\left(\frac{\partial V}{\partial \dot{q}}\right) \tag{5a}$$

$$= P$$

where

$$P = -\frac{\partial V}{\partial q} + \frac{d}{dt}\left(\frac{\partial V}{\partial \dot{q}}\right)$$

In most cases, since V is not a function of velocity,

$$P = -\frac{\partial V}{\partial q} \tag{6}$$

It is noted that if only some of the forces acting on the system are *derivable from a potential*, the Lagrange equation can be written in the form

$$\frac{d}{dt}\left(\frac{\partial T}{\partial \dot{q}}\right) - \frac{\partial T}{\partial q} = P + F \tag{7}$$

where F represents the force not arising from a potential. This is the case if there is friction, which makes it a nonconservative system:

$$F = -B\dot{q} \tag{8}$$

Rayleigh developed a dissipation function D from which frictional force, F, can be derived:

$$D = \tfrac{1}{2} B \dot{q}^2 \tag{9}$$

and

$$F = -\frac{\partial D}{\partial \dot{q}} \tag{8a}$$

Combining (7) and (8a) gives

$$\frac{d}{dt}\left(\frac{\partial T}{\partial \dot{q}}\right) - \frac{\partial}{\partial q}(T - V) = F$$

or

$$\frac{d}{dt}\left(\frac{\partial T}{\partial \dot{q}}\right) - \frac{\partial}{\partial q}(T - V) + \frac{\partial D}{\partial \dot{q}} = 0 \tag{10}$$

or

$$\frac{d}{dt}\left(\frac{\partial T}{\partial \dot{q}}\right) - \frac{\partial T}{\partial q} + \frac{\partial V}{\partial q} + \frac{\partial D}{\partial \dot{q}} = 0 \qquad (10a)$$

which is Lagrange's equation for a nonconservative system.

7. Application to a Simple Pendulum

Derive the differential equation for a simple pendulum suspended in air (as shown in Fig. 1-22) from Lagrange's equations.

The kinetic energy is

$$T = \tfrac{1}{2}m(l\dot{\theta})^2 \qquad (11)$$

Fig. 1-22. A simple pendulum.

The potential energy is

$$V = -mgl \cos \theta \qquad (12)$$

The dissipation function is

$$D = \tfrac{1}{2}B(l\dot{\theta})^2 \qquad (13)$$

Substituting (11), (12), and (13) into (10a) yields

$$\frac{d}{dt}\left(\frac{\partial T}{\partial \dot{\theta}}\right) - \frac{\partial T}{\partial \theta} + \frac{\partial V}{\partial \theta} + \frac{\partial D}{\partial \dot{\theta}} = 0$$

or

$$ml^2\ddot{\theta} - 0 + mgl \sin \theta + Bl^2\dot{\theta} = 0$$

$$\ddot{\theta} + \frac{B}{m}\dot{\theta} + \frac{g}{l}\sin \theta = 0 \qquad (14)$$

8. Applications to a Simple Circuit

Derive the differential equation for the *RLC* circuit described in Fig. 1-23 from Lagrange's equations.

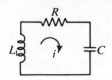

Fig. 1-23. An RLC circuit.

The kinetic energy of the circuit is

$$T = \tfrac{1}{2} L \dot{q}^2 \tag{15}$$

The potential energy of the circuit is

$$V = \frac{1}{2} C v^2 = \frac{1}{2} C \frac{q^2}{C^2} = \frac{1}{2C} q^2 \tag{16}$$

The dissipation function of the circuit is

$$D = \tfrac{1}{2} R \dot{q}^2 \tag{17}$$

Substituting into Lagrange's equation, we have

$$\frac{d}{dt}\left(\frac{\partial T}{\partial \dot{q}}\right) - \frac{\partial T}{\partial q} + \frac{\partial V}{\partial q} + \frac{\partial D}{\partial \dot{q}} = 0$$

or

$$L \frac{di}{dt} + \frac{1}{C} q + R i = 0 \tag{18}$$

9. Lagrange's Equations for a System with n Degrees of Freedom

In the last two sections, examples were restricted to problems with only one degree of freedom. When there are two or more degrees of freedom, Eq. (10a) must be generalized. This is easily done:

$$\frac{d}{dt}\left(\frac{\partial T}{\partial \dot{q}_n}\right) - \frac{\partial T}{\partial q_n} + \frac{\partial V}{\partial q_n} + \frac{\partial D}{\partial \dot{q}_n} = Q_n \tag{19}$$

where n is the number of degrees of freedom and Q_n are the generalized forces which can be considered as a further generalization of P. It can be seen that when $n = 2$, (19) represents two simultaneous equations.

10. Application to an Elastic Pendulum

An elastic pendulum is shown Fig. 1-24. The mass of the bob is m; the length of the pendulum is l_0 in the absence of motion; the spring constant is K, and the coefficient of viscous friction is B.

The energy functions and the dissipation function of the entire system are

$$T = \tfrac{1}{2} m \dot{r}^2 + \tfrac{1}{2} m (r \dot{\theta})^2 \tag{20}$$

$$D = \tfrac{1}{2} B \dot{r}^2 + \tfrac{1}{2} B (r \dot{\theta})^2 \tag{21}$$

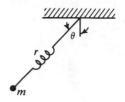

Fig. 1-24. An elastic pendulum.

$$V = \tfrac{1}{2}K(r - l_0)^2 + mg(l_0 - r\cos\theta) \tag{22}$$

For this two degrees of freedom problem, we have:

$$\frac{d}{dt}\left(\frac{\partial T}{\partial \dot{r}}\right) - \frac{\partial T}{\partial r} + \frac{\partial D}{\partial \dot{r}} + \frac{\partial V}{\partial r} = 0$$

or

$$m\ddot{r} - mr\dot{\theta}^2 + B\dot{r} + K(r - l_0) - mg\cos\theta = 0 \tag{23}$$

and

$$\frac{d}{dt}\left(\frac{\partial}{\partial \dot{\theta}}T\right) - \frac{\partial T}{\partial \theta} + \frac{\partial D}{\partial \dot{\theta}} + \frac{\partial V}{\partial \theta} = 0$$

or

$$mr^2\ddot{\theta} + 2m\dot{\theta}r\dot{r} + Br^2\dot{\theta} + mgr\sin\theta = 0$$

Simplifying gives

$$\ddot{\theta} + \left(2\frac{\dot{r}}{r} + \frac{B}{m}\right)\dot{\theta} + \frac{g}{r}\sin\theta = 0 \tag{24}$$

It is interesting to note that (24) reduces to (14) if we consider the pendulum as nonelastic.

11. Application to a Capacitor Microphone

In Fig. 1-25, b represents the movable plate of a capacitor microphone. At equilibrium, when no external force is applied to b, there is a charge q_0 on the capacitor which produces a force of attraction between the plates so that the spring is stretched by an amount x_1 and the space between plates is x_0. When sound waves apply force to plate b, there will be a resulting displacement x. The distance between the plates will then be $x_0 - x$ and the charge on the plates will be $q_0 + q$. The capacitance is known as

$$C_0 = \frac{\epsilon A}{x_0} \tag{25}$$

when the moving plate is at the initial position, and it becomes

$$C = \frac{\epsilon A}{x_0 - x} \tag{25a}$$

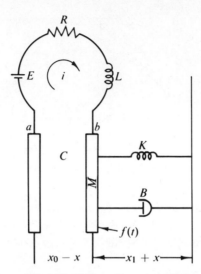

Fig. 1-25. A capacitor microphone.

when the moving plate is at the new postion $x_0 - x$; A is the area of the plates and ϵ is the dielectric constant of air.

The energy functions for the system are

$$T = \frac{1}{2} L\dot{q}^2 + \frac{1}{2} M\dot{x}^2 \tag{26}$$

$$V = \frac{1}{2C}(q_0 + q)^2 + \frac{1}{2} K(x_1 + x)^2$$
$$= \frac{1}{2\epsilon A}(x_0 - x)(q_0 + q)^2 + \frac{1}{2} K(x_1 + x)^2 \tag{27}$$

and the dissipation function of the system is

$$D = \tfrac{1}{2} R\dot{q}^2 + \tfrac{1}{2} B\dot{x}^2 \tag{28}$$

The two degrees of freedom are the displacement x and the charge q. Substituting in Lagrange's equations gives

$$M\ddot{x} + B\dot{x} - \frac{1}{2\epsilon A}(q_0 + q)^2 + K(x_1 + x) = f(t) \tag{29}$$

$$L\ddot{q} + R\dot{q} + \frac{1}{\epsilon A}(x_0 - x)(q_0 + q) = E \tag{30}$$

12. Three Degrees of Freedom

Equation (19) is for several degrees of freedom. If $n = 3$, there will be three sets of variables and three dynamic equations.

$$\frac{d}{dt}\left(\frac{\partial T}{\partial \dot{q}_1}\right) - \frac{\partial T}{\partial q_1} + \frac{\partial D}{\partial \dot{q}_1} + \frac{\partial V}{\partial q_1} = Q_1 \tag{31}$$

$$\frac{d}{dt}\left(\frac{\partial T}{\partial \dot{q}_2}\right) - \frac{\partial T}{\partial q_2} + \frac{\partial D}{\partial \dot{q}_2} + \frac{\partial V}{\partial q_2} = Q_2 \tag{32}$$

$$\frac{d}{dt}\left(\frac{\partial T}{\partial \dot{q}_3}\right) - \frac{\partial T}{\partial q_3} + \frac{\partial D}{\partial \dot{q}_3} + \frac{\partial V}{\partial q_3} = Q_3 \tag{33}$$

We will consider two examples, one electrical and the other mechanical, in the following two sections.

13. Application to a Circuit with Mutual Inductances

There are three loop currents; $\dot{q}_1$, $\dot{q}_2$, and $\dot{q}_3$ represent the three variables (Fig. 1-26). The kinetic energy of the entire system includes not only that resulting from self-inductance but also that from mutual inductance.

The kinetic energy is

$$T = \tfrac{1}{2}L_1\dot{q}_1^2 + \tfrac{1}{2}L_2\dot{q}_2^2 + \tfrac{1}{2}L_3\dot{q}_3^2 + M_{12}\dot{q}_1\dot{q}_2 - M_{13}\dot{q}_1\dot{q}_2 - M_{23}\dot{q}_2\dot{q}_3 \tag{34}$$

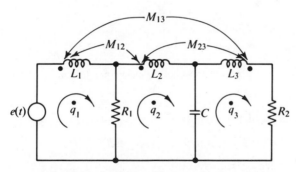

Fig. 1-26. A circuit with mutual inductances.

The potential energy is

$$V = \frac{1}{2C}(q_2 - q_3)^2 \tag{35}$$

The dissipation function is

$$D = \tfrac{1}{2}R_1(\dot{q}_1 - \dot{q}_2)^2 + \tfrac{1}{2}R_2\dot{q}_3^2 \tag{36}$$

Substituting (34) through (36) into (31), (32), and (33) leads to

$$L_1\ddot{q}_1 + M_{12}\ddot{q}_2 - M_{13}\ddot{q}_3 + R_1(\dot{q}_1 - \dot{q}_2) = e(t) \tag{37}$$

$$L_2\ddot{q}_2 + M_{12}\ddot{q}_1 - M_{23}\ddot{q}_3 + \frac{1}{C}(q_2 - q_3) - R_1(\dot{q}_1 - \dot{q}_2) = 0 \tag{38}$$

$$L_3\ddot{q}_3 - M_{13}\ddot{q}_1 - M_{23}\ddot{q}_2 - \frac{1}{C}(\dot{q}_2 - q_3) + R_2\dot{q}_3 = 0 \qquad (39)$$

This example illustrates the advantages of using Lagrange's equations.

14. Application to a Spherical Elastic Pendulum

A spherical elastic pendulum is shown in Fig. 1-27; m is the mass of the bob; l, the length of the pendulum in its initial condition; K, the spring constant; B, the coefficient of viscous friction.

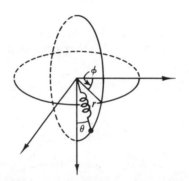

Fig. 1-27. A spherical elastic pendulum.

The kinetic energy of the system is

$$T = \tfrac{1}{2}m[(r\dot{\theta})^2 + (r\dot{\phi}\sin\theta)^2 + \dot{r}^2] \qquad (40)$$

The potential energy of the system is

$$V = mg(l - r\cos\theta) + \tfrac{1}{2}K(r - l)^2 \qquad (41)$$

and the dissipation function is

$$D = \tfrac{1}{2}B[(r\dot{\theta})^2 + (r\dot{\phi}\sin\theta)^2 + \dot{r}^2] \qquad (42)$$

Substituting (40) through (42) into (31), (32), and (33) with the variables being θ, ϕ, and r, we obtain the following three equations:

$$mr^2\ddot{\theta} + 2mr\dot{\theta}\dot{r} - mr^2\dot{\phi}^2\sin\theta\cos\theta + Br^2\dot{\theta} + mgr\sin\theta = 0 \quad (43)$$

$$mr^2\sin^2\theta\,\ddot{\phi} + 2m\sin^2\theta\,\dot{\phi}r\dot{r} + 2mr^2\dot{\phi}\sin\theta\cos\theta\,\dot{\theta} + Br^2\dot{\phi}\sin^2\theta = 0 \quad (44)$$

$$\ddot{r} + \frac{B}{m}\dot{r} - \left(-\frac{K}{m} + \dot{\theta}^2 + \dot{\phi}^2\sin^2\theta\right)r - g\cos\theta - \frac{K}{m}l = 0 \quad (45)$$

III. Block Diagrams

1. Block and Arrows

In the preceding sections, we have learned to represent a system by a set of differential equations derived by applying cause-effect reasoning or by applying Lagrange's equations. A system can also be represented by a diagram consisting of blocks and arrows.

For a simple system which can be described by an algebraic equation, the diagram can consist simply of a block and two arrows, as shown in Fig. 1-28. However, when a system is represented by a set of differential equations, no corresponding diagram is possible that will adequately symbolize the derivative operation.

Fig. 1-28. A block.

But there is a way to convert the differential equations to equivalent algebraic equations. This is known as Laplace transformation.

In this section we will explain how to use block diagrams to express a complicated system. Then we will develop conventional symbols for the block diagram procedure. Finally, we will apply the technique to some examples.

2. Elements of Block Diagrams

Terms generally used in block diagram expressions are as follows:

A. SUMMING SYMBOL. When two or more excitation sources are acting simultaneously, they should be summed up and applied to the system. This process is indicated by a summer.

Consider the simple circuit shown in Fig. 1-29(a), in which two voltages e_1 and e_2 are connected in series opposition across a resistance R. To express it graphically, the block diagram Fig. 1-29(b) is used. The symbol to the left of the system function, $1/R$, is called a summer, which is a symbol for summing up the input signals with their proper signs; in the present case the minus sign is used, since e_2 is of different polarity from e_1 and thereby requires subtraction. The input signals referred to above are general and are not limited to primary exciting functions.

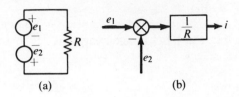

Fig. 1-29. A circuit with two inputs: (a) schematic diagram; (b) block diagram.

B. TAKE-OFF SYMBOL. When one excitation source is applied to two or more systems, its action is indicated by a take-off symbol.

In Fig. 1-30(a) a voltage source v is applied to two resistances, R_1 and R_2, in parallel, producing currents i_1 and i_2, respectively. The corresponding block diagram is shown in Fig. 1-30(b). The junction point shows that v acts on both the system functions $1/R_1$ and $1/R_2$ simultaneously.

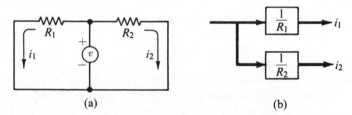

Fig. 1-30. A circuit with two outputs: (a) schematic diagram; (b) block diagram.

C. TRANSFER SYMBOL. Frequently an excitation source is to be multiplied by a constant factor k before being applied to a system. If it is reduced to a fraction of its original value or only a portion of it is applied, k is less than unity; if it is amplified, k is greater than unity. This process is indicated by a transfer symbol (or block).

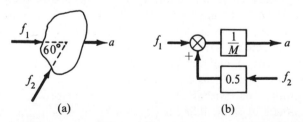

Fig. 1-31. A mechanical system with two inputs: (a) schematic diagram; (b) block diagram.

In Fig. 1-31(a) a mass M is under the action of two forces f_1 and f_2. It is desired to find its acceleration if the motion is limited to the direction of f_1. In constructing the block diagram, Fig. 1-31(b), two exciting functions have to be represented: f_1 and the component of f_2 along f_1. The latter is

$$f_2 \cos 60° = 0.5 f_2$$

To indicate the effect of this second force f_2 (in the direction of f_1) in a block diagram, a new block is inserted. The effect of this block is to change it by the constant factor $k = 0.5$.

3. Feedback

In the preceding examples several independent excitation sources are applied to the system. However, a part of the excitation function may be derived directly or indirectly from the response. This is known as feedback. Figure 1-32 shows a typical feedback system in which R is the independent excita-

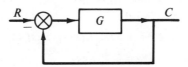

Fig. 1-32. A typical feedback system.

tion function, G is the system function, and C is the response. The entire C is fed back and subtracted from R. If a portion of C is fed back, then a transfer block should be used in the feedback path. It should be noted that this block diagram is different from a circuit diagram, as the feedback path does not divert a part of C to the input side; the output is still C.

As an example, consider a direct-current motor, Fig. 1-33(a). The motor is excited from a separate source and the field current is kept constant. If the effect of armature reaction is neglected, then the developed torque T will be proportional to the armature current i_a; the back emf, e_b, will be proportional to the speed $d\theta/dt$. The motor equations may be written as follows:

$$e_i - e_b = R_a i_a + L_a \frac{di_a}{dt} \tag{1}$$

$$T = K_t i_a \tag{2}$$

$$T = J \frac{d^2\theta}{dt^2} + f \frac{d\theta}{dt} \tag{3}$$

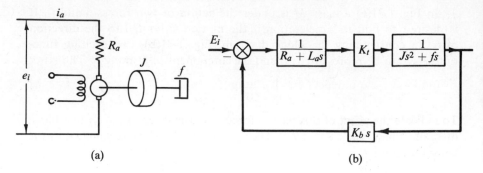

(a)   (b)

Fig. 1-33. A direct current motor: (a) schematic diagram; (b) block diagram.

$$K_b \frac{d\theta}{dt} = e_b \tag{4}$$

where $J, f,$ and θ are the combined moment of inertia of the rotor of the motor and the load, damping coefficient, and angular displacement of the motor respectively. Taking the Laplace transforms of the above equations with initially quiescent state, we obtain

$$E_i(s) - E_b(s) = R_a I_a(s) + L_a s I_a(s) \tag{1a}$$

$$\Gamma(s) = K_t I_a(s) \tag{2a}$$

$$\Gamma(s) = Js^2\,\Theta\,(s) + fs\,\Theta\,(s) \tag{3a}$$

$$K_b s\,\Theta\,(s) = E_b(s) \tag{4a}$$

The block diagram for the above four equations is shown in Fig. 1-33(b). It can be easily seen that the back emf forms the feedback path. The back emf is due to motor velocity, a direct function of the response, and becomes a source modifying the performance of the entire system.

As a further example, consider the triode circuit of Fig. 1-34(a), with its equivalent circuit shown in Fig. 1-34(b). The following circuit equations may be readily obtained:

$$e_o = iR_k \tag{5}$$

$$e_{gk} = e_s - iR_k = e_s - e_o \tag{6}$$

$$\mu e_{gk} = i(r_p + R_k) \tag{7}$$

where e_s, e_o, and i are the input voltage, output voltage developed on the load resistor R_k, and the plate current respectively; and μ and r_p are two constants. Eliminate e_{gk} from (6) and (7).

$$\mu(e_s - e_o) = ir_p + iR_k \tag{8}$$

or

$$\mu e_s - (\mu + 1)e_o = ir_p \tag{8a}$$

Equation (8a) may be represented by the block diagram shown in Fig. 1-34(c)

From Fig. 1-34(c) it can be readily seen that the system is not only under the action of the independent excitation function e_s, but also the feedback

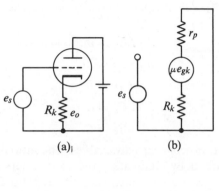

(a)   (b)

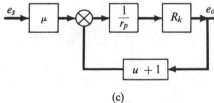

(c)

Fig. 1-34. A triode circuit: (a) schematic diagram; (b) equivalent circuit; (c) block diagram.

signal e_o as well. Since the signal travels forward in the upper part of the diagram, it is designated as the forward path; the lower part is the feedback path.

4. Block Diagram Identities

Let Fig. 1-34(c) be now redrawn with more general symbols, as in Fig. 1-35. Find the relationship between the cause R and the result C by means of the equation satisfied by the various parts of the block diagram. The procedure is as follows:

1. Label each variable of the intermediate stages, such as R, m_1, m_2, etc.

2. Write down the equations between these variables, immediately obtainable from the block diagram:

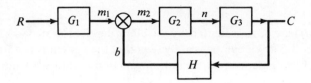

Fig. 1-35. A typical block diagram example.

$$RG_1 = m_1 \tag{9}$$

$$m_1 - b = m_2 \tag{10}$$

$$CH = b \tag{11}$$

$$m_2 G_2 = n \tag{12}$$

$$nG_3 = C \tag{13}$$

The above five equations connect six variables. The ratio of any two of these variables may be found by eliminating the four others among these five equations. However, we are most interested in the ratio C/R, which can be obtained readily as:

$$\frac{C}{R} = \frac{G_1 G_2 G_3}{1 + G_2 G_3 H} \tag{14}$$

Equation (14) can be represented by means of a combined block diagram, Fig. 1-36, which shows the same relation between C and R, so they can be considered as identical.

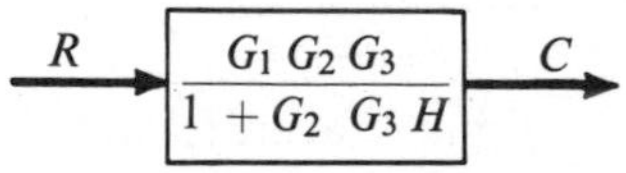

$$R \longrightarrow \boxed{\frac{G_1 G_2 G_3}{1 + G_2 G_3 H}} \longrightarrow C$$

Fig. 1-36. A single block diagram equivalent to Fig. 1-35.

Various identities between block diagrams can be established so that one may be replaced by the other without necessity of changing the corresponding algebraic equations. Those commonly used are given in Fig. 1-37 and can all be proved by algebraic procedures similar to preceding examples.

The identities in Fig. 1-37 may be applied to the system represented by the block diagram in Fig. 1-35. The steps taken to accomplish this are shown in the sequence in Fig. 1-38. The last one agrees with Fig. 1-36. The process of transforming the block diagrams by rules of identity is, like solving most problems, not unique. The alternative procedure shown in Fig. 1-39 may be followed. The final result is, of course, the same.

Rule 1

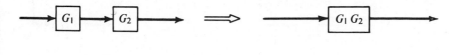

Rule 2

Rule 3

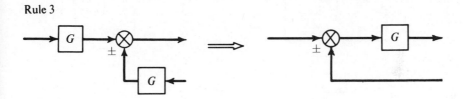

Rule 5

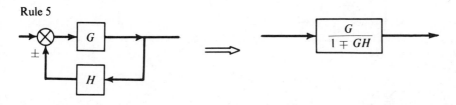

Rule 4

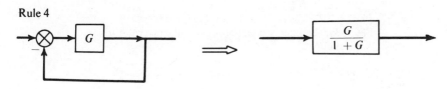

Rule 6

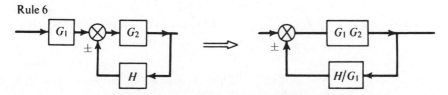

Fig. 1-37. Useful rules for transforming block diagrams.

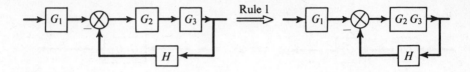

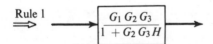

Fig. 1-38. One application of the identity rules to the example.

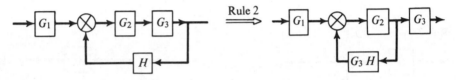

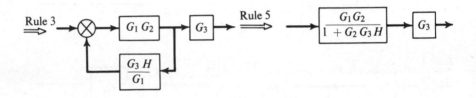

Fig. 1-39. Another application of the identity rules to the same example.

5. A Circuit

Consider a network as shown in Fig. 1-40. It is desired to set up the corresponding block diagram. The loop equations may be easily obtained:

$$e_i = R_1 i_1 + \frac{1}{C_1} \int (i_1 - i_2) \, dt$$

$$0 = \frac{1}{C_1} \int (i_2 - i_1) \, dt + R_2 i_2 + \frac{1}{C_2} \int i_2 \, dt \qquad (15)$$

$$e_o = \frac{1}{C_2} \int i_2 \, dt$$

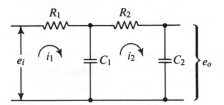

Fig. 1-40. A circuit example.

Let all the initial charges be zero. Then the Laplace transforms of the above equations will be

$$E_i = R_1 I_1 + \frac{1}{C_1 s}(I_1 - I_2)$$

$$0 = \frac{1}{C_1 s}(I_2 - I_1) + \left(R_2 + \frac{1}{C_2 s}\right) I_2 \tag{15a}$$

$$E_o = \frac{I_2}{C_2 s}$$

Figure 1-41 shows the block diagram for the network of Fig. 1-40. It is seen that the over-all transfer function or the system function is as follows.

$$\frac{E_o}{E_i} = \frac{\left(\frac{1}{C_1 s}\right)\left(\frac{1}{C_2 s}\right)}{\left(R_1 + \frac{1}{C_1 s}\right)\left(R_2 + \frac{1}{C_1 s} + \frac{1}{C_2 s}\right) - \frac{1}{C_1^2 s^2}} \tag{16}$$

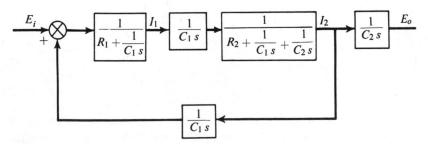

Fig. 1-41. The block diagram for the example.

Sometimes the system function can be simplified as below.

$$\frac{E_o}{E_i} \simeq \frac{\frac{1}{C_1 s}}{\left(R_1 + \frac{1}{C_1 s}\right)} \cdot \frac{\frac{1}{C_2 s}}{\left(R_2 + \frac{1}{C_2 s}\right)} \tag{17}$$

which is the product of the transfer functions of the two individual sections.

6. Controller and Controlled Element

On the basis of the foregoing, a simple electromechanical device such as a direct-current motor will now be represented by its block diagram and transfer function. Figure 1-42 shows the diagrammatic representation of the physical structure of a direct-current motor. The armature is connected to a constant current source I_a. The excitation function of input is a voltage across the field circuit. The sequence of action is as follows:

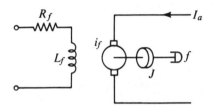

Fig. 1-42. A direct current motor.

1. The input voltage e_i produces a field current i_f.
2. i_f produces a proportional air gap flux ϕ in the motor if the magnetic saturation is low.
3. ϕ produces a torque T proportional to the product ϕI_a.
4. The torque produces a rotation of the motor shaft.

By applying the various basic laws of mechanics and electricity, we obtain the following equations:

$$e_i = R_f i_f + L_f \frac{di_f}{dt} \tag{18}$$

$$\phi = K_1 i_f \tag{19}$$

$$T = K_2 \phi \tag{20}$$

$$T = J \frac{d^2\theta}{dt^2} + f \frac{d\theta}{dt} \tag{21}$$

where J is the sum of the moments of inertia of the motor armature and the load and f is the sum of the friction coefficients. With all initial conditions at zero, the corresponding Laplace transforms of the above equations are

$$E_i = R_f I_f + L_f s I_f \tag{18a}$$

$$\Phi = K_1 I_f \tag{19a}$$

$$\Gamma = K_2 \Phi \tag{20a}$$

$$\Gamma = J s^2 \Theta + f s \Theta \tag{21a}$$

The block diagram is shown in Fig. 1-43(a). It can also be represented as in Fig. 1-43(b).

In the above system, the input voltage controls the motor field and thereby

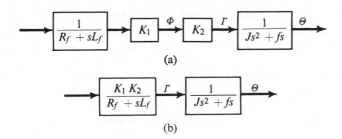

(a)

(b)

Fig. 1-43. Block diagram of the d-c motor: (a) showing every element; (b) with some elements combined.

the torque developed. The rotation of the motor armature generates a back electromotive force in the armature winding only; therefore, it has no reaction on the input circuit. The power level is amplified through a power source supplying the armature current I_a. The presence of a separate power source is quite common in a control system. The first block of Fig. 1-43(a) shows the development of torque and is designated as the controller, whereas the second block, representing the mechanical part of the rotating system, is the controlled element.

In this simple motor drive, the input E_i is entirely arbitrary. This type of system without feedback is known as open loop control.

7. Voltage Regulator—Feedback System

A typical block diagram for a control mechanism is shown in Fig. 1-44. The blocks represent the components; and the arrowheads, the variables. In addition there is a summer.

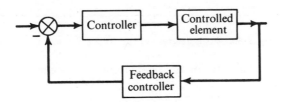

Fig. 1-44. A control mechanism.

If the feedback controller has a coefficient of unity, then the output signal is directly fed into the input; hence this is known as a unity feedback system. If the coefficient is zero, there is no feedback and it is called "open loop."

The following examples show what is meant by a controller, a controlled element, and a summing device.

Figure 1-45 contains a voltage regulator of the type suitable for a direct current generator. The output voltage e_g is fed back to appear between the input terminal of the amplifier and one of the terminals of the reference battery. The potential difference between the reference battery and the output voltage is the input to the amplifier. Those parts enclosed in the dotted lines

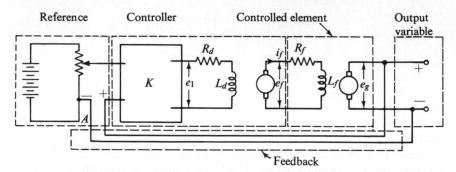

Fig. 1-45. The schematic diagram of a voltage regulator.

can be symbolized in a block diagram as a summing device. In Fig. 1-46, R corresponds to the reference voltage, C to the generator voltage. The output of the summing device is fed into K. The output of this amplifier is a voltage e_1, which establishes a field current in the exciter of a generator. This in turn excites the field of the main generator with a voltage e_f. The voltage output of the main generator, which is e_g, then appears at the output terminals.

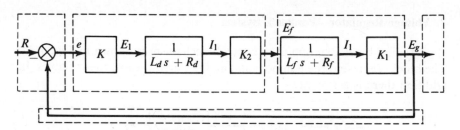

Fig. 1-46. Block diagram of the voltage regulator.

The differential equation for the field circuit of the exciter is

$$e_1 = L_d \frac{di_1}{dt} + R_d i_1 \qquad (22)$$

and the differential equation for the field of the main generator is

$$e_f = R_f i_f + L_f \frac{di_f}{dt} \qquad (23)$$

The generator voltage e_g is assumed to be proportional to i_f; i.e.,

$$e_g = K_1 i_f \qquad (24)$$

where K_1 is a proportionality constant.

The exciter voltage e_f is assumed to be proportional to i_1:

$$e_f = K_2 i_f \tag{25}$$

Taking the Laplace transform of (22), (23), (24), and (25), we have

$$E_1(s) = L_d s I_1(s) + R_d I_1(s) \tag{22a}$$

$$E_f(s) = R_f I_f(s) + L_f s I_f(s) \tag{23a}$$

$$E_g(s) = K_1 I_f(s) \tag{24a}$$

$$E_f(s) = K_2 I_1(s) \tag{25a}$$

A block diagram is constructed as shown in Fig. 1-46. The four dotted-line boxes correspond to the four boxes in Fig. 1-45, denoting the summer, controller, controlled element, and the feedback path.

Several intermediate variables, E_1, I_1, E_f, I_f, appear in the diagram. The input variable is R and the output variable is $C = E_g$. If R is set to a desired value, there will result a difference between R and C. Then E_1, I_1, E_f, I_f, and E_g will change successively in a sort of "chain reaction." The new value of E_g will be such as to reduce the difference between it and R. This new difference causes a new chain reaction. Finally, when $R - E_g = 0$, the output will be at the desired value and regulation will have been accomplished.

With this system, E_g can be set and maintained at any desired value by adjusting the terminal voltage of the battery.

8. Engine Governor—Another Feedback System

The flyball governor was perhaps the first automatic control which was rigorously analyzed by Maxwell. A simplified version will be used to describe this feedback system.

The engine is the controlled element; it is represented simply by a flywheel having a moment of inertia J.

$$J\alpha = \tau + \tau_n \tag{26}$$

where α is the angular acceleration, $\alpha = \dot{\omega}$. τ is the torque developed by the engine, and τ_n is the noise torque, some undesirable disturbance.

Equation (26) can also be written in terms of the angular velocity:

$$J\dot{\omega} = \tau + \tau_n \tag{26a}$$

Centrifugal force F is proportional to the square of the angular velocity; or

$$f = K_1(\omega - \omega_r)^2 \tag{27}$$

where ω_r is the reference speed, or the speed desired.

For simplicity, assume a linear relationship:

$$f \cong K_1(\omega - \omega_r) \tag{28}$$

and that the torque τ is proportional to the displacement of the governor x:

$$\tau = K_2 x \qquad (29)$$

The governor itself is assumed to have mass M, damping B, and a spring constant K. Therefore,

$$M\ddot{x} + B\dot{x} + K = f \qquad (30)$$

Performing Laplace transformation of (26a), (28), (29), and (30), we have

$$Js\Omega = \tau + \tau_n \qquad (26a)$$

$$F = K_1(\Omega - \Omega_r) \qquad (28a)$$

$$T = K_2 X \qquad (29a)$$

$$(Ms^2 + Bs + K)X = F \qquad (30a)$$

A block diagram is then constructed as shown in Fig. 1-47.

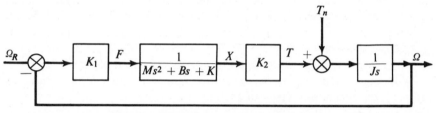

Fig. 1-47. Block diagram of an engine governor.

The over-all, or closed loop, function can be obtained by combining the four equations:

$$\frac{\Omega_c}{\Omega_R} = \frac{K_1 K_2}{(Ms^2 + Bs + K)Js + K_1 K_2}\bigg|_{\text{closed loop}}$$

The block diagram shows the cause and effect relationships among the components of the system.

What corresponds to the open-loop operation of the engine-governor system?

Set $\omega = 0$ in Eq. (28). The forward or open-loop function then is

$$\frac{\Omega_0}{\Omega_r} = \frac{K_1 K_2}{(Ms^2 + Bs + K)Js}\bigg|_{\text{open loop}} \qquad (31)$$

9. Summary

In this chapter we have discussed two approaches to the representation problem.

The first is the cause-effect or input-output viewpoint, which originated in electrical engineering circuit and communication theory. Basic techniques, the loop method and the node method, are presented, and then these are extended from an electric circuit into the formulation of a mechanical system.

The second section deals with the same kind of problem, but the approach is more mechanically orientated. The Lagrange equation and its terminology are introduced; then it is applied, first to a mechanical, then to an electrical, and finally to a mechanical-electrical system.

The advantage of the cause and effect approach is its simplicity, if only linear systems are considered. However, many of the techniques developed will not apply to a nonlinear case.

On the other hand, the energy approach is advantageous because it is unifying. As we have seen, many nonlinear system formulation problems can be handled easily with this approach. It does not require either graphic constructions or geometric deductions, but only algebraic operations proceeding after a regular and uniform plan.

The third section of this chapter is for the purpose of familiarizing the reader with a modern language—the block diagram. As a result of writing the Laplace transformations of the differential equations, we have produced a set of algebraic equations whose elements can be clearly represented by blocks and arrows. The operations, such as simplification and reduction, have become routine algebra.

In the final section of this chapter the feedback idea was developed with a mechanical and an electrical system as two illustrative examples.

This chapter lays the foundation for the construction of mathemacial models of physical systems.

The models used in the examples were high order differential equations. These are not the only possible models. In a more modern technique the model consists of a set of first order equations. This will be discussed in Chapter 6.

PROBLEMS

1-1. Indicate which of the following differential equations are linear and which are nonlinear:

(a) $\dfrac{d^3x}{dt^3} + 5\dfrac{d^2x}{dt^2} + 8\dfrac{dx}{dt} + 10x = e^{-t}$.

(b) $\dfrac{d^2x}{dt^2} + \sin t\dfrac{dx}{dt} + 9x = 0$.

(c) $\dfrac{d^2x}{dt^2} + x^2 = 5$.

(d) $\dfrac{d^2x}{dt^2} + t\left(\dfrac{dx}{dt}\right)^2 + x = e^{-t}$.

(e) $\dfrac{d^2x}{dt^2} + \dfrac{1}{t}\left(\dfrac{dx}{dt}\right) + \left(1 - \dfrac{1}{t^2}\right)x = 0.$

(f) $\dfrac{d^3x}{dt^3} + t\dfrac{d^2x}{dt^2} + 2x\left(\dfrac{dx}{dt}\right) + x = 1 + t^2.$

1-2. State which of the following systems are linear and which are nonlinear:
  (a) The simple pendulum (Eq. 14).
  (b) The simple circuit (Eq. 18).
  (c) The elastic pendulum (Eqs. 23–24).
  (d) The capacitor microphone (Eqs. 29–30).
  (e) The circuit with mutual inductances (Eqs. 37–39).
  (f) The spherical elastic pendulum (Eqs. 43–45).

1-3. Use Lagrange's equations to derive the differential equations for each of the circuits shown in Fig. P1-3. Then verify the results by applying the loop method or the node method.

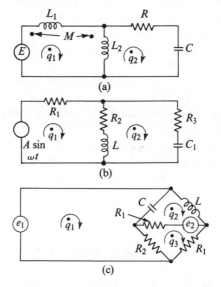

(a)

(b)

(c)

Fig. P1-3

1-4. For the mechanical systems shown in Fig. P1-4, (1) draw the mechanical circuit diagrams and then give the differential equations; (2) use Lagrange's equations to verify the results obtained in (1).

1-5. Derive the differential equations for each of the systems shown in Fig. P1-5, using Lagrange's equations.

1-6. Verify that the three networks in Fig. P1-6 have the same transfer function:

$$\dfrac{E_o}{E_i} = \dfrac{RCs}{(RC)^2s^2 + 3RCs + 1}.$$

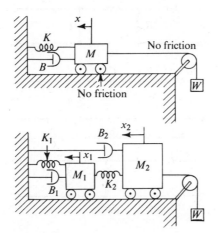

Fig. P1-4

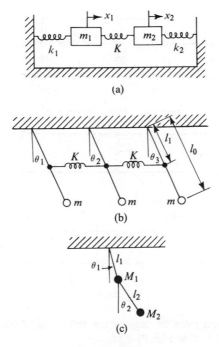

(a)

(b)

(c)

Fig. P1-5

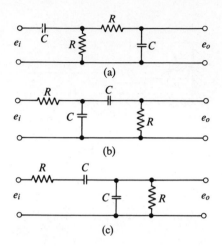

Fig. P1-6

1-7. For Fig. P1-7 find the transfer function for each circuit.

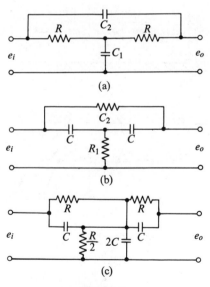

Fig. P1-7

1-8. Apply the reduction rules to simplify the following block diagrams:
(a) Find C/R and C/V in Fig. P1-8(a).

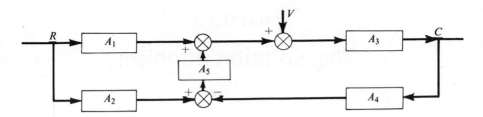

Fig. P1-8(a)

(b) Find C/R and E/R in Fig. P1-8(b).

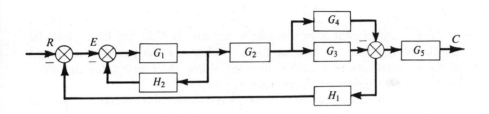

Fig. P1-8(b)

(c) Find C/R in Fig. P1-8(c).

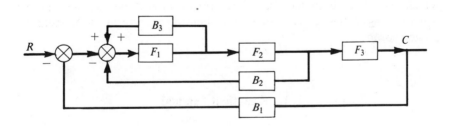

Fig. P1-8(c)

CHAPTER 2

The Solution Problem

The mathematical representations or models that we have thus far developed in Chapter 1 have all been differential equations. A complicated dynamic system can be described by a set of low order differential equations or by a single high order differential equation.

But the dynamic behavior and physical meaning pertaining to the system is not easily seen from equations containing derivatives. What we wish to know are the relationships among the variables; in other words, the solutions of the equations.

Finding the relationships among the variables, free of derivatives, which satisfy the equations identically, is what is known as the solution problem.

There are many techniques useful for the solution problem, among which the classical method and the transform method are well known. We will review these topics briefly and then introduce a machine method—analog computation.

I. Classical Method

1. A Simple Differential Equation

Consider an ordinary differential equation of the first order:

$$\frac{dx}{dt} = f(t) \tag{1}$$

Applying calculus, we can write (1) as follows:

$$x = \int f(t)\, dt + C = \varphi(t) + C \tag{1a}$$

(1) and (1a) are the same except that one has the differentiation symbol and the other has the integral symbol.

One might describe the solution problem as that of removing the differentiation symbol from an equation. The result gives a direct relationship between the independent variable t and the dependent variable x; (1a) is the solution of (1).

For example, given:

$$\frac{dx}{dt} = gt + v_0 \tag{3}$$

where g and v_0 are constants, the solution of (3) is

$$x = g\frac{t^2}{2} + v_0 t + c_0 \tag{4}$$

Equation (3) describes a physical phenomenon, such as a falling body having an initial velocity of v_0 in a gravitational field g. The relationship between the displacement of the body and the time it has been falling is shown in (4), in which c_0 is the initial displacement from its reference position.

2. Linear nth Order Differential Equations

A linear, nth order, differential equation is one which is linear in the dependent variable and its derivatives. It has the following form:

$$a_0(t)x^{(n)} + a_1(t)x^{(n-1)} + \cdots + a_{n-1}(t)x' + a_n(t)x = f(t) \tag{5}$$

If $f(t) = 0$, (5) is called a homogeneous linear equation, since it is homogeneous in the dependent variable x and its derivatives.

We are interested in the particular case in which all the a_i's are constants and $f(t)$ is zero.

$$a_0 x^{(n)} + a_1 x^{(n-1)} + \cdots + a_{n-1}x' + a_n x = 0 \tag{6}$$

For simplicity it is customary to write (6) as

$$H[x] = 0 \tag{7}$$

where $H[x]$ is called the linear differential operator, or

$$H[x] = a_0 x^{(n)} + a_1 x^{(n-1)} + \cdots + a_n x \tag{8}$$

If such a linear differential operator is applied to the sum of two variables, x_1 and x_2, the result is the same as the sum of the results obtained in applying the operator to each variable separately; i.e.,

$$H[x_1 + x_2] = H[x_1] + H[x_2]$$

This is known as the superposition theorem. It can be expressed in a more general form:

$$H\left[\sum_{i=1}^{n} c_i x_i\right] = \sum_{i=1}^{n} c_i H[x_i] \tag{9}$$

where the c_i's are constants.

3. Homogeneous Equations with Distinct Roots

A homogeneous linear differential equation with constant coefficients has the form

$$a_0 x^{(n)} + a_1 x^{(n-1)} + \cdots a_{n-1} x' + a_n x = 0 \tag{10}$$

Assume a particular solution:

$$x = e^{\lambda t} \tag{11}$$

The reason for making this assumption is that the derivative of an exponential curve is also an exponential, and therefore (11) is a proper solution with the correct value for the constant λ.

Substituting (11) into (10), we have

$$a_0 \lambda^n e^{\lambda t} + a_1 \lambda^{n-1} e^{\lambda t} + a_2 \lambda^{n-2} e^{\lambda t} + \cdots + a_n e^{\lambda t} = 0 \tag{12}$$

or

$$a_0 \lambda^n + a_1 \lambda^{n-1} + \cdots + a_n = 0 \tag{12a}$$

This is called the characteristic equation of (10). The values of λ must be determined.

The polynomial (12a) can be factored. This will give a group of roots:

$$\lambda_1, \lambda_2, \cdots, \lambda_n.$$

The solution of (10) is then

$$x_1 = c_1 e^{\lambda_1 t}$$
$$x_2 = c_2 e^{\lambda_2 t}$$
$$\cdot$$
$$\cdot \tag{13}$$
$$\cdot$$
$$x_n = c_n e^{\lambda_n t}$$

where c_i's are arbitrary constants. However, according to the principle of superposition, a linear combination with arbitrary, constant coefficients, $\sum_{i=1}^{n} c_i x_i$, of *solutions* $x_1, x_2, \ldots, x_n$ of the homogeneous linear equation $H[x] = 0$ is *also* a *solution of the equation.* Thus

$$x = c_1 e^{\lambda_1 t} + c_2 e^{\lambda_2 t} + \cdots + c_n e^{\lambda_n t} \tag{14}$$

is the general solution of Eq. (10).

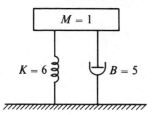

Fig. 2-1. Mechanical interpretation of $\dfrac{d^2x}{dt^2} + 5\dfrac{dx}{dt} + 6x = 0$.

Example. The differential equation for the system in Fig. 2-1 is

$$\frac{d^2x}{dt^2} + 5\frac{dx}{dt} + 6x = 0 \tag{15}$$

Assume

$$x = e^{\lambda t} \tag{16}$$

Substitution of (16) into (15) gives

$$\lambda^2 + 5\lambda + 6 = 0 \tag{17}$$

which is the characteristic equation of (15).

The roots are $\lambda_1 = -2$, $\lambda_2 = -3$. Hence the general solution of the given equation has the form

$$x = c_1 e^{-2t} + c_2 e^{-3t}$$

4. Homogeneous Equations with Repeated Roots

If the characteristic equation has a root λ_1 repeated m times, then not only is $e^{\lambda_1 t}$ a solution, but so are $t e^{\lambda_1 t}$, $t^2 e^{\lambda_1 t}$, ... , $t^{m-1} e^{\lambda_1 t}$.

Suppose a characteristic equation has a root $\lambda_1 = 0$, of multiplicity m. Then the characteristic equation has a common factor λ^m, or the coefficients

$$a_n = a_{n-1} = a_{n-m+1} = 0 \tag{18}$$

and the characteristic equation has the form

$$a_0\lambda^n + a_1\lambda^{n-1} + \cdots + a_{n-m}\lambda^m = 0$$

The corresponding linear homogeneous equations would be

$$a_0 x^n + a_1 x^{n-1} + \cdots + a_{n-m}x^m = 0$$

and it has particular solutions $1, t, t^2, \ldots, t^{m-1}$, since the equation does not contain derivatives of order lower than m.

Following similar reasoning, we can directly write a particular solution for a differential equation whose characteristic equation has a root $\lambda_1 \neq 0$ of multiplicity m.

For example, in the equation

$$\frac{d^3x}{dt^3} + 3\frac{d^2x}{dt^2} + 3\frac{dx}{dt} + x = 0 \tag{19}$$

the characteristic equation is $(\lambda + 1)^3 = 0$.
The general solution is

$$x = c_1 e^{-t} + c_2 t e^{-t} + c_3 t^2 e^{-t} \tag{20}$$

5. Nonhomogeneous Equations

A nonhomogeneous linear equation has the form

$$a_0 x^{(n)} + a_1 x^{(n-1)} + \cdots + a_n x = f(t) \tag{21}$$

For simplicity we can normalize the equation, making $a_0 = 1$. It can also be written

$$H[x] = f(t) \tag{22}$$

The general solution is equal to the sum of the solution $\sum_{i=1}^{n} c_i x_i$ of the corresponding homogeneous equation and a particular integral X of the nonhomogeneous solution.

In other words, the complete solution of (22) has the form

$$x = \sum_{i=1}^{n} c_i x_i + X \tag{23}$$

where $\sum_{i=1}^{n} c_i x_i$ is usually called the complementary function and X, the particular integral.

We find the complementary function by merely solving the homogeneous equation (21), or by solving

$$H[x] = 0 \tag{24}$$

However, finding the particular integral is usually by inspection. Fortunately, there is a general method for finding the solution of the nonhomogeneous equation once the solution of the corresponding homogeneous equation is known. The method is called Lagrange's *variation of parameters*.

6. Variation of Parameters

The general solution of (24), or the complementary function of (22), has the form:

$$x = u_1 x_1 + u_2 x_2 + \cdots + u_n x_n \tag{25}$$

Lagrange's variation of parameters method consists in seeking to modify the u's in such a way that (25) becomes a solution of (22).

In doing this, think of the u's as the functions "to be determined" of the independent variable.

Differentiating (25) gives

$$\dot{x} = \dot{u}_1 x_1 + u_1 \dot{x}_1 + \dot{u}_2 x_2 + u_2 \dot{x}_2 + \cdots + \dot{u}_n x_n + u_n \dot{x}_n$$
$$= (u_1 \dot{x}_1 + u_2 \dot{x}_2 + \cdots + u_n \dot{x}_n) + (\dot{u}_1 x_1 + \dot{u}_2 x_2 + \cdots + \dot{u}_n x_n)$$

Imposing as a first condition upon the u's that

$$\dot{u}_1 x_1 + \dot{u}_2 x_2 + \cdots + \dot{u}_n x_n = 0 \tag{26}$$

so that

$$\dot{x} = u_1 \dot{x}_1 + u_2 \dot{x}_2 + \cdots + u_n \dot{x}_n \tag{27}$$

and hence

$$\ddot{x} = (u_1 \ddot{x}_1 + u_2 \ddot{x}_2 + \cdots + u_n \ddot{x}_n) + (\dot{u}_1 \dot{x}_1 + \dot{u}_2 \dot{x}_2 + \cdots + \dot{u}_n \dot{x}_n)$$

and imposing a second condition

$$\dot{u}_1 \dot{x}_1 + \dot{u}_2 \dot{x}_2 + \cdots + \dot{u}_n \dot{x}_n = 0 \tag{28}$$

gives

$$\ddot{x} = u_1 \ddot{x}_1 + u_2 \ddot{x}_2 + \cdots + u_n \ddot{x}_n \tag{29}$$

Differentiating again

$$\dddot{x} = (u_1 \dddot{x}_1 + u_2 \dddot{x}_2 + \cdots + u_n \dddot{x}_n) + (\dot{u}_1 \ddot{x}_1 + \dot{u}_2 \ddot{x}_2 + \cdots + \dot{u}_n \ddot{x}_n)$$

and again imposing a third condition

$$\dot{u}_1 \ddot{x}_1 + \dot{u}_2 \ddot{x}_2 + \cdots + \dot{u}_n \ddot{x}_n = 0 \tag{30}$$

gives

$$\dddot{x} = u_1 \dddot{x}_1 + u_2 \dddot{x}_2 + \cdots + u_n \dddot{x}_n = 0 \tag{31}$$

If we continue the process n times the result is

$$x^{(n)} = (u_1 x_1^{(n)} + u_2 x_2^{(n)} + \cdots + u_n x_n^{(n)})$$
$$+ (\dot{u}_1 x_1^{(n-1)} + \dot{u}_2 x_2^{(n-1)} + \cdots + \dot{u}_n x_n^{(n-1)})$$

Now let

$$\dot{u}_1 x_1^{(n-1)} + \dot{u}_2 x_2^{(n-1)} + \cdots + \dot{u}_n x^{(n-1)} = f(t) \tag{32}$$

The conditions thus imposed on the u's are rewritten as follows:

$$
\begin{aligned}
\dot{u}_1 x_1 \quad &+ \dot{u}_2 x_2 \quad &&+ \cdots + \dot{u}_n x_n \quad &&= 0 \\
\dot{u}_1 \dot{x}_1 \quad &+ \dot{u}_2 \dot{x}_2 \quad &&+ \cdots + \dot{u}_n \dot{x}_n \quad &&= 0 \\
\dot{u}_1 \ddot{x}_1 \quad &+ \dot{u}_2 \ddot{x}_2 \quad &&+ \cdots + \dot{u}_n \ddot{x}_n \quad &&= 0 \\
&\cdots\cdots\cdots\cdots\cdots\cdots\cdots\cdots\cdots\cdots\cdots \\
\dot{u}_1 x_1^{(n-1)} \quad &+ \dot{u}_2 x^{(n-1)} \quad &&+ \cdots + \dot{u}_n x^{(n-1)} \quad &&= f(t)
\end{aligned}
\tag{33}
$$

Solving (33) for $\dot{u}_1$, we obtain

$$
\dot{u}_1 = \frac{\begin{vmatrix}
0 & x_2 & \cdots & x_n \\
0 & \dot{x}_2 & \cdots & \dot{x}_n \\
0 & \ddot{x}_2 & \cdots & \ddot{x}_n \\
\multicolumn{4}{c}{\cdots\cdots\cdots\cdots\cdots\cdots} \\
f(t) & x_2^{(n-1)} & \cdots & x_n^{(n-1)}
\end{vmatrix}}{\begin{vmatrix}
x_1 & x_2 & \cdots & x_n \\
\dot{x}_1 & \dot{x}_2 & \cdots & \dot{x}_n \\
\ddot{x}_1 & \ddot{x}_2 & \cdots & \ddot{x}_n \\
\multicolumn{4}{c}{\cdots\cdots\cdots\cdots\cdots\cdots} \\
x_1^{(n-1)} & x_2^{(n-1)} & \cdots & x_n^{(n-1)}
\end{vmatrix}}
\tag{34}
$$

Now u_1 can be obtained from the integration of (34). The general solution of (22) will be found by substituting the u_i values into (25).

It must now be shown that (25) is a solution of (22), if the u's satisfy (31). Collecting Eqs. (25), (27), (29), (31), etc. together, we have

$$
\begin{aligned}
x &= u_1 x_1 + u_2 x_2 + \cdots + u_n x_n \\
\dot{x} &= u_1 \dot{x}_1 + u_2 \dot{x}_2 + \cdots + u_n \dot{x}_n \\
\ddot{x} &= u_1 \ddot{x}_1 + u_2 \ddot{x}_2 + \cdots + u_n \ddot{x}_n \\
&\cdots\cdots\cdots\cdots\cdots\cdots\cdots\cdots\cdots \\
x^{(n)} &= u_1 x_1^{(n)} + u_2 x_2^{(n)} + \cdots + u_n x_n^{(n)} + f(t)
\end{aligned}
$$

and substituting them into (22), we obtain the following equations:

$$
\begin{aligned}
&u_1 x_1^{(n)} + u_2 x_2^{(n)} + \cdots + u_n x_n^{(n)} + f(t) \\
&\quad + a_1(u_1 x^{(n-1)} + u_2 x^{(n-1)} + \cdots + u_n x^{(n-1)}) + \cdots \\
&\quad + a_{n-1}(u_1 \dot{x}_1 + u_2 \dot{x}_2 + \cdots + u_n \dot{x}_n) \\
&\quad + a_n(u_1 x_1 + u_2 x_2 + \cdots + u_n x_n) = f(t)
\end{aligned}
$$

or

$$
\begin{aligned}
&u_1(x_1^{(n)} + a_1 x_1^{(n-1)} + \cdots + a_{n-1} \dot{x}_1 + a_n x_1) \\
&\quad + u_2(x_2^{(n)} + a_1 x_2^{(n-1)} + \cdots + a_{n-1} \dot{x}_2 + a_n x_2) \\
&\quad + u_3(\cdots) + \cdots + u_n(\cdots) + \cdots = 0
\end{aligned}
$$

Each pair of parentheses has to be equal to zero, because x_i is a solution of Eq. (24). Thus (25), with the u's in the form determined, is indeed a solution of (21).

Example 1. Consider the following second order system as the first example.

$$
\ddot{x} + 5\dot{x} + 6x = 10
\tag{35}
$$

The solution of the corresponding homogeneous equation is

$$x = u_1 e^{-2t} + u_2 e^{-3t} \tag{36}$$

Following the given procedures, we differentiate (36) and obtain the following equation:

$$\dot{x} = u_1(-2)e^{-2t} + u_2(-3)e^{-3t} + e^{-2t}\dot{u}_1 + e^{-3t}\dot{u}_2$$

The first condition is imposed as follows. Set

$$e^{-2t}\dot{u}_1 + e^{-3t}\dot{u}_2 = 0 \tag{37}$$

so that

$$\dot{x} = u_1(-2)e^{-2t} + u_2(-3)e^{-3t} \tag{38}$$

Differentiating again gives

$$\ddot{x} = u_1(4)e^{-2t} + u_2(9)e^{-3t} + (-2)e^{-2t}\dot{u}_1 + (-3)e^{-3t}\dot{u}_2$$

The second condition is as follows:

$$-2e^{-2t}\dot{u}_1 + (-3)e^{-3t}\dot{u}_2 = 10 \tag{39}$$

Rewriting (37) and (39) as

$$\begin{cases} e^{-2t}\dot{u}_1 + e^{-3t}\dot{u}_2 = 0 \\ -2e^{-2t}\dot{u}_1 + (-3)e^{-3t}\dot{u}_2 = 10 \end{cases}$$

and solving for $\dot{u}_1$, we obtain the following equation:

$$\dot{u}_1 = \frac{\begin{vmatrix} 0 & e^{-3t} \\ 10 & -3e^{-3t} \end{vmatrix}}{\begin{vmatrix} e^{-2t} & e^{-3t} \\ -2e^{-2t} & -3e^{-3t} \end{vmatrix}} = 10e^{2t}$$

Equation (40) is the result of performing the integration of u_1,

$$u_1 = \int 10e^{2t}\,dt = 5e^{2t} + c_1 \tag{40}$$

By repeating the procedures illustrated with u_1, we can obtain u_2.

$$\dot{u}_2 = \frac{\begin{vmatrix} e^{-2t} & 0 \\ -2e^{-2t} & 10 \end{vmatrix}}{\begin{vmatrix} e^{-2t} & e^{-3t} \\ -2e^{-2t} & -3e^{-3t} \end{vmatrix}} = -10e^{3t}$$

$$u_2 = \int -10e^{3t}\,dt = -\tfrac{10}{3}e^{3t} + c_2 \tag{41}$$

The solution is then

$$x = u_1 x_1 + u_2 x_2 = (5e^{2t} + c_1)(e^{-2t}) + (-\tfrac{10}{3}e^{3t} + c_2)(e^{-3t})$$

$$= \tfrac{5}{3} + c_1 e^{-2t} + c_2 e^{-3t} \tag{42}$$

which is just what was expected.

Example 2.

$$\ddot{x} + 5\dot{x} + 6x = f(t) \tag{43}$$

Following the procedure used in the last example, we can determine u_1.

$$\dot{u}_1 = \frac{\begin{vmatrix} 0 & e^{-3t} \\ f(t) & -3e^{-3t} \end{vmatrix}}{\begin{vmatrix} e^{-2t} & e^{-3t} \\ -2e^{-2t} & -3e^{-3t} \end{vmatrix}} = f(t)e^{2t}$$

$$u_1 = \int f(t)e^{2t}\,dt + c_1 \tag{44}$$

Similarly, we can find u_2.

$$\dot{u}_2 = \frac{\begin{vmatrix} e^{-2t} & 0 \\ -2e^{-2t} & f(t) \end{vmatrix}}{\begin{vmatrix} e^{-2t} & e^{-3t} \\ -2e^{-2t} & -3e^{-3t} \end{vmatrix}} = -f(t)e^{3t}$$

$$u_2 = -\int f(t)e^{3t}\,dt + c_2 \tag{45}$$

The solution of the problem is then

$$x = \left[\int f(\tau)e^{2\tau}\,d\tau + c_1\right]e^{-2t} + \left[\int -f(\tau)e^{3\tau}\,d\tau + c_2\right]e^{-3t}$$

$$= \int f(\tau)e^{-2(t-\tau)}\,d\tau - \int f(\tau)e^{-3(t-\tau)}\,d\tau + c_1e^{-2t} + c_2e^{-3t} \tag{46}$$

II.   Laplace Transformations

1. Substitution and Transformation

Some mathematical operations can be greatly simplified by a substitution method. For example, in solving the algebraic equation

$$x^2 + 2x + 5 = 0 \tag{1}$$

Let

$$y - 1 = x \tag{2}$$

Substituting (2) in (1) gives

$$y^2 - 2y + 1 + 2(y - 1) + 5 = 0$$
$$y^2 = -4$$
$$y = \pm j\sqrt{2} \tag{3}$$

This gives the answer in terms of the variable y.

To get the answer in terms of x, one has to rewrite (2) as

$$y = x + 1 \tag{2a}$$

Then

$$x = -1 \pm j\sqrt{2} \tag{4}$$

Note that the quadratic formula was not used; instead a new variable was substituted into the equation.

One pays for the substitution method in having to establish a relationship between the variables and in having to substitute back into the equation.

This process can be called the "transformation method." Relationship (2) is the direct transformation formula by which the variable x of (1) is transformed to the variable y. Relationship (2a) is called the inverse transformation formula by which the result in terms of y is transformed into a solution in terms of x.

2. Transformation Table

The relationship between x and y is fixed by (2); however, from this a table can be produced showing specific values of x and the corresponding values of y. It can be used somewhat like a dictionary: when one has a specific value of x, he can look up the corresponding value of y.

x	y
1	2
-1	0
$-1 + j\sqrt{2}$	$+j\sqrt{2}$
$-1 - j\sqrt{2}$	$-j\sqrt{2}$

Similarly, another kind of dictionary can be generated. For example:

x domain	y domain
x	$y - 1$
x^2	$y^2 - 2y + 1$
$2x$	$2y - 2$

The latter consists of algebraic relationships between the variables whereas the former contained only numbers. These tables can be generated in a straightforward manner from the basic relationship in (2).

3. Laplace Transformation

One of the methods of solving differential equations is by a table of Laplace transformations. The basic Laplace transform is defined as follows:

$$F(s) = \int_0^\infty f(t)e^{-st}\, dt \tag{4}$$

4. Function Transform Table

We shall construct a few members of the Laplace table. To find the Laplace transform of the exponential function e^{at}, substitute the function into (4).

$$F(s) = \int_0^\infty e^{at}e^{-st}\, dt \overset{\Delta}{=} \mathscr{L}\, e^{at}$$
$$= \frac{1}{s-a} \tag{5}$$

Case 1. $a = 0$.

$$\mathscr{L}\, 1 = \frac{1}{s}$$

Case 2. $a = j\omega$.

$$\mathscr{L}\, e^{j\omega t} = \frac{1}{s - j\omega} \tag{6}$$

Expanding the left-hand side into its trigonometric identity and rationalizing the right-hand side yield

$$\mathscr{L}(\cos \omega t + j \sin \omega t) = \frac{s + j\omega}{s^2 + \omega^2}$$

or

$$\mathscr{L}\, \cos \omega t = \frac{s}{s^2 + \omega^2} \tag{7}$$

$$\mathscr{L}\, \sin \omega t = \frac{\omega}{s^2 + \omega^2} \tag{8}$$

Case 3. $a = -\alpha + j\beta$

$$\mathscr{L}\, e^{(-\alpha + j\beta)t} = \frac{1}{s - (-\alpha + j\beta)} = \frac{1}{(s + \alpha) - j\beta}$$

$$\mathscr{L}\, e^{-\alpha t}(\cos \beta t + j \sin \beta t) = \frac{(s + \alpha) + j\beta}{(s + \alpha)^2 + \beta^2}$$

Equating the real parts, we have

$$\mathscr{L}\, e^{-\alpha t} \cos \beta t = \frac{s + \alpha}{(s + \alpha)^2 + \beta^2} \tag{9}$$

Then equating the imaginary parts, we obtain

$$\mathscr{L} e^{-\alpha t} \sin \beta t = \frac{\beta}{(s + \alpha)^2 + \beta^2} \tag{10}$$

Case 4. a = 1.

$$\mathscr{L} e^t = \frac{1}{s - 1} \tag{11}$$

Develop the left-hand side into a power series:

$$e^t = 1 + t + \frac{t^2}{2!} + \frac{t^3}{3!} + \frac{t^4}{4!} + \cdots \tag{12}$$

Expand the right-hand side by using long division:

$$\frac{1}{s - 1} = \frac{1}{s} + \frac{1}{s^2} + \frac{1}{s^3} + \frac{1}{s^4} + \cdots \tag{13}$$

Equating corresponding terms of (12) and (13), we obtain the following pairs simultaneously:

$$\mathscr{L} 1 = \frac{1}{s}$$

$$\mathscr{L} t = \frac{1}{s^2}$$

$$\mathscr{L} \frac{t^2}{2!} = \frac{1}{s^3} \tag{14}$$

$$\mathscr{L} \frac{t^n}{n!} = \frac{1}{s^{n+1}}$$

Collecting these pairs produces the accompanying table.

$f(t)$ (t domain)	$F(s)$ (s domain)
e^{at}	$\dfrac{1}{s - a}$
1 or $u(t)$	$\dfrac{1}{s}$
$\cos \omega t$	$\dfrac{s}{s^2 + \omega^2}$
$\sin \omega t$	$\dfrac{\omega}{s^2 + \omega^2}$
t^n	$\dfrac{n!}{s^{n+1}}$
$e^{-\alpha t} \sin \beta t$	$\dfrac{\beta}{(s + \alpha)^2 + \beta^2}$
$e^{-\alpha t} \cos \beta t$	$\dfrac{s + \alpha}{(s + \alpha)^2 + \beta^2}$

If a differential equation can be recognized as being the same as one of the forms in the table, then the transform can be read directly from it. This is much easier than substituting the function into (4) and simplifying the result, as is being done here to make up the table.

5. Operation Transform Table

If the Laplace transform of $f(t)$ is $F(s)$, what is the transform of $\frac{d}{dt}f(t)$?

By definition,

$$\mathscr{L}\frac{d}{dt}f(t) \overset{\Delta}{=} \int_0^\infty \frac{d}{dt}f(t)e^{-st}\,dt$$

Before evaluating the integral, consider the following:

$$\int_0^\infty f(t)e^{-st}\,dt = \frac{-1}{s}f(t)e^{-st}\Big|_0^\infty + \frac{1}{s}\int_0^\infty \left[\frac{df(t)}{dt}\right]e^{-st}\,dt$$

$$= \frac{f(0)}{s} + \frac{1}{s}\int_0^\infty \left[\frac{df(t)}{dt}\right]e^{-st}\,dt$$

Rearrangement gives

$$\int_0^\infty \left[\frac{df(t)}{dt}\right]e^{-st}\,dt = s\int_0^\infty f(t)e^{-st} - f(0)$$

or

$$\mathscr{L}\frac{df(t)}{dt} = sF(s) - f(0) \tag{15}$$

where $f(0)$ is the value of $f(t)$ when $t = 0+$.

The next question is: If the Laplace transform of $f(t)$ is $F(s)$, what is the transform of $\int f(t)\,dt$?

Let $u = e^{-st}$ and $dv = f(t)\,dt$
Evaluate:

$$\int u\,dv = uv - \int v\,du$$

or

$$\int_0^\infty f(t)e^{-st}\,dt = e^{-st}\int f(t)\,dt\Big|_0^\infty + s\int_0^\infty \left[\int f(t)\,dt\right]e^{-st}\,dt$$

$$= -f^{-1}(0) + s\int_0^\infty \left[\int f(t)\,dt\right]e^{-st}\,dt$$

Rearrangement yields

$$\int_0^\infty \left[\int f(t)\,dt\right]e^{-st}\,dt = \frac{F(s)}{s} + \frac{f^{-1}(0)}{s} \tag{16}$$

where $f^{-1}(0)$ represents the value of the integration of $f(t)$ at $t = 0+$.
The accompanying table of operation transforms can be made.

$f(t)$ (t domain)	$F(s)$ (s domain)
$f'(t)$	$sF(s) - f(0)$
$f''(t)$	$s[sF(s) - f(0)] - f'(0)$
$f^{(n)}(t)$	$s^n F(s) - \sum_{k=1}^{n} f^{(k-1)}(0)s^{n-k}$

in which $f^{(k)}(t) = \dfrac{d^k f(t)}{dt^k}$

and $f^{(0)}(t) \overset{\Delta}{=} f(t)$

$f^{(-1)}(t)$	$\dfrac{F(s)}{s} + \dfrac{f^{(-1)}(0)}{s}$
$f^{(-2)}(t)$	$\dfrac{F(s)}{s^2} + \dfrac{f^{(-1)}(0)}{s^2} + \dfrac{f^{(-2)}(0)}{s}$
$f^{(-n)}(t)$	$\dfrac{F(s)}{s^n} + \sum_{k=1}^{n} \dfrac{f^{(-k)}(0)}{s^{n-k+1}}$

6. Initial Value Theorem

Consider Eq. (15) in the last section.

$$\int_0^\infty f'(t)e^{-st}\,dt = sF(s) - f(0) \tag{15}$$

Allowing s to approach infinity as a limit:

$$\lim_{s\to\infty} \int_0^\infty f'(t)e^{-st}\,dt = \lim_{s\to\infty}\,[sF(s) - f(0)]$$

we obtain

$$\lim_{s\to\infty}\,[sF(s) - f(0)] = 0$$

but $f(0)$ is the $\lim_{t\to 0} f(t)$, from which we conclude that

$$\lim_{s\to\infty} sF(s) = \lim_{t\to 0} f(t) \tag{17}$$

Equation (17) shows that the value of $f(t)$ at $t = 0$ is equal to the limit of the product of $sF(s)$ as s approaches infinity.

7. Final Value Theorem

From Eq. (15) again, letting s approach zero, we have

$$\lim_{s\to 0} \int_0^\infty f'(t)e^{-st}\,dt = \lim_{s\to 0}\,[sF(s) - f(0)]$$

The left-hand side can also be written as

$$\int_0^\infty f'(t)\,dt = f(\infty) - f(0)$$

Therefore, we have

$$\lim_{s \to 0} [sF(s)] - f(0) = \lim_{t \to \infty} f(t) - f(0)$$

from which we conclude that

$$\lim_{s \to 0} [sF(s)] = \lim_{t \to \infty} [f(t)] \qquad (18)$$

This is called the final value theorem.

8. Real Translation Theorem

The translation theorem states that a translation in the t domain corresponds to multiplication by an exponential in the s domain, i.e.,

$$\mathscr{L}[f(t - a)] = e^{as}F(s) \qquad (19)$$

If $f(t)$ is as shown in Fig. 2-2, then $f(t - a)$ will be as shown in Fig. 2-3. Note that the whole curve is translated a distance a.

Fig. 2-2. The function $f(t)$.   Fig. 2-3. The function $f(t - a)$.

The translation theorem is easily proved from the basic definition:

$$\int_a^\infty f(t - a)e^{-st}\, dt = e^{-as}F(s) \qquad (19a)$$

9. Complex Translation Theorem

The complex translation theorem states that multiplication by an exponential in the t domain corresponds to translation of the function in the s domain; i.e.,

$$\mathscr{L}[e^{-at} f(t)] = F(s + a)$$

We can prove the theorem as follows. By definition

$$\int_0^\infty f(t)e^{-qt}\, dt = F(q)$$

Let $q = s + a$; then

$$\int_0^\infty f(t)e^{-(s+a)t}\, dt = \int_0^\infty [f(t)e^{-at}]e^{-st}\, dt = F(s + a) \qquad (20)$$

10. Scaling Theorem

The scaling theorem states that the division of the variable by a constant in the t domain corresponds to the multiplication of both the transform and its variable by this same constant in the s domain; i.e.,

$$\mathscr{L}\left[f\left(\frac{t}{a}\right)\right] = aF(as) \tag{21}$$

The proof of the theorem starts from the basic definition

$$\int_0^\infty f(\tau)e^{-q\tau}\,d\tau = F(q) \tag{22}$$

in which q is a complex variable. Let $\tau = a\tau/a$ and substitute into (22):

$$\int_0^\infty f\left(\frac{a\tau}{a}\right)e^{-(a\tau/a)q}\,d\tau = F(q)$$

Then let $t = a\tau$ and $q/a = s$

$$\int_0^\infty f\left(\frac{t}{a}\right)e^{-(t/a)us}\,d\left(\frac{t}{a}\right) = F(as) \quad \text{or} \quad \int_0^\infty f\left(\frac{t}{a}\right)e^{-st}\,dt = aF(as)$$

The theorem is proven.

III.   The Transform Method

1. Classical Method vs. Transform Method

It is well-known that the process of multiplication can be simplified by a transformation to logarithms. In a similar sense Laplace transformation simplifies the solution of linear differential equations. The charts in Figs. 2-4 and 2-5 illustrate this analogy.

In either case the transformation method converts a complicated operation into a simpler one, but requires additional steps.

2. The Solution of a Differential Equation

Given a differential equation:

$$\frac{dx}{dt} + 4x = 10 \tag{1}$$

$$x(0) = 0 \tag{2}$$

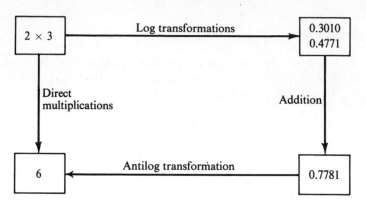

Fig. 2-4. Direct and indirect procedures for multiplying two numbers.

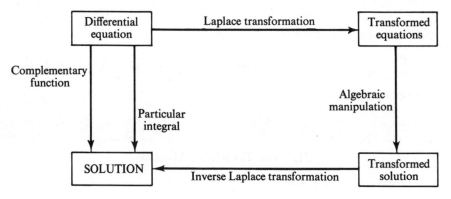

Fig. 2-5. Direct and transform methods for solving a differential equation.

where $x(0)$ is the initial condition. Transform both sides of Eq. (1):

$$\mathscr{L}\left\{\frac{dx}{dt}\right\} + \mathscr{L}\{4x\} = \mathscr{L}\{10\} \qquad (1a)$$

Using the table transforms in the last section, we obtain

$$sX(s) - x(0) + 4X(s) = \frac{10}{s} \qquad (3)$$

Equation (3) is called the Laplace-transformed equation of (1). It contains the unknown function $X(s)$ and the known values $x(0)$, $10/s$, etc. Next, solve (3) for $X(s)$ after substituting the initial condition.

$$(s + 4)X(s) = \frac{10}{s} + 2$$

$$X(s) = \frac{\dfrac{10}{s} + 2}{s + 4} \qquad (4)$$

$$X(s) = \frac{10 + 2s}{s(s + 4)} = \frac{\frac{5}{2}}{s} + \frac{-\frac{1}{2}}{s + 4}$$

Equation (4) is the solution of the differential equation, but it is in the transformed form.

Consult table of transforms to convert the solution back to the time domain:

$$x(t) = \tfrac{5}{2} - \tfrac{1}{2}e^{-4t}$$

which is the desired solution of (1).

Note that the initial condition is carried along throughout the solution.

3. Solution of a Second Order Differential Equation: Overdamped Case

Given a second order differential equation

$$\frac{d^2x}{dt^2} + 5\frac{dx}{dt} + 6x = 0 \tag{5}$$

with the initial conditions

$$\begin{cases} x(0) = 0 \\ x'(0) = 1 \end{cases}$$

Laplace transformation yields

$$s^2 X(s) - sx(\cancel{0}) - x'(0) + 5sX(s) - 5x(\cancel{0}) + 6X(s) = 0$$

Solving for $X(s)$ gives

$$X(s) = \frac{1}{s^2 + 5s + 6} \tag{6}$$

Factor the denominator:

$$X(s) = \frac{1}{(s + 2)(s + 3)} \tag{7}$$

Assume

$$X(s) = \frac{A}{(s + 2)} + \frac{B}{(s + 3)} = \frac{As + 3A + Bs + 2B}{(s + 2)(s + 3)}$$

Equate the two numerators to determine A and B.

$$As + 3A + Bs + 2B = 1$$

or

$$\begin{cases} A + B = 0 \\ 3A + 2B = 1 \end{cases}$$

therefore,

$$A = 1, \quad B = -1 \tag{8}$$

Substitute (8) into (7).

$$X(s) = \frac{1}{s+2} + \frac{-1}{s+3} \tag{9}$$

Obtain the inverse Laplace transform by consulting the table.

$$x(t) = e^{-2t} - e^{-3t} \tag{10}$$

Equation (5) could be the differential equation of a system such as that shown in Fig. 2-6. The solution (10) gives the relationship between the displacement x and time t. If the roots of the denominator of the transfer function turn out to be real, as they did in this case, the second order mechanical system which the equation represents is said to be overdamped.

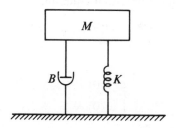

Fig. 2-6. Mechanical interpretation of

$$M\frac{d^2x}{dt^2} + B\frac{dx}{dt} + Kx = 0$$

4. Solution of a Second Order Differential Equation: Underdamped Case

Consider the following differential equation:

$$\frac{d^2x}{dt^2} + 2\frac{dx}{dt} + 5x = 0 \tag{11}$$

$$x(0) = 0, \quad x'(0) = 1$$

The corresponding transformed equation is

$$s^2 X(s) + 2s X(s) + 5X(s) = 1 \tag{12}$$

The transfer function is

$$X(s) = \frac{1}{s^2 + 2s + 5}$$

Factoring gives

$$X(s) = \frac{1}{(s + 1 + j2)(s + 1 - j2)} \qquad (13)$$

Again assume

$$\frac{1}{(s + 1 + j2)(s + 1 - j2)} = \frac{A}{(s + 1 + j2)} + \frac{B}{(s + 1 - j2)} \qquad (14)$$

Multiply both sides by the function $(s + 1 + j2)$.

$$\frac{1}{(s + 1 + j2)(s + 1 - j2)}(s + 1 + j2)$$

$$= \frac{A}{s + 1 + j2}(s + 1 + j2) + \frac{B}{s + 1 - j2}(s + 1 + j2) \qquad (15)$$

Since (15) is an identity, we can set s to any value. Let $s = -1 - j2$; then (15) becomes

$$\frac{1}{s + 1 - j2}\bigg|_{s \to (-1 - j2)} = A$$

or

$$A = \frac{1}{-j4} \qquad (16)$$

Similarly, determine B.

$$B = \frac{1}{j4} \qquad (17)$$

Substitute these values for A and B into (14) to obtain:

$$X(s) = \frac{\frac{1}{-j4}}{s + 1 + j2} + \frac{\frac{1}{j4}}{s + 1 - j2}$$

Then perform the inverse Laplace transform

$$x(t) = \frac{1}{-j4}e^{-(1+j2)t} + \frac{1}{j4}e^{-(1-j2)t} \qquad (18)$$

Equation (18) is the solution. However, it is not a conventional form. It can be converted into a trigonometric function by the following steps:

$$x(t) = \frac{1}{-j4}e^{-t}(\cos 2t - j \sin 2t) + \frac{1}{j4}e^{-t}(\cos 2t + j \sin 2t)$$

$$= \frac{je^{-t}(\cos 2t - j \sin 2t)}{4} + \frac{-je^{-t}(\cos 2t + j \sin 2t)}{4}$$

$$= \frac{je^{-t}\cos 2t + e^{-t} \sin 2t - je^{-t}\cos 2t + e^{-t} \sin 2t}{4} \qquad (19)$$

$$= \frac{1}{2}e^{-t} \sin 2t$$

If the two roots of a second order equation are complex numbers then the latter can represent a mechanical system which is said to be underdamped.

In this section the constants A and B were determined by the *Heaviside technique of partial fraction expansion*, which is simpler than solving a set of simultaneous equations as done in Sec. 3.

5. Solution of a Second Order Differential Equation: Critically Damped Case

Consider the following differential equation:

$$\frac{d^2x}{dt^2} + 6\frac{dx}{dt} + 9x = 1$$

$$x(0) = 0, \qquad x'(0) = 0 \tag{20}$$

The corresponding transformed equation is

$$X(s) = \frac{1}{s^2 + 6s + 9} \cdot \frac{1}{s} \tag{21}$$

or

$$X(s) = \frac{1}{s(s+3)^2}$$

Write the partial fraction expansion

$$X(s) = \frac{A}{(s+3)^2} + \frac{B}{(s+3)} + \frac{C}{s} \tag{22}$$

$$\frac{1}{s(s+3)^2}(s+3)^2 = \frac{A}{(s+3)^2}(s+3)^2 + \frac{B}{(s+3)}(s+3)^2 + \frac{C}{s}(s+3)^2 \tag{23}$$

Then let $s = -3$. Therefore,

$$A = \frac{-1}{3}$$

To find B, multiply both sides of (22) by $(s+3)$.

$$\frac{1}{s(s+3)^2}(s+3) = \frac{A}{(s+3)^2}(s+3) + \frac{B}{(s+3)}(s+3) + \frac{C}{s}(s+3)$$

If we now let s equal -3, we cannot obtain B, because the first term on the right side is not zero.

Differentiate both sides of (23) with respect to s:

$$\frac{-1}{s^2} = B + \frac{[s2(s+3) - (s+3)^2]C}{s^2}$$

Then let s equal -3:

$$-\tfrac{1}{9} = B$$

To find C:

$$\frac{1}{\not{s}(s+3)^2}\Bigg|_{s\to 0} = \frac{A}{(s+3)^2}s + \frac{B}{(s+3)}s + \frac{C}{\not{s}}\not{s}$$

Therefore,

$$C = \tfrac{1}{9}$$

Substitute these constants into (22):

$$X(s) = \frac{-\tfrac{1}{3}}{(s+3)^2} + \frac{-\tfrac{1}{9}}{(s+3)} + \frac{\tfrac{1}{9}}{s} \tag{24}$$

$$x(t) = \tfrac{1}{9} - \tfrac{1}{9}e^{-3t} - \tfrac{1}{3}te^{-3t} \tag{25}$$

which is the solution of (20).

6. A General Inverse Laplace Transform Formula

If

$$F(s) = \frac{Q(s)}{P(s)} \tag{26}$$

where $Q(s)$ and $P(s)$ are polynomials of s with constant coefficients, by using "partial fraction expansion," (26) can be written as

$$F(s) = \frac{Q(s)}{s^n + a_{n-1}s^{n-1} + \cdots + a_1 s + a_0}$$

$$= \frac{Q(s)}{(s - s_1)(s - s_2) \ldots (s - s_k)^m \ldots (s - s_{n-m})}$$

The denominator has a multiple root of order m.

By making a partial fraction expansion, we have

$$F(s) = \frac{A_1}{s - s_1} + \frac{A_2}{s - s_2} + \cdots + \frac{A_{k-1}}{s - s_{k-1}}$$

$$+ \frac{A_k(s)}{(s - s_k)^m} + \frac{A_{k+1}}{s - s_{k+1}} + \cdots + \frac{A_{n-m+1}}{s - s_{n-m+1}} + \frac{A_{n-m}}{s - s_{n-m}} \tag{27}$$

The contribution to the inverse transform of each single root is readily obtained. For example, the first term of (27) contributes $A_1 e^{s_1 t}$, where

$$A_1 = (s - s_1)F(s)|_{s\to s_1}$$

The contribution made by a term with multiple roots, s_k, should be considered separately. Expand as follows:

$$\frac{A_k(s)}{(s - s_k)^m} = \frac{A_{k1}}{(s - s_k)^m} + \frac{A_{k2}}{(s - s_k)^{m-1}} + \cdots + \frac{A_{km}}{(s - s_k)} \tag{28}$$

where the coefficients A_{km} are constants given by

$$A_{k1} = [(s - s_k)^m F(s)]_{s \to s_k}$$

$$A_{k2} = \left\{ \frac{d}{ds} [(s - s_k)^m F(s)] \right\} \Big|_{s \to s_k}$$

$$A_{k3} = \left\{ \frac{1}{2!} \frac{d^2}{ds^2} [(s - s_k)^m F(s)] \right\} \Big|_{s \to s_k} \tag{29}$$

$$A_{km} = \left\{ \frac{1}{(m-1)!} \frac{d^{m-1}}{ds^{m-1}} [(s - s_k)^m F(s)] \right\} \Big|_{s \to s_k}$$

7. A General Example

A differential equation

$$x^v + 12x^{iv} + 55\ddot{x} + 120\ddot{x} + 124\dot{x} + 48x = 1 \tag{30}$$

with a set of initial conditions

$$x(0) = 0, \quad x'(0) = 0, \quad x''(0) = 0, \quad x'''(0) = 0, \quad x^{iv}(0) = 6 \tag{31}$$

is given. The complete solution of the differential equation is desired.

First of all, performing Laplace transformation on (30) yields

$$s^5 X(s) - s^4 x(0) - s^3 x'(0) - s^2 x''(0) - sx'''(0) - x^{iv}(0)$$
$$+ 12[s^4 X(s) - s^3 x(0) - s^2 x'(0) - sx''(0) - x'''(0)]$$
$$+ 55[s^3 X(s) - s^2 x(0) - sx'(0) - x''(0)]$$
$$+ 120[s^2 X(s) - sx(0) - x'(0)]$$
$$+ 124[sX(s) - x(0)]$$
$$+ 48[X(s)] = \frac{1}{s}$$

And then substitution of the initial conditions gives

$$(s^5 + 12s^4 + 55s^3 + 120s^2 + 124s + 48)X(s) = \frac{1}{s} + 6$$

$$X(s) = \frac{6s + 1}{s(s^5 + 12s^4 + 55s^3 + 120s^2 + 124s + 48)} \tag{32}$$

Factoring the denominator, we have

$$X(s) = \frac{6s + 1}{s(s + 1)(s + 2)^2(s + 3)(s + 4)} \tag{33}$$

For taking the partial fraction expansion, we write

$$X(s) = \frac{A_1}{s} + \frac{A_2}{(s + 1)} + \frac{A_3(s)}{(s + 2)^2} + \frac{A_4}{(s + 3)} + \frac{A_5}{(s + 4)} \tag{34}$$

where the middle term of the right-hand side can be written as

$$\frac{A_3(s)}{(s+2)^2} = \frac{A_{31}}{(s+2)^2} + \frac{A_{32}}{s+2} \tag{35}$$

Substituting (35) into (34) gives

$$X(s) = \frac{A_1}{s} + \frac{A_2}{s+1} + \frac{A_{31}}{(s+2)^2} + \frac{A_{32}}{(s+2)} + \frac{A_4}{s+3} + \frac{A_5}{s+4} \tag{36}$$

For determining A_1, multiply both sides of (34) by s.

$$X(s)(s) = \frac{A_1}{s}s + \frac{A_2}{s+1}s + \frac{A_3(s)}{(s+2)^2}s + \frac{A_4}{s+3}s + \frac{A_5}{s+4}s$$

The value of A is found by letting $s \to 0$.

$$A_1 = \frac{1}{1 \cdot 2^2 \cdot 3 \cdot 4} = \frac{1}{48}$$

Similarly,

$$A_2 = \frac{-5}{(-1)(1)^2(2)(3)} = \frac{-5}{-6} = \frac{5}{6}$$

$$A_4 = \frac{-18+1}{(-3)(-2)(-1)^2(1)} = \frac{-17}{6}$$

$$A_5 = \frac{-24+1}{(-4)(-3)(-2)^2(1)} = \frac{-23}{48}$$

$$A_{31} = \frac{-12+1}{(-2)(-1)(1)(2)} = \frac{-11}{4}$$

A_{32} must be obtained by differentiation, or by applying the second formula of (29).

$$
\begin{aligned}
A_{32} &= \left\{ \frac{d}{ds}[(s+2)^2 X(s)] \right\}\Big|_{s \to -2} \\
&= \left\{ \frac{d}{ds}\left[\frac{6s+1}{s(s+1)(s+3)(s+4)} \right] \right\}\Big|_{s \to -2} \\
&= \left\{ \frac{d}{ds}\left(\frac{6s+1}{s^4+8s^3+19s^2+12s} \right) \right\}\Big|_{s \to -2} \\
&= \frac{(s^4+8s^3+19s^2+12s)(6) - (6s+1)(4s^3+24s^2+38s+12)}{(s^4+8s^3+19s^2+12s)^2}\Big|_{s \to -2} \\
&= \frac{(4)(6)-(-11)(6)}{(4)^2} \\
&= \frac{3}{2}
\end{aligned}
$$

The solution of (30) is then

$$x(t) = \tfrac{1}{48} + \tfrac{5}{6}e^{-t} - \tfrac{11}{4}te^{-2t} + \tfrac{3}{2}e^{-2t} - \tfrac{17}{6}e^{-3t} - \tfrac{23}{48}e^{-4t} \tag{37}$$

8. Convolution Integral

Partial fractions expansion is one way to decompose a transfer function into several simple components in preparation for using inverse Laplace transforms to arrive at a solution. However, there is another way which is illustrated by the following example. The transfer function

$$F(s) = \frac{1}{(s+1)(s+2)} \tag{38}$$

can be decomposed in two ways:

$$F(s) = \frac{1}{(s+1)} + \frac{-1}{(s+2)}, \quad \text{which is the partial fraction form} \tag{38a}$$

$$F(s) = \frac{1}{(s+1)} \cdot \frac{1}{(s+2)}, \quad \text{which is the product form} \tag{38b}$$

We cannot immediately evaluate the inverse Laplace of $F(s)$ through (38b) because, even though we know the transforms for $1/(s+1)$ and $1/(s+2)$ in (38b), we do not know how to write the time function for $F(s)$. Here one of the most useful deductions in transform theory comes into play. We will prove that

$$\mathscr{L}^{-1}[F_1(s)F_2(s)] = \int_0^t f_1(\tau)f_2(t-\tau)\,d\tau \tag{39}$$

The integral on the right-hand side is called a convolution integral. If for $t < 0$, $f_1(t) = 0$ and $f_2(t) = 0$, we can write

$$\int_0^t f_1(\tau)f_2(t-\tau)d\tau = \int_0^\infty f_1(\tau)f_2(t-\tau)\,d\tau \tag{40}$$

Now we transform both sides of (39).

$$F_1(s)F_2(s) = \int_0^\infty e^{-st}\,dt \int_0^\infty f_1(\tau)f_2(t-\tau)\,d\tau$$

Changing the order of integration yields

$$F_1(s)F_2(s) = \int_0^\infty f_1(\tau)\,d\tau \int_0^\infty e^{-st}f_2(t-\tau)\,dt$$

Since $f_2(t-\tau) = 0$ for $t < \tau$, we can change the limit on the second integral:

$$F_1(s)F_2(s) = \int_0^\infty f_1(\tau)\,d\tau \int_\tau^\infty e^{-st}f_2(t-\tau)\,dt$$

Let $t - \tau = x$ and rewrite the above equation:

$$F_1(s)F_2(s) = \int_0^\infty f_1(\tau)\,d\tau \int_0^\infty e^{-(x+\tau)s}f_2(x)\,dx$$

$$= \left[\int_0^\infty f_1(\tau)e^{-s\tau}\,d\tau\right]\left[\int_0^\infty f_2(x)e^{-sx}\,dx\right] \tag{41}$$

Since (41) is an identity, the original equation, (39), is proven.

Example. Take the inverse Laplace transform of Eq. (21) through the convolution integral technique.

Rewrite (21):

$$X(s) = \frac{1}{s(s+3)^2}$$

Instead of decomposing the function by partial fraction expansion, write it in a product form:

$$X(s) = \frac{1}{s} \cdot \frac{1}{s+3} \cdot \frac{1}{s+3} \tag{42}$$

$$= F_1(s)F_2(s)F_3(s)$$

Find $\mathscr{L}^{-1}\, F_2(s)F_3(s)$ first:

$$F_2(s) = \frac{1}{s+3}$$

$$F_3(s) = \frac{1}{s+3}$$

$$f_2(t) = e^{-3t}$$

$$f_3(t) = e^{-3t}$$

and

$$\mathscr{L}^{-1}\left[\frac{1}{(s+3)^2}\right] = \int_0^t e^{-3\tau} e^{-3(t-\tau)}\, d\tau$$

$$= e^{-3t} \int_0^t d\tau$$

$$= e^{-3t}(\tau)\Big|_0^t$$

$$= te^{-3t} \tag{43}$$

Then find $x(t)$.

$$\mathscr{L}^{-1}\frac{1}{s(s+3)^2} = \int_0^t \tau e^{-3\tau}\, d\tau$$

$$= \tau \frac{e^{-3\tau}}{-3}\Big|_0^t - \int_0^t \frac{e^{-3\tau}}{-3}\, d\tau$$

$$= t\frac{e^{-3t}}{-3} + \frac{1}{3}\left(\frac{e^{-3\tau}}{-3}\right)\Big|_0^t$$

$$= -\frac{1}{3}te^{-3t} - \frac{1}{9}e^{-3t} + \frac{1}{9} \tag{44}$$

This answer coincides with (25).

Since (44) is a pure integration of the function te^{-3t}, we conclude that multiplying a transfer function $G(s)$ by $1/s$ corresponds to integrating $g(t)$ once.

9. Factorization

Finding the roots of the characteristic equation is an essential step in the solution of a closed-form differential equation.

There are formulae for finding the roots of characteristic equations of the second, third, and fourth order; above this one must resort to a numerical method. The formula for solving a quadratic is easily applied, but the cubic solution is rather long and the quartic is very complicated. It turns out that we usually use a numerical method for all equations above the quadratic.

One well-known method, attributed to Lin, is carried out as follows: Consider the general characteristic equation:

$$s^n + a_{n-1}s^{n-1} + a_{n-2}s^{n-2} + \cdots + a_2s^2 + a_1s + a_0 = 0 \qquad (45)$$

A trial divisor is chosen:

$$s^2 + \frac{a_1}{a_2}s + \frac{a_0}{a_2} \qquad (46)$$

which is the last three terms after normalization. The original equation is divided by this.

$$
\begin{array}{r}
s^{n-2} + \cdots \\
\hline
s^2 + \dfrac{a_1}{a_2}s + \dfrac{a_0}{a_2} \Big) \; s^n + a_{n-1}s^{n-1} + a_{n-2}s^{n-2} + \cdots + a_2s^2 + a_1s + a_0 \\
\end{array}
$$

$$
\begin{array}{r}
b_2s^2 + b_1s + a_0 \;\leftarrow \\
b_2s^2 + c_1s + c_0 \\
\hline
d_1s + d_0
\end{array}
$$

If the remainder is appreciable, then use the next trial divisor and repeat the long division.

$$s^2 + \frac{b_1}{b_2}s + \frac{a_0}{b_2} \qquad (47)$$

Finally, the remainder becomes negligible and we have extracted a quadratic factor from (45).

The following example illustrates the details of Lin's method. Given:

$$s^4 + 15s^3 + 270s^2 + 1600s + 2000 = 0 \qquad (48)$$

The first trial divisor is

$$s^2 + \tfrac{1600}{270}s + \tfrac{2000}{270} = s^2 + 5.93s + 7.4 \qquad (49)$$

Perform the long division:

$$
s^2 + 5.93s + 7.4 \overline{\smash{\big)}\, s^4 + 15s^3 + 270s^2 + 1600s + 2000}
$$

$$
\begin{array}{r}
s^2 + 9.07s + 208.82 \\
\hline
s^4 + 5.93s^3 + 7.4s^2 \\
\hline
9.07s^3 + 262.6s^2 + 1600s \\
9.07s^3 + 53.78s^2 + 67.1s \\
\hline
208.82s^2 + 1532.9s + 2000 \quad \leftarrow \\
208.82s^2 + 1238.3s + 1545.3 \\
\hline
294.6s + 454.7
\end{array}
$$

The second trial division is found.

$$\frac{208.82s^2 + 1532.9s + 2000}{208.82} = s^2 + 7.34s + 9.57 \tag{50}$$

$$
s^2 + 7.34s + 9.57 \overline{\smash{\big)}\, s^4 + 15s^3 + 270s^2 + 1600s + 2000}
$$

$$
\begin{array}{r}
s^2 + 7.66s + 204.21 \\
\hline
s^4 + 7.34s^3 + 9.57s^2 \\
\hline
7.66s^3 + 260.43s^2 + 1600s \\
7.66s^3 + 56.22s^2 + 73.3s \\
\hline
204.21s^2 + 1526.7s + 2000 \quad \leftarrow \\
204.21s^2 + 1498.9s + 1954.3 \\
\hline
27.8s + 45.7
\end{array}
$$

The third trial division is then

$$s^2 + 7.47s + 9.79 \tag{51}$$

Following the operation, finally, we find that (47) is factorized as follows:

$$(s^2 + 7.48s + 9.805)(s^2 + 7.52s + 203.945) = 0 \tag{52}$$

By using the quadratic formula, we get

$$(s + 1.69)(s + 5.79)(s + 3.76 + j13.78)(s + 3.76 - j13.78) = 0 \tag{53}$$

Based on Lin's method, a general digital computer program and its flow-chart are written as follows:

Program 1. FINDING ROOTS OF A POLYNOMIAL BY LIN'S METHOD.

```
C   C    ROOTS OF A POLYNOMIAL...LIN S METHOD

C   FIRST DATA CARD CONTAINS N, THE DEGREE OF THE
C   POLYNOMIAL AND THE TOLERANCE
C   NEXT DATA CARDS CONTAIN THE COEFFICIENTS IN
C   E10.3
        DIMENSION A(11),D(11)
        READ101,N,TOL
101 FORMAT(I4,E10.3)
        NP1=N+1
        READ201,(A(I),I=1,NP1)
201 FORMAT(E10.3)
        PUNCH 300,N,(A(I),I=1,NP1)
 18 K=0
        B=A(N)/A(N-1)
        C=A(N+1)/A(N-1)
        NN=N-1
  5 D(1)=A(1)
        D(2)=A(2)-B*D(1)
        DO 1 I=3,NN
```

```
   1  D(I)=A(I)-B*D(I-1)-C*D(I-2)
      K=K+1
      B1=(A(N)-C*D(N-2))/D(N-1)
      C1=A(N+1)/D(N-1)
      BCHEK=B-B1
      CCHEK=C-C1
      IF(ABS(BCHEK)-TOL)2,2,3
   2  IF(ABS(CCHEK)-TOL)4,4,3
   3  IF(K-100)20,21,21
  20  B=B1
      C=C1
      GO TO 5
  21  PUNCH 60
      PUNCH 75,B,C,B1,C1
      PUNCH 80
      GO TO 22
   4  PUNCH 50,K
  22  DISCR=B*B-4.*C
      IF(DISCR)6,7,7
   7  R1=(-B+DISCR**.5)/2.
      R2=(-B-DISCR**.5)/2.
      PUNCH 100,R1,R2
      GO TO 8
   6  RR=-B/2.
      RI=ABS(DISCR)**.5/2.
      PUNCH 200,RR,RI
   8  N=N-2
      IF (N-1)9,10,11
  11  IF(N-2)9,12,13
  10  FR=-D(2)/D(1)
      PUNCH 100,FR
      GO TO 9
  13  NP1=N+1
      DO 14 I=1,NP1
  14  A(I)=D(I)
      GO TO 18
  12  DISCR=D(2)*D(2)-4.*D(1)*D(3)
      IF (DISCR)15,16,16
  15  RR=-D(2)/(2.*D(1))
      RI=(ABS(DISCR)**.5)/(2.*D(1))
      PUNCH 200,RR,RI
   9  STOP
  16  R1=(-D(2)+DISCR**.5)/(2.*D(1))
      R2=(-D(2)-DISCR**.5)/(2.*D(1))
      PUNCH 100,R1,R2
      STOP
  50  FORMAT(/4HK = I3)
  60  FORMAT(7HK = 1005X14HQUADRATICS ARE)
  75  FORMAT(6HX**2 +E12.5,3HX +E12.5/6HX**2 +E12.5
     13HX +E12.5)
  80  FORMAT(22HROOTS OF FIRST EQ. ARE)
 100  FORMAT(E12.5,6XE12.5)
 200  FORMAT(E12.5,4X1HJE12.5)
 300  FORMAT(9HDEGREE = I2/9HCOEF. ARE/4X5E12.5)
      END
```

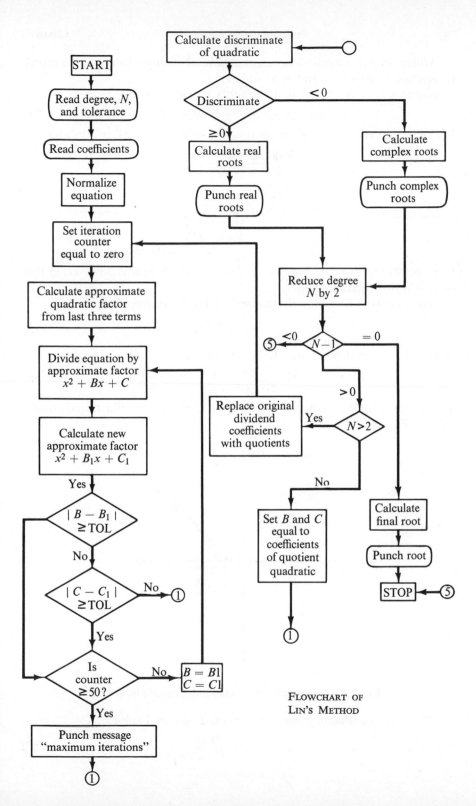

START

Read degree, N, and tolerance

Read coefficients

Normalize equation

Set iteration counter equal to zero

Calculate approximate quadratic factor from last three terms

Divide equation by approximate factor $x^2 + Bx + C$

Calculate new approximate factor $x^2 + B_1x + C_1$

Yes

$|B - B_1| \geq \text{TOL}$

No

$|C - C_1| \geq \text{TOL}$ No   ①

Yes

Is counter ≥ 50?   No $B = B1$ $C = C1$

Yes

Punch message "maximum iterations"

①

Calculate discriminate of quadratic

Discriminate < 0

≥ 0

Calculate real roots

Punch real roots

Calculate complex roots

Punch complex roots

Reduce degree N by 2

⑤ < 0 $N-1$ $= 0$

> 0

Replace original dividend coefficients with quotients   Yes $N > 2$

No

Set B and C equal to coefficients of quotient quadratic

①

Calculate final root

Punch root

STOP   ⑤

FLOWCHART OF LIN'S METHOD

73

Whenever Lin's method does not give a satisfactory solution, we resort to Newton's method, which is as follows:

Suppose it is desired to find the roots of $F(s)$, where

$$F(s) = s^3 + 20s^2 + 600s + 1200 \tag{54}$$

We know that at least one of the roots is real, so we assume $(s + 2)$ as a factor.

$$
\require{enclose}
\begin{array}{r}
s^2 + 18s + 564 \\
s + 2 \enclose{longdiv}{s^3 + 20s^2 + 600s + 1200} \\
\underline{s^3 + 2s^2} \\
18s^2 + 600s \\
\underline{18s^2 + 36s} \\
564s + 1200 \\
\underline{564s + 1128} \\
72 \text{ (remainder)}
\end{array}
$$

If the remainder is zero, then $(s = -2)$ is one of the roots; however, in this case, it is not equal to zero.

The next trial divisor is determined by the following formula:

$$s_1 = s_0 - \frac{F(s_0)}{F'(s_0)} \tag{55}$$

where s_0 is the trial value used before. We have

$$s_1 = (-2) - \frac{(-2)^3 + (20)(-2)^2 + 600(-2) + 1200}{3(-2)^2 + 40(-2) + 600} \tag{56}$$

The reason why (55) is a proper choice for a trial divisor is illustrated graphically in Fig. 2-7.

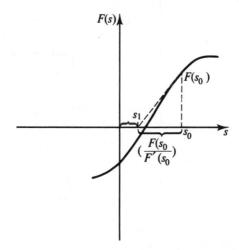

Fig. 2-7. Geometrical interpretation of Newton's formula.

If it is not satisfactory, use s_2 to obtain a new trial value:

$$s_2 = s_1 - \frac{F(s_1)}{F'(s_1)} \tag{57}$$

In general,

$$s_{i+1} = s_i - \frac{F(s_i)}{F'(s_i)} \tag{58}$$

The iterative procedure is illustrated in Fig. 2-8.

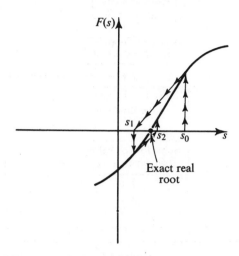

Fig. 2-8. Geometrical interpretation of the iteration procedure of Newton's method.

Newton's method is particularly suitable for programing on a digital computer. A typical program that will solve the above case is as follows:

Program 2. SOLUTION OF AN EQUATION BY NEWTON'S METHOD

```
C   C   SOLUTION OF AN EQUATION BY NEWTON S METHOD
        READ,A,B,C,D
        READ,S
        N=0
      1 FS=A*S*S*S + B*S*S + C*S + D
        DFS = 3.*A*S*S + 2.*B*S + C
        FS=ABS(FS)
        IF(FS-.0000001)3,3,2
      2 S=S-FS/DFS
        N=N+1
        GO TO 1
      3 PUNCH,S,FS,N
        END
     1.   20.   600.   1200.
    -2.
```

```
C   C   SOLUTION OF AN EQUATION BY NEWTON S METHOD
   -2.1358        0.0000              3
```

For this example we have the program and the flow chart.
Figure 2-9 is the flow chart of the simple program.

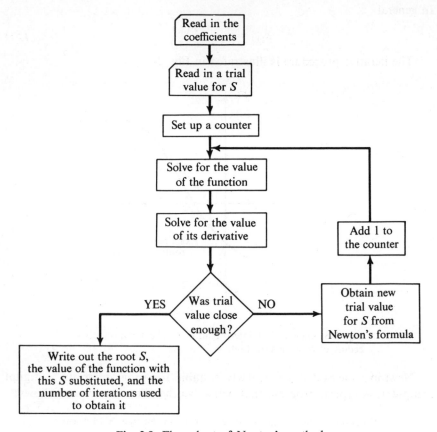

Fig. 2-9. Flow chart of Newton's method.

10. Poles and Zeros

When a transfer function

$$T(s) = \frac{b_0 s^m + b_1 s^{m-1} + \cdots + b_m}{a_0 s^n + a_1 s^{n-1} + a_2 s^{n-2} + \cdots + a_n} \tag{59}$$

is factored as follows:

$$T(s) = K \frac{(s - z_1)(s - z_2)\ldots(s - z_m)}{(s - p_1)(s - p_2)\ldots(s - p_n)} \tag{60}$$

we say that $T(s)$ has a pole at $(s = p_i)$ because

$$T(s)|_{s \to p_i} = \infty, \text{ where } i = 1, 2, \ldots, n$$

and we say that $T(s)$ has a zero at $(s = z_j)$ because

$$T(s)|_{s \to z_j} = 0, \text{ where } j = 1, 2, \ldots, m$$

The values of p_i, and z_j are complex in general and can be located on the s plane. For example, the transfer function

$$T_1(s) = \frac{K(s+3)}{(s+4)(s+1+j2)(s+1-j2)} \tag{61}$$

can be graphically represented by Fig. 2-10. The symbols × and o represent the poles and zeros, respectively. The coordinates of each are equal to its value.

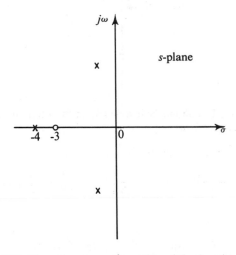

Fig. 2-10. The pole-zero configuration of the function

$$\frac{k(s+3)}{(s+4)\,(s+1+j2)\,(s+1-j2)}.$$

Poles and zeros on the s plane can also be used to represent the inverse Laplace transformation of a system. For example,

$$T_2(s) = \frac{s+A}{s(s^2+B^2)} \tag{62}$$

$$= \frac{s+A}{s(s+jB)(s-jB)} \tag{62a}$$

The pole, zero configuration of $T_2(s)$ is shown in Fig. 2-11. The inverse Laplace transformation of (62a) can be written

$$\mathscr{L}^{-1}T_2(s) = \frac{A}{B^2} - \frac{C}{B^2}\cos{(Bt+\phi)} \tag{63}$$

where ϕ and C are defined in Fig. 2-11.

Another example:

$$T_3(s) = \frac{s+a_0}{(s+\gamma)[(s+\alpha)^2+\beta^2]} \tag{64}$$

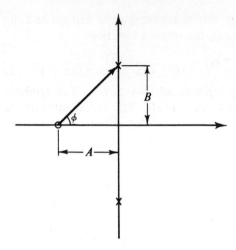

Fig. 2-11. Evaluating the inverse Laplace transform from the pole-zero configuration.

Its inverse Laplace transformation is

$$\mathscr{L}^{-1}T_3(s) = \frac{a_0 - \gamma}{(\alpha - \gamma)^2 + \beta^2}e^{-\gamma t} + \frac{1}{\beta}\left[\frac{(a_0 - \alpha)^2 + \beta^2}{(\gamma - \alpha)^2 + \beta^2}\right]^{1/2} e^{-\alpha t} \sin{(\beta t + \phi)}$$

(65)

where

$$\phi = \tan^{-1}\frac{\beta}{a_0 - \alpha} - \tan^{-1}\frac{\beta}{\gamma - \alpha}$$

The pole-zero configuration is shown in Fig. 2-12 along with the relative distances between them and the angles.

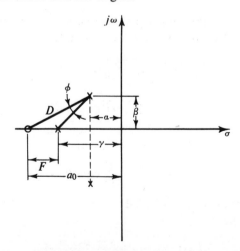

Fig. 2-12. Obtaining the inverse Laplace transform from the distances and angles among poles and zeros.

The inverse Laplace transform can be written as

$$\mathscr{L}^{-1}T_3(s) = \frac{F}{E^2}e^{-\gamma t} + \frac{D}{\beta E}e^{-\alpha t}\sin(\beta t + \phi) \qquad (66)$$

In fact, this representation of the inverse Laplace transformation in terms of β, D, E, F, and ϕ is a graphical interpretation of Heaviside partial fraction expansion. It is not surprising that (65) and (66) are equivalent.

A digital computer program which is based on pole-zero configurations to evaluate inverse Laplace transforms is shown as follows.

Program 3. EVALUATING TRANSIENT RESPONSE FROM MEASUREMENTS
OF POLE AND ZERO LOCATIONS.

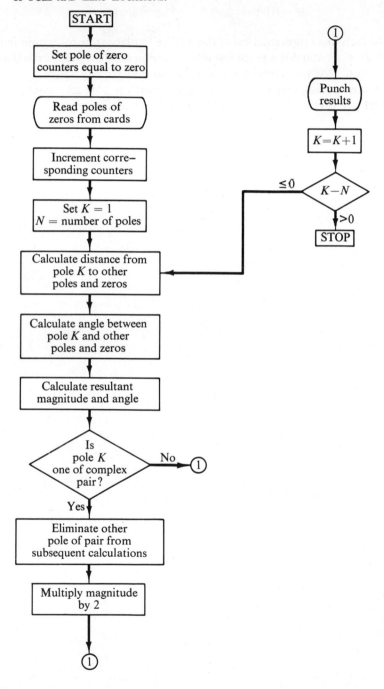

```
C   C TRANSIENT RESPONSE FROM POLE-ZERO LOCATIONS

C       PUNCH EITHER THE WORD POLE OR THE WORD ZERO START
C       ING IN COL. 11.
C   THEN PUNCH THE REAL PART STARTING IN COL. 21 AND THE
C       IMAGINARY IN 41.
C   ON THE LAST CARD PUNCH THE WORD LAST IN 11 AND ZEROS
C       IN 21 AND 41.
        DIMENSION XRP(20),XIP(20),XRZ(20),XIZ(20)
    1   IPOLE = 0
        IZERO=0
        THETA = 0.
        ZMAG = 0
        FACTN = 1.0
        DO 1000 I = 1,20
        XRP(I)=0
        XIP(I) = 0
        XRZ(I) = 0
 1000   XIZ(I) = 0
    2   READ 904, ALPHA, REAL, YIMAG
        IF(ALPHA -4.16263E2) 3,10,3
    3   IF(ALPHA-5.65345E6) 5,4,5
    4   IPOLE = IPOLE+1
        XRP(IPOLE) = REAL
        XIP(IPOLE) = YIMAG
        GO TO 2
    5   IF(ALPHA -4.55956E18)2,6,2
    6   IZERO = IZERO+1
        XRZ(IZERO) = REAL
        XIZ(IZERO) = YIMAG
        GO TO 2
   10   DO 900 J=1,IPOLE
        IF (XIP(J)) 900,15,15
   15   JOINT =0
        TFACT = 1.0
        BFACT = 1.0
        BETA = 0
        PHI = 0
        DO 20 IP = 1,IPOLE
        JP = IP
        IF (J-IP) 16,20,16
   16   IF (XRP(J)-XRP(IP)) 165,160,165
  160   IF (XIP(J)-XIP(IP)) 165, 161, 165
  161   JOINT = JOINT + 1
        GO TO 20
  165   XREAL = XRP(J)-XRP(IP)
        XIMAG = XIP(J)-XIP(IP)
        CALL POLAR
        BFACT = ZMAG*BFACT
        BETA = THETA + BETA
   20   CONTINUE
        IF (J-JP) 23,25,23
   23   IF (JOINT) 25,25,24
   24   CALL MULTI
   25   IF(IZERO) 40,60,40
```

Program 3 (continued)

```
 40 DO 50   IZ = 1, IZERO
    XREAL = XRP(J)-XRZ(IZ)
    XIMAG = XIP(J)-XIZ(IZ)
    CALL POLAR
    TFACT = TFACT * ZMAG
 50 PHI=PHI+THETA
    GO TO 65
 60 TFACT=1.0
    PHI=0.0
 65 IF(XIP(J))   900,200,400
200 ZMAG=TFACT/BFACT
    THETA=PHI-BETA
    CALL RECTI
    IF(XRP(J)) 250,260,250
250 IF (JOINT) 255,255,254
254 LOOK = 1
    GO TO 800
255 PUNCH 901,J,XREAL,XRP(J)
    GO TO 900
260 IF(JOINT) 265,265,261
261 LOOK = 0
    GO TO 800
265 PUNCH 905, J,XREAL
    GO TO 900
400 ZAGK=2.0*TFACT/BFACT
    THETA=PHI-BETA
    IF(JOINT) 410,410,409
409 LOOK=-1
    GO TO 800
410 PUNCH 9025,XIP(J),THETA
    PUNCH 902,J,ZAGK,XRP(J)
    CALL DANG
    PUNCH 903, THETA
    GO TO 900
800 N=JOINT
    CALL FACT
    NFACT=FACTN
    IF (LOOK) 803,804,805
805 PUNCH 907,J,NFACT,FACTN,XRP(J)
    GO TO 900
804 PUNCH 908,J,NFACT,FACTN
    GO TO 900
803 PUNCH 9025, XIP(J), THETA
    PUNCH 909,J,NFACT, FACTN
    CALL DANG
    PUNCH 903, THETA
    GO TO 900
900 CONTINUE
    PUNCH 906
    GO TO 1
901 FORMAT  (6X,3HRF(,I2,5H) =  ,E14.6,7H*H*EXP(,E14.6
   13H*T
902 FORMAT(6X,3HRF(,I2,5H)   =  ,E12.4,8H*H*(EXP(,E12.4
   19H*T))*REMD)
9025 FORMAT(6X,11HREMD = COS(,E12.4,4H*T+(,E12.4,2H)))
903 FORMAT(1HC,5X,8HTHETA=  F10.5,8H DEGREES)
904 FORMAT(10X,A4,6X,F20.8,F20.8)
```

Program 3 (continued)

```
  905 FORMAT(6X,3HRF(,I2,5H)    = ,E12.4,2H*H)
  906 FORMAT(1HC,5X,14HNEXT ANALYSIS.)
  907 FORMAT    (6X,3HRF(, I2, 10H) = H*T**,I2,1H/
     1E12.4,5H*EXP(E12.4,3M*T))
  908 FORMAT    (6X, 3HRF(, I2, 10H)   = H*T**,I2,1H/
     1E12.4
  909 FORMAT    (6X, 3HRF(, I2, 10H)   = H*T**,I2,1H/
     1E12.4,5H*REMD)
C                    SUBPROGRAMS FOR COMPLEX ALGEBRA
     SUBROUTINE FACT
     FACTN = 1.0
     DO 9990  I = 1,N
     F = N
 9990 FACTN=FACTN*F
     RETURN
     SUBROUTINE MULTI
     PUNCH9900,JOINT
 9900 FORMAT (1HC, 5X, 18HORDER OF POLES     I2)
     RETURN
     SUBROUTINE DANG
     THETA = THETA *57.298
     RETURN
     SUBROUTINE POLAR
     IF (ABSF(XREAL) -1.0E-50) 9000,9000,9006
 9000 IF (ABSF(XIMAG) -1.0E-50) 9001, 9001, 9002
 9001 PUNCH 9019
     GO TO 9004
 9002 IF (XIMAG) 9003, 9004, 9005
 9003 THETA = -1.5707963
     GO TO 9017
 9004 THETA = 0.0
     GO TO 9017
 9005 THETA = 1.5707963
     GO TO 9017
 9006 IF (XREAL) 9007, 9008, 9008
 9007 RCHEK   = -1.0
     GO TO 9009
 9008 RCHEK   = 1.0
 9009 IF (XIMAG) 9010, 9011, 9011
 9010 YCHEK = -1.0
     GO TO 9012
 9011 YCHEK = 1.0
 9012 IF(RCHEK + YCHEK)9013,9015,9014
 9013 OMEGA = 3.1415927
     GO TO 9016
 9014 OMEGA = 0.0
     GO TO 9016
 9015 IF(RCHEK)9013,9014,9014
 9016 THETA = ATNF(XIMAG/XREAL) +OMEGA
 9017 ZMAG =SQRTF(XREAL*XREAL   +XIMAG*XIMAG)
 9019 FORMAT (22HINDETERMINATE IN POLAR   )
     RETURN
     SUBROUTINE RECTI
     XREAL  = ZMAG*COS(THETA)
     XIMAG = ZMAG*SIN(THETA)
     RETURN
     END
```

11. Analysis of a Meter.

A voltmeter shown in Fig. 2-13 has moving parts with inertia: $J = 1.25 \cdot 10^{-4}$ newton-meter-sec². The movement is spring restrained with a spring constant $K_s = 0.15$ newton-meter. The back emf constant, $K_v = 0.1$ volt-sec and the torque constant $K_t = 0.1$ newton-meter/amp.

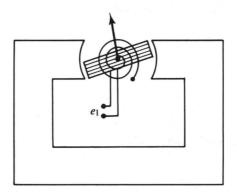

Fig. 2-13. A meter.

Find the equation of meter deflection, θ, when a unit step function, 1v, is applied to the input terminals.

We have to use several physical laws to find the differential equation of the system.

1. The spring must provide a torque proportional to deflection θ:

$$\tau_s = -K_s\theta \tag{67}$$

where $K_s = 0.015$.

2. The back emf is proportional to the angular velocity $\omega = d\theta/dt$:

$$e_b = K_v\frac{d\theta}{dt} \tag{68}$$

where $K_v = 0.1$.

3. The torque developed by the system is proportional to the current.

$$\tau_t = K_t i \tag{69}$$

where $K_t = 0.1$.

The sum of the torques is equal to the inertial force. We may write

$$\tau_t + \tau_s = J\frac{d^2\theta}{dt^2} \tag{70}$$

where $J = 1.25 \cdot 10^{-4}$.

Substitute (67) and (69) into (70).

$$K_t i = J\frac{d^2\theta}{dt^2} + K_s\theta \tag{71}$$

4. For the input part, we write

$$e_1 = Ri + e_b \tag{72}$$

After performing the Laplace transformations on (72), (68), and (71), we simplify to obtain

$$E_1(s) = RI(s) + K_v s\Theta(s) \tag{73}$$

$$K_t I(s) = Js^2\Theta(s) + K_s s\Theta(s) \tag{74}$$

Solving (74) for $I(s)$ and substituting into (73), we have

$$E_1(s) = \left(\frac{RJ}{K_t}s^2 + \frac{RK_s}{K_t} + K_v s\right)\Theta(s)$$

from which

$$\frac{\Theta(s)}{E_1(s)} = \frac{1}{\dfrac{RJ}{K_t}s^2 + K_v s + \dfrac{RK_s}{K_t}} \tag{75}$$

Since $E_1(s) = 1/s$, we have

$$\Theta(s) = \frac{K_t}{RJ} \cdot \frac{1}{s\left(s^2 + \dfrac{K_t K_v}{RJ}s + \dfrac{K_s}{J}\right)} \tag{76}$$

Substituting the given numerical values, we obtain

$$\Theta(s) = \frac{100}{s(s^2 + 10s + 120)} \tag{77}$$

Partial fraction expansion gives

$$\Theta(s) = \frac{\frac{5}{6}}{s} + \frac{0.469\underline{/152.85°}}{s + 5 - j\sqrt{95}} + \frac{0.469\underline{/152.85°}}{s + 5 + j\sqrt{95}} \tag{78}$$

Inverse Laplace transformation yields

$$\theta(t) = 0.833 + 0.93e^{-5t}\cos(9.75t - 2.68) \tag{79}$$

The deflection curve of the meter is shown in Fig. 2-14.

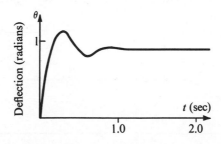

Fig. 2-14. The unit step response of the meter.

IV. Analog Computation

1. Mechanization of Mathematical Operations

Solving an ordinary differential equation with constant coefficients involves the following mathematical operations:
1. Multiplication (or division) by a constant.
2. Addition (or subtraction).
3. Differentiation (or integration).

Is it possible to make a machine which will perform these operations and give some kind of output that will be analogous to the solution of an equation?

Such machines have been built, a gear train being used to perform the operation of multiplication, a differential gear box for subtraction, and a "disk and plate" unit for integration. The last device was much like a planimeter, used by draftsmen to measure areas, since the graphic interpretation of an integral is an area. These machines were called differential analyzers. They were very accurate, but they were also bulky and slow. Since World War II they have been nearly completely replaced by electronic analog computers. These instruments can perform all the foregoing mathematical operations, using high gain amplifiers, resistors, and capacitors.

2. Multiplication By a Constant and Sign Inversion

Consider the circuit shown in Fig. 2-15. The triangle represents a high gain, direct-coupled amplifier. The external resistors have been so connected as to perform the two-fold operation of multiplication by a constant and sign inversion.

Because this is a high gain amplifier, the input potential e_g is very small. (If the output were 100v and the gain were 10^6, the input would be 10^{-4} v.)

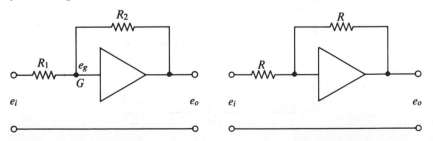

Fig. 2-15. A typical electronic analog circuit.

Fig. 2-16. The sign inverter.

In the following analyses it will be considered zero, making G a virtual ground. It will also be assumed that the input lead to the amplifier is connected to the grid of a vacuum tube or to the gate of a field effect transistor. Then it will be correct to assume that the current flow into the amplifier is zero, and a nodal equation describing the currents at G can be written.

$$\frac{e_g - e_i}{R_1} + \frac{e_g - e_o}{R_2} = 0 \tag{1}$$

which means that $\sum i = 0$ at G (Kirchhoff's first law). Since $e_g = 0$, this can be simplified:

$$\frac{e_o}{e_i} = -\frac{R_2}{R_1} \tag{2}$$

In other words, the output voltage is equal to the input voltage multiplied by a constant $(-R_2/R_1)$. Note the change of sign. The amplifiers are ordinarily so constructed that the output voltage is opposite in sign to the input.

If $R_2 = R_1$, as shown in Fig. 2-16, then Eq. (2) is simplified and

$$e_o = -e_i \tag{2a}$$

This is appropriately called a sign inverter and can be used to restore any sign changes which are undesirable.

3. Addition and Subtraction

If there are two voltages put in, as shown in Fig. 2-17, a nodal equation can be written as follows:

$$\frac{e_1 - e_g}{R_1} + \frac{e_2 - e_g}{R_1} + \frac{e_o - e_g}{R_1} = 0$$

Assuming $e_g = 0$ gives

$$e_1 + e_2 = -e_o \tag{3}$$

This circuit is called an adder. The output voltage is the sum of the two input voltages (with the sign changed). These voltages can take on any values within the range of the amplifiers. They can vary in time in such a way as to

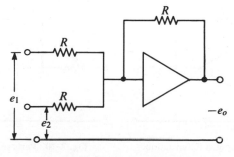

Fig. 2-17. Performing addition.

be analogous to some function whose value varies with respect to some independent variable. Hence, when the voltages are properly scaled, they can be used for analog computation.

A combination of sign inverter and adder will perform subtraction, as shown in Fig. 2-18.

$$e_1 - e_2 = -e_o \tag{4}$$

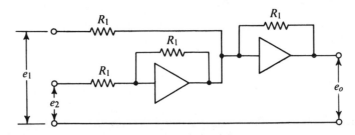

Fig. 2-18. Performing subtraction.

4. Integration and Differentiation

Replace one external resistor by a capacitor, as shown in Fig. 2-19. Again write the nodal equation at G.

$$C\frac{d(e_i - e_g)}{dt} + \frac{e_o - e_g}{R} = 0$$

or

$$e_o = -RC\frac{de_i}{dt} \tag{5}$$

Equation (5) shows that the output voltage is the derivative of the input, multiplied by a constant and with sign changed. (The constant could be unity.) The circuit is called a differentiator and is seldom used when the amplifiers are vacuum tube devices, because of the large amount of noise that is normally present in the output. In some high-quality transistorized analog computers a couple of differentiations in a problem setup are permissible.

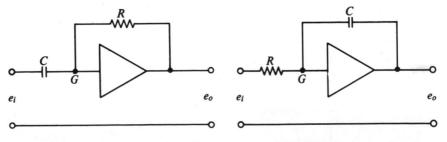

Fig. 2-19. Performing differentiation.   **Fig. 2-20.** Performing integration.

Normally we prefer to rearrange the differential equation so that integration is called for instead of differentiation. Then circuits, called integrators, are employed which do not have the noise problem that is present in the differentiators. Such a circuit is produced by the configuration shown in Fig. 2-20. The nodal equation at G is

$$\frac{e_i - e_g}{R} + C\frac{d(e_o - e_g)}{dt} = 0$$

After neglecting e_g, we have

$$\frac{e_i}{R} = -C\frac{de_o}{dt}$$

or

$$e_o = -\frac{1}{RC}\int e_i\, dt \tag{6}$$

Equation (6) shows that the output voltage of this circuit is the integral of the input voltage, multiplied by a constant (which can be made unity) and with sign reversed.

5. Programming a First Order Differential Equation

As an example, all the necessary steps for solving the equation

$$\frac{dx}{dt} + 4x = 10 \tag{7}$$

will be described.

Step 1. Solve the equation for its highest derivative.

$$\frac{dx}{dt} = 10 - 4x \tag{8}$$

Step 2. Generate the term $-4x$ by integrating dx/dt with the proper constant multiplier: Note that the values of R and C shown in Fig. 2-21 do make the term equal to $-4x$ if the input is dx/dt.

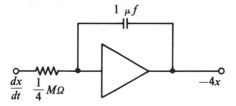

Fig. 2-21. Generating $-4x$ from dx/dt.

Step 3. Complete the right-hand side of (8) by using a summer and a battery. The output of the circuit shown in Fig. 2-22 is $(4x - 10)$.

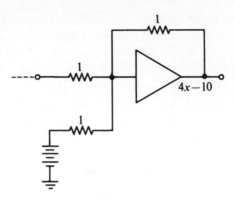

Fig. 2-22. Generating $4x - 10$.

Step 4. A sign inverter is needed.

Step 5. The "equals" sign of (8) is simulated by connecting the output of Fig. 2-23 to the input of Fig. 2-21. The complete setup is shown in Fig. 2-24.

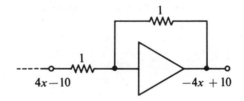

Fig. 2-23. Inverting the sign.

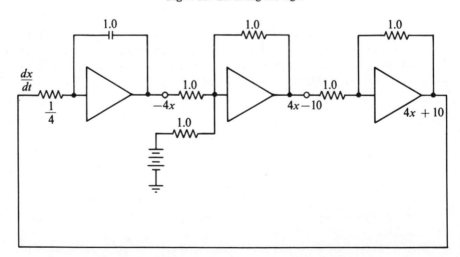

Fig. 2-24. The complete setup for $dx/dt + 4x = 10$.

6. Programming a Higher Order Differential Equation

Consider the solution of the following well-known equation.

$$\frac{d^2y}{dx^2} + B\frac{dy}{dx} + Ky = f(t) \tag{9}$$

Step 1. Solve the equation for the highest order derivative.

$$\frac{d^2y}{dt^2} = f(t) - B\frac{dy}{dt} - Ky \tag{9a}$$

Step 2. Assume that d^2y/dt^2 exists and is located at our starting point. Send it through an integrator to produce a dy/dt term; send this through another integrator to produce a y term. These are shown in Fig. 2-25.

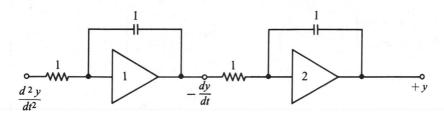

Fig. 2-25. Generating y from d^2y/dt^2.

Step 3. Modify the generated variables by inserting the proper constants (Fig. 2-26).

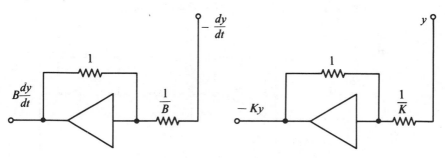

Fig. 2-26. Multiplying variables by constants.

Step 4. Correct the signs by inserting sign inverters where necessary. Figure 2-27 is for this purpose.

Step 5. Obtain $f(t)$ from an appropriate function generator. Machines are available which will produce such functions.

Step 6. Now that every term on the right side of (9a) is available at some point in the circuit and since the sum of these terms is equal to d^2y/dt^2,

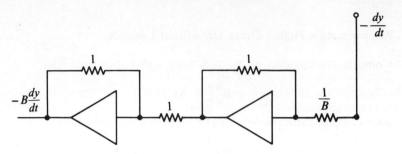

Fig. 2-27. Inverting the sign and modifying the constant.

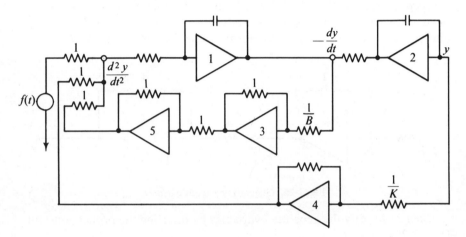

Fig. 2-28. Complete setup for $d^2y/dt^2 + B(dy/dt) + Ky = f(t)$.

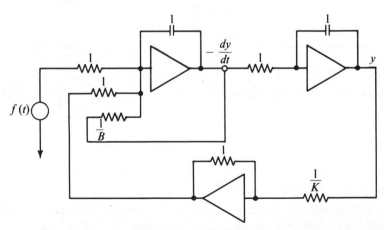

Fig. 2-29. Simplifying the last setup in order to use fewer amplifiers.

we connect these into a summer at the starting point where we originally assumed the second derivative to be. The complete circuit is shown in Fig. 2-28.

Step 7. Examine the circuit and note that (as is usually the case) some sign changers and multipliers can be eliminated without changing the result. In this case, amplifiers 3 and 5 can be eliminated, giving the final program, as shown in Fig. 2-29.

7. Programming a Set of Simultaneous Differential Equations

The method of directly simulating a set of equations will be illustrated with the following example.

$$\frac{d^2x}{dt^2} - 10\frac{dx}{dt} + 3y = 20$$

$$\frac{d^2y}{dt^2} + 5y + 2x = 0 \tag{10}$$

Step 1. Solve for the highest order derivatives for both equations.

$$\frac{d^2x}{dt^2} = 20 + 10\frac{dx}{dt} - 3y$$

$$\frac{d^2y}{dt^2} = -2x - 5y \tag{10a}$$

Step 2. Generate the lower order derivatives by using integrators (Fig. 2-30).

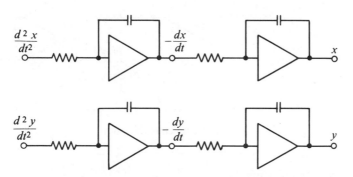

Fig. 2-30. Simulating the two equations separately.

Step 3. Synthesize the right-hand side of (10a), term by term (Fig. 2-31).

Step 4. Complete the circuit by connecting the loops. A final program is shown in Fig. 2-32.

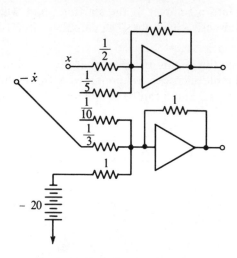

Fig. 2-31. Including some summers.

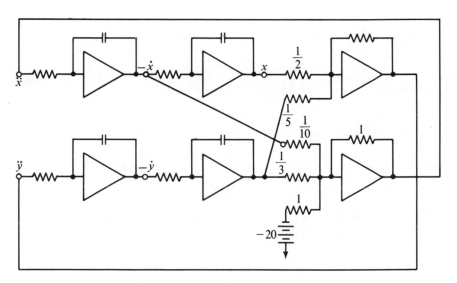

Fig. 2-32. Completing the setup.

8. Introducing Initial Conditions

In all cases so far, we have assumed that the initial conditions were zero. If there are nonzero initial conditions, they will necessitate the application of a specified voltage across the capacitors in the integrators. This can be done with a circuit which is essentially like that shown in Fig. 2-33. The source

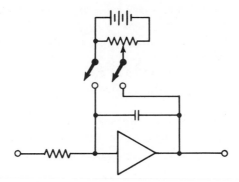

Fig. 2-33. Simulating an initial condition.

voltage is set at the value of the initial condition. The switches are closed and the capacitor is charged to this value. When the problem is "started" the switches are opened. The output of the integrator is then

$$x(t) = -\left[\int_0^t x \, dt + x(0)\right]$$

where $x(0)$ is the initial condition.

9. Summary

In this chapter we deal with the problem of solving ordinary differential equations with constant coefficients.

In the first section some of the basic classical methods are treated. The procedure for finding the general solution of a homogeneous equation is (1) assume an exponential function as a particular solution; (2) solve the characteristic equation; (3) linearly combine the two to produce the required solution.

For finding the general solution of a nonhomogeneous equation, we emphasized the "variation of parameters" by Lagrange. There are several reasons for that.

1. It can be extended quite easily to solving differential equations with time varying coefficients.

2. It can be used as a starting point for a quite natural development of the concept of the convolution integral, one of the most important in analysis.

3. Based entirely on Lagrange's variation of parameters technique, it is possible to evaluate the time response, almost automatically, of any linear system which is described by a vector-matrix differential equation form. This technique will be developed in Chapter 7.

Section 2 of this chapter reviewed Laplace transformations. Relevant theorems were treated. A general Heaviside expansion formula was given and many particular cases were illustrated by examples.

Evaluation of a time response from a pole-zero confiugration was explained by examples.

The last section presented the basic electronic circuits and programing techniques for analog computation. This provides a sound foundation for flow graphs in Chapter 7.

This chapter was written with state space techniques in mind; Chapter 7 might be considered as a further development or natural extension of it.

PROBLEMS

2-1. Using the Laplace transform method, solve each of the following:

(a) $d^2x/dt^2 + x = t$:
$$\begin{cases} x(0) = 1 \\ x'(0) = 2 \end{cases}$$

(b) $d^2x/dt^2 + 3(dx/dt) + 2x = \sin t$:
$$\begin{cases} x(0) = 2 \\ x'(0) = 0 \end{cases}$$

(c) $d^3x/dt^3 + 3d^2x/dt^2 + 3dx/dt + x = t^2e^{-t}$:
$$\begin{cases} x(0) = 1 \\ x'(0) = 0 \\ x''(0) = 0 \end{cases}$$

(d) $d^2x/dt^2 + x = f(t)$:
$$\begin{cases} x(0) = 1 \\ x'(0) = 2 \end{cases}$$

2-2. Solve the above problems, again using Lagrange's method of variation of parameters.

2-3. Find the general solutions of the following differential equations:

(a) $d^2x/dt^2 + 4dx/dt + 4x = e^{-2t}$.

(b) $d^2x/dt^2 = x - y$
$d^2y/dt^2 = y - x + 10$.

(c) $dx/dt = y - 5$
$dy/dt = x + \sin t$.

2-4. Draw an analog computer program for solving each of the equations in Problem 2-3, assuming all initial conditions equal zero.

2-5. Give the inverse Laplace transformations by directly measuring the relative distances and angles of the pole-zero configurations in Fig. P2-5.

2-6. In Fig. P2-6(a) the system is initially at rest. The scissors cuts the string at $t = 0$. Find $x(t)$ and plot the curve. The values of the parameters are as follows: $M = 1$ slug; $W = 50$ lb; $K = 40$ lb/ft; $B = 10$ lb-sec/ft.

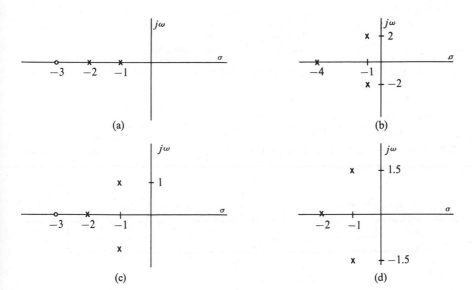

Fig. P2-5

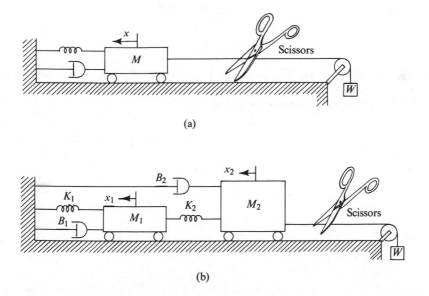

Fig. P2-6

For the system shown in Fig. P2-6(b), find $x_1(t)$ and $x_2(t)$. At $t = 0$ the scissors cuts the string.

$$M_1 = 1 \text{ slug} \qquad M_2 = 2 \text{ slugs}$$
$$B_1 = 20 \text{ lb-sec/ft} \qquad B_2 = 5$$
$$W = 100 \text{ lb} \qquad K_2 = 80 \text{ lb/ft}$$

2-7. Using Newton's or Lin's method solve each of the following.
  (a) $F_1(s) = s^3 + 9.55s^2 + 24s + 18$.
  (b) $F_2(s) = s^4 + 13.5s^3 + 62s^2 + 114s + 72$.
  (c) $F_3(s) = s^4 + 3s^3 + 12s^2 + 13s + 21$.
  (d) $F_4(s) = s^5 + 5.4s^4 + 11.64s^3 + 12.52s^2 + 6.72s + 1.44$.

2-8. Given the feedback system in Fig. P2-8(a) where the twin T network is as shown in Fig. P2-8(b):

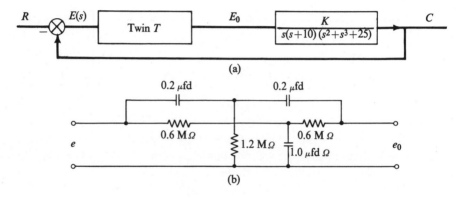

(a)

(b)

Fig. P2-8

  (a) Assume $K = 10$. If a unit step is applied as an input, what would be the corresponding response, $c(t)$?
  (b) What is $e(t)$, the response at the summer?

CHAPTER 3

The Stability Problem

Most control systems that require the attention of control engineers are large, complex, and involve many variables. These factors are related to one another in complicated mathematical models; quantitative answers to problems are obtained by numerical solutions of the model equations.

Besides this approach to the analysis of control systems, engineers need a simplified, qualitative investigation which will yield a "yes or no" type of answer and serve as a guide to their thinking in analysis as well as design.

Will the ship float or sink? Will the missile stay in orbit? Will the machine run or run away? Will the bridge stand or fall? Will the kettle boil or blow up? All these systems can be examined in the light of a generalized and unified theory known as *stability analysis*. All these questions then become one: is the system stable?

I. Basic Concepts

1. Definition of Stability in the Time Domain

A linear system is said to be stable if, and only if, any bounded input produces a bounded output. This definition of stability is based on the cause-effect or input-output point of view in system study.

As an illustration of the definition, Fig. 3-1 shows a system to which a pulse function has been applied and a response has been produced. Both are bounded, so the system is said to be stable.

In Fig. 3-2, the same bounded pulse function is applied to M_2, but the response is seen to be unbounded; therefore, the system is unstable.

99

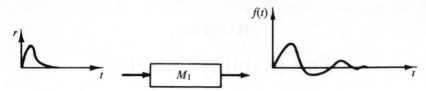

Fig. 3-1. A bounded input and a bounded output.

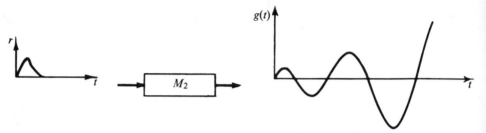

Fig. 3-2. A bounded input and an unbounded output.

2. Unit Impulse and Weighting Function

A special function which is often used as an input in analysis is the "unit impulse." It is rectangular with width a and height $1/a$; therefore, the product or area is unity; the dimension a approaches zero as a limit.

When a unit impulse shown in Fig. 3-3 noted by $\delta(t)$ is applied to a system, the corresponding response function is called $W(t)$, or the weighting function of the system.

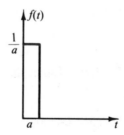

Fig. 3-3. Definition of a unit impulse function; $a \to 0$.

In the Laplace transform domain, this means that

$$[\mathscr{L}\,\delta(t)][M(s)] = [W(s)] \tag{1}$$

Evaluation of $\mathscr{L}\,\delta(t)$ yields

$$\int_0^\infty \delta(t)e^{-st}\,dt = \frac{1}{a}\int_0^a e^{-st}\,dt$$

$$= \frac{1}{-as}\left(1 - as + \frac{a^2 s^2}{2!} - \cdots - 1\right) \doteq 1 \tag{2}$$

Substitution of (2) into (1) gives

$$M(s) = W(s) \qquad (3)$$

Equation (3) shows that the Laplace transform of a weighting function is the transfer function of the system.

Any function multiplied by "unity" is not changed in value. Thus the stability study of a system carried out by applying a unit impulse reduces to a study of the transfer function itself.

3. Definition of Stability in the s Domain

In general, a transfer function such as

$$M(s) = \frac{s^q + b_1 s^{q-1} + b_2 s^{q-2} + \cdots + b_q}{s^n + a_0 s^{n-1} + a_1 s^{n-2} + \cdots + a_n}, \qquad n > q \qquad (4)$$

may be decomposed by Heaviside expansion into the following form:

$$M(s) = \frac{k_1}{s - \gamma_1} + \frac{k_2}{s - \gamma_2} + \cdots + \frac{k_n}{s - \gamma_n} \qquad (5)$$

where γ_i are roots of the characteristic equation or the denominator of (4).

Inverse Laplace transformation gives

$$m(t) = k_1 e^{\gamma_1 t} + k_2 e^{\gamma_2 t} + \cdots + k_n e^{\gamma_n t} \qquad (6)$$

It is obvious that the stability of $M(s)$ depends on the boundedness of $m(t)$, which in turn depends upon the signs of γ_i.

In general, γ_i are complex numbers. If all their real parts are less than zero, the response will die away in time, thereby exhibiting a bounded output and a corresponding stable system. For example,

$$m_1(t) = k_1 e^{-t} + k_2 e^{(-2+j5)t} + k_3 e^{(-2-j5)t} + k_4 e^{-6t} \qquad (7)$$

is a stable response; therefore, $M_1(s)$ is a stable system.

But if one (or more) γ_i has a positive real part, as in (8),

$$m_2(t) = k_a e^{-t} + k_b e^{(2+j8)t} + k_c e^{(2-j8)t} + k_d e^{3t} \qquad (8)$$

the response, as t goes to infinity, will have no bound and M_2 represents a system that is unstable.

Therefore, the study of stability in a linear system has become a study of the sign of the roots of the characteristic equation. Geometrically speaking, this means investigating the location of roots to determine whether all the roots lie in the left half of the s-plane.

4. D(s) plane and s-plane

The roots of a characteristic equation can be found directly, Lin's or Newton's method being used, as discussed in the previous chapter. However, a different and simpler procedure will be presented in this section, which involves the examination of a *certain area* of the *s*-plane to determine if it contains any roots. This is done with the aid of the $D(s)$ plane. When *s* is considered as an independent complex variable, then $D(s)$ is a dependent complex variable.

By way of analogy, consider the relationship between a *real* independent variable *x* and a dependent variable $f(x)$ which can be expressed graphically on a single plane as in Fig. 3-4. On the other hand, a *complex* independent

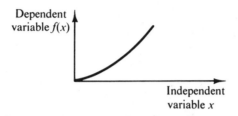

Fig. 3-4. Graphical expression of a real variable.

variable and a complex dependent variable can be expressed graphically also, but the relationship will require two planes: an *s*-plane for the independent variable with its real and imaginary components and a $D(s)$ plane for the dependent variable with its two components.

For every point in the *s*-plane, there will be a corresponding point in the $D(s)$ plane, as in Fig. 3-5(a) and (b).

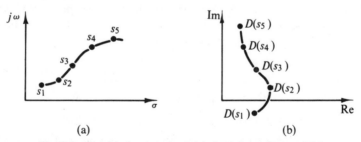

(a)   (b)

Fig. 3-5. Graphical expression: (a) the independent variable plane; (b) the dependent variable plane.

For example, consider $D(s) = s^2 + 3s + 6$. Substitute specific values for *s* and obtain $D(s)$, as shown in the following table and in Fig. 3-5.

s	$0.5 + j0.5$	$0.79 + j1$	$0.97 + j1.5$	$0.85 + j2.0$	$0.35 + j2.5$
$D(s)$	$7.5 + j2$	$7.99 + j4.58$	$7.6 + j7.4$	$5.27 + j9.4$	$0.92 + j9.25$

If $D(s)$ is any *n*th degree polynomial of *s*, it can be factored:

$$D(s) = (s - \gamma_1)(s - \gamma_2)(s - \gamma_3) \ldots (s - \gamma_n) \qquad (9)$$

in which $\gamma_1, \gamma_2, \gamma_3, \ldots, \gamma_n$ are the roots of $D(s) = 0$.

Each factor in (9) can be expressed by a vector as in Fig. 3-6. For each particular value of s, $D(s)$ can be evaluated by using the following graphical measurements and calculations:

$$D(s) = (A_1\underline{/\theta_1})\,(A_2\underline{/\theta_2}) \ldots (A_n\underline{/\theta_n})$$

$$= A_1 A_2 \ldots A_n\underline{/\theta_1 + \theta_2 + \theta_n}$$

$$= A\,\underline{/\theta}$$

where $A_i\underline{/\theta_i}$ = the magnitude and phase angle of the vector $s - \gamma_i$ and
$\quad\quad A$ = the magnitude of the resultant vector
$\quad\quad \theta$ = the resultant phase angle
$A\underline{/\theta}$ gives one point in the $D(s)$ plane.

Now take successive complex values of s as would be the case if s were a point moving in the s-plane along the contour Γ. A corresponding contour C would result in the $D(s)$ plane, as in Fig. 3-7.

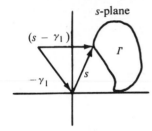

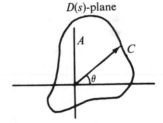

Fig. 3-6. A typical vector $(s - \gamma_1)$ in the independent variable plane.

Fig. 3-7. A corresponding vector $A\angle\theta$ in the dependent variable plane.

5. The Cauchy Theorem

Cauchy established a method of determining how many roots lie in any given closed area of the s-plane. *"The number of roots contained in the contour Γ in the s-plane is equal to the number of encirclements of the origin by the line C in the D(s) plane."*

In Fig. 3-8(a) the values of s substituted in a certain equation trace the closed figure Γ, which happens to contain one root. The corresponding figure traced by $D(s)$ in Fig. 3-8(b) encircles the origin once. This is the meaning of Cauchy's theorem. The contour Γ is said to be mapped onto the $D(s)$ plane as the contour C.

For example, for the equation $D(s) = (s + 1.5)\,[(s + 0.4)^2 + 1]$, substitution of the proper values for s has resulted in the contour Γ in Fig. 3-9. When this is mapped onto the $D(s)$ plane, it results in the contour C of Fig. 3-10.

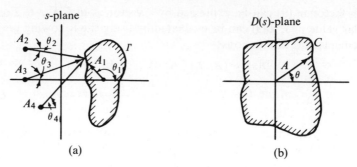

Fig. 3-8. Contour Γ in the s-plane is mapped onto the $D(s)$ plane: (a) the s-plane; (b) the $D(s)$ plane.

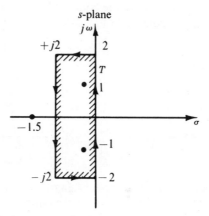

Fig. 3-9. A contour Γ in the s-plane and three roots of $D(s) = (s + 1.5)[(s + 0.4)^2 + 1]$.

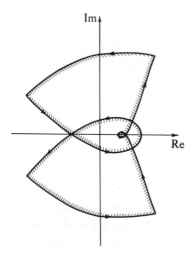

Fig. 3-10. A corresponding contour C in the $D(s)$ plane.

Although the derivation of the Cauchy theorem is based on the complex variable theory, its validity can be established geometrically by noting in Fig. 3-8(a) that the net argument was contributed by A_1/θ_1, with the other vectors contributing nothing. Hence the number of roots in a given region of the s-plane may be counted by finding the number of encirclements of the origin by C in the $D(s)$ plane. In Fig. 3-10 there are two such encirclements; therefore, the equation has two roots within the contour in Fig. 3-9.

6. A Special Contour

The Cauchy theorem implies no restriction on the location of the contour Γ in the s-plane; however, in stability studies the particular contour Γ_0 shown in Fig. 3-11 is of particular interest. Γ_0 extends from $-j\infty$ to $+j\infty$ in a semicircle in the right half of the s-plane. If the characteristic equation of a system has no roots in this region, the system under consideration is stable.

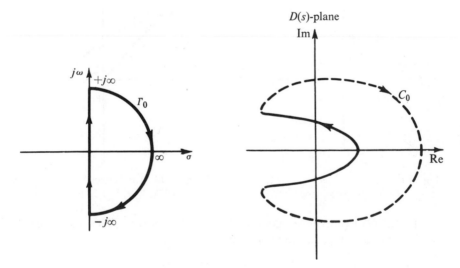

Fig. 3-11. A particular contour Γ_0 is of interest in stability studies.

Fig. 3-12. The corresponding contour C_0 in the $D(s)$ plane. $D(s) = s^2 + 3s + 2$.

As an example, for a system with the following characteristic equation:

$$D(s) = s^2 + 3s + 2$$

the C_0 contour map for the Γ_0 of Fig. 3-11 has been plotted in Fig. 3-12. Since there is no encirclement of the origin, this function, according to the Cauchy theorem, has no root in the right half of the s-plane. Thus the system is stable.

II. The Direct Method of Routh-Hurwitz

1. Hurwitz Polynomial

A polynomial, $D(s)$, with all its roots in the left half of the s-plane is called a Hurwitz polynomial. It has the following property: Let $D_n(s)$ be a Hurwitz polynomial of order n. If $D_n(j\omega)$ is plotted on polar coordinates, the plot will start from a point on the real axis and turn in a counterclockwise direction around the origin, $n\pi/2$ radians or n quadrants.

Example 1. $D_2(s) = s^2 + 3s + 2$ is a Hurwitz polynomial whose roots are -1 and -2, as shown in Fig. 3-13. The corresponding $D_2(j\omega)$ plot is shown in Fig. 3-14.

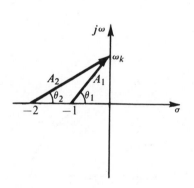

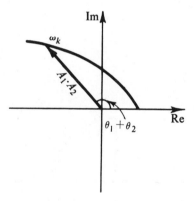

Fig. 3-13. Vectors $A_1\underline{/\theta_1} = j\omega + 1$,
$A_2\underline{/\theta_2} = j\omega + 2$ in the s-plane.

Fig. 3-14. The vector $A_1 \cdot A_2\underline{/\theta_1 + \theta_2}$
$= (j\omega)^2 + 3(j\omega) + 2$ in the $D(s)$ plane.

The $D_2(j\omega)$ plot never goes to the third quadrant. This fact is easily demonstrated if the equation is rewritten as

$$D_2(s) = (s + 1)(s + 2)$$

which becomes

$$D_2(j\omega) = (j\omega + 1)(j\omega + 2)$$

$$= A_1\underline{/\theta_1} \cdot A_2\underline{/\theta_2} \quad \text{in polar form}$$

Choose a particular frequency ω_k at the imaginary axis of the s-plane. Draw two vectors from -1 and -2 to ω_k. $D(j\omega_k)$ is then plotted in the $D(s)$ plane. The magnitude is $A_1 A_2$ and the angle is $\theta_1 + \theta_2$. When ω is varied from 0 to $+\infty$, the corresponding plot $D_2(j\omega)$ is produced. It is evident that θ_1 and θ_2 increase as ω_k increases. However, even when $\omega_k = \infty$, neither θ_1 nor θ_2 is ever greater than 90 deg. In the $D_2(j\omega)$ plane this means that the argument of the plot never exceeds $2(\pi/2)$. In other words, the $D_2(j\omega)$ vector moves around the origin $2\pi/2$ radians or two quadrants at most, as shown in Fig. 3-14.

Example 2. $D_3(s) = s^3 + 1.7s^2 + 0.8s + 0.1$ is a Hurwitz polynomial. The three roots are -0.2, -0.5, and -1.0, as shown in Fig. 3-15. The angle of the $D_3(j\omega)$ vector in the $D_3(s)$ plane $\theta_1 + \theta_2 + \theta_3 \leq 3\pi/2$. The $D_3(s)$ vector never goes into the fourth quadrant, as shown in Fig. 3-16.

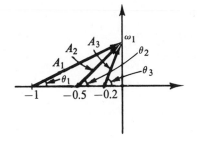

Fig. 3-15. Vectors $(0.2 + j\omega)$, $(0.5 + j\omega,)$ and $(1 + j\omega)$ in the s-plane.

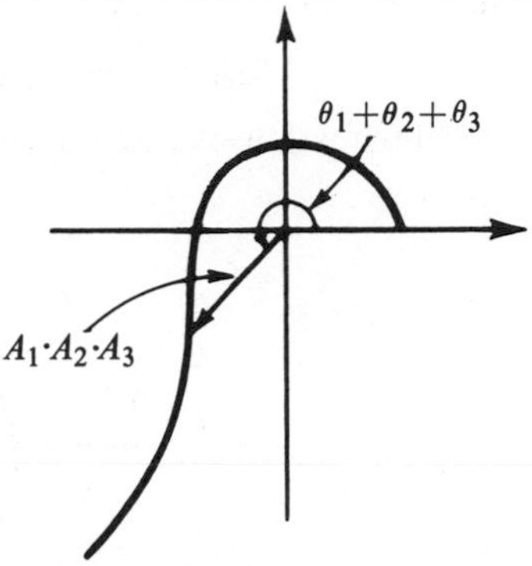

Fig. 3-16. The vector $A_1 \cdot A_2 \cdot A_3 / \theta_1 + \theta_2 + \theta_3$ $= (j\omega)^3 + 1.7(j\omega)^2 + 0.8(j\omega) + 1$.

The following theorem is thus established: *If a Hurwitz polynomial, $D_n(s)$, of order n is plotted with pure imaginary values of s from 0 to $j\infty$, the $D_n(s)$ plane mapping will move around the origin in a counterclockwise direction for a total angle of $n\pi/2$ radians.*

This can be considered as another version of the Cauchy theorem, but the trace for $-j\omega$ is ignored in this one.

The converse of the above theorem is also true: If the plot of a polynomial $D_n(j\omega)$ of degree n is made from 0 to $j\omega$ and it encircles the origin for $n\pi/2$ radians, then $D_n(s)$ is a Hurwitz polynomial.

2. The Mikhailov Criterion

Mikhailov converted this theorem into the nomenclature of control engineers: A linear control system is characterized by

$$D_n(s) = \sum_{n=0}^{n} a_n s^n = 0 \qquad (1)$$

where a_n are real numbers and $s = \sigma + j\omega$ is a complex variable. If this system is to be stable, it is necessary and sufficient that the following condition be satisfied.

$$\text{Arg of } D_n(j\omega) = n\pi/2 \quad \text{when } \omega \text{ varies from 0 to } \infty$$

3. The Stability of a Third Order System

Given a third order system whose characteristic equation is

$$a_0 s^3 + a_1 s^2 + a_2 s + a_3 = 0 \tag{2}$$

the Mikhailov plot of the system is based on

$$a_0(j\omega)^3 + a_1(j\omega)^2 + a_2(j\omega) + a_3 = 0 \tag{3}$$

If it is stable, it will look like the curve in Fig. 3-17.

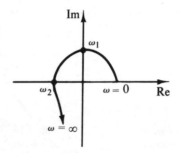

Fig. 3-17. Mikhailov's criterion for a third order system.

The plot will start on the real axis and travel three successive quadrants as ω increases from 0 to ∞.

Assume that the plot crosses the positive imaginary axis when $\omega = \omega_1$ and the negative real axis when $\omega = \omega_2$. The necessary condition for the plot to intersect the positive imaginary axis is

$$- a_0\omega_1^3 + a_2\omega_1 > 0 \tag{4}$$

when

$$a_1(j\omega_1)^2 + a_3 = 0 \tag{5}$$

Solving (5), one obtains

$$\omega_1^2 = \frac{a_3}{a_1} \tag{5a}$$

Substituting (5a) into (4) gives

$$a_2 - a_0 \frac{a_3}{a_1} > 0$$

or

$$a_1 a_2 - a_0 a_3 > 0 \tag{6}$$

A corresponding determinant form is

$$\begin{vmatrix} a_1 & a_0 \\ a_3 & a_2 \end{vmatrix} > 0 \tag{7}$$

with the condition that

$$a_0 > 0, \quad a_1 > 0, \quad a_2 > 0, \quad a_3 > 0 \tag{8}$$

Next, we consider the necessary condition for the plot to cross the negative real axis at ω_2:

$$- a_0 \omega_2^3 + a_2 \omega_2 = 0 \tag{9}$$

$$a_1(j\omega_2)^2 + a_3 < 0 \tag{10}$$

Combining (9) and (10) gives

$$a_1\left(\frac{-a_2}{a_0}\right) + a_3 < 0$$

or

$$\frac{a_1 a_2}{a_0} - a_3 > 0 \tag{11}$$

In determinant form:

$$\begin{vmatrix} a_1 & a_0 \\ a_3 & a_2 \end{vmatrix} > 0 \tag{12}$$

with the following condition:

$$a_0 > 0, \quad a_1 > 0, \quad a_2 > 0, \quad a_3 > 0 \tag{13}$$

(7) and (12) are identical. Therefore, the necessary and sufficient conditions for the stability of a third order system are that Δ_1 and Δ_2 must be greater than zero, where

$$\Delta_1 = a_1$$

$$\Delta_2 = \begin{vmatrix} a_1 & a_0 \\ a_3 & a_2 \end{vmatrix}$$

4. Stability of a Fourth Order System

The characteristic equation of a fourth order system is

$$a_0 s^4 + a_1 s^3 + a_2 s^2 + a_3 s + a_4 = 0 \tag{14}$$

If the system is stable, the Mikhailov plot will look like the curve shown in Fig. 3-18. Call the points of intersection with the axes ω_1, ω_2, ω_3, respectively. By definition, the plot in Fig. 3-18 is produced from

$$a_0(j\omega)^4 + a_1(j\omega)^3 + a_2(j\omega)^2 + a_3(j\omega) + a_4 = 0 \tag{15}$$

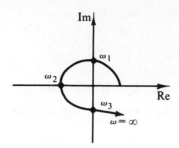

Fig. 3-18. Mikhailov's criterion for a fourth order system.

or by letting

$$a_0\omega_{1,3}^4 - a_2\omega_{1,3}^2 + a_4 = 0 \tag{16}$$

one can solve for ω_1 and ω_3 as follows:

$$\omega_{1,3}^2 = \frac{a_2 \pm \sqrt{a_2^2 - 4a_0a_4}}{2a_0} \tag{17}$$

The necessary condition for stability is

$$a_1(-\omega_1^2) + a_3 > 0 \quad \text{and} \quad a_1(-\omega_3^2) + a_3 < 0 \tag{18}$$

The combination of (17) and (18) gives final condition in determinant form.

$$\begin{vmatrix} a_1 & a_0 & 0 \\ a_3 & a_2 & a_1 \\ 0 & a_4 & a_3 \end{vmatrix} > 0 \tag{19}$$

and

$$\begin{vmatrix} a_1 & a_0 \\ a_3 & a_2 \end{vmatrix} > 0 \tag{20}$$

If ω_2 is used, one obtains the same results shown in (19) and (20). Thus we have the stability criterion of a fourth order system: The quantities $\Delta_1, \Delta_2,$ and Δ_3 must be greater than zero, where

$$\Delta_1 = |a_1|$$

$$\Delta_2 = \begin{vmatrix} a_1 & a_0 \\ a_3 & a_2 \end{vmatrix}$$

$$\Delta_3 = \begin{vmatrix} a_1 & a_0 & 0 \\ a_3 & a_2 & a_1 \\ 0 & a_4 & a_3 \end{vmatrix}$$

5. Hurwitz's Criterion

On the basis of the Mikhailov plot, which is a particular version of the Cauchy theorem, stability criteria for a third and a fourth order system have been derived. The determinant form can easily be extended to higher order systems. Hurwitz derived such a general formula for an nth order system in a manner which can be summarized as follows:

The characteristic equation of an nth order system is

$$a_0 s^n + a_1 s^{n-1} + \cdots + a_{n-1} s + a_n = 0 \tag{21}$$

For the system to be stable, it is necessary and sufficient that the n determinants be positive, where these determinants are taken as the principal minors of the following arrangement.

$$
\begin{array}{cccccc}
a_1 & a_0 & 0 & 0 & 0 & 0 \\
a_3 & a_2 & a_1 & a_0 & 0 & 0 \\
a_5 & a_4 & a_3 & a_2 & a_1 & a_0 \\
\cdot & \cdot & \cdot & \cdot & \cdot & \cdot \\
\cdot & \cdot & \cdot & \cdot & \cdot & \cdot
\end{array}
\tag{22}
$$

In other words, the necessary and sufficient conditions are

$$\Delta_1 = a_1 > 0$$

$$\Delta_2 = \begin{vmatrix} a_1 & a_0 \\ a_3 & a_2 \end{vmatrix} > 0$$

$$\Delta_3 = \begin{vmatrix} a_1 & a_0 & a \\ a_3 & a_2 & a_1 \\ a_5 & a_4 & a_3 \end{vmatrix} > 0, \cdots \tag{22a}$$

and $\Delta_n > 0$, where Δ_n is the entire arrangement of (22).

Examining the conditions of $a_0, a_1, \ldots, a_n$ as Hurwitz did to predicate the stability of a system is sometimes referred to as a *direct* method.

6. Routh's Criterion

An alternate form, derivable from Hurwitz, is called Routh's criterion. Routh formulated a series of $(n + 1)$ polynomials which he derived from the characteristic equation, (21).

$$f_1(s) = a_0 s^n + a_2 s^{n-2} + a_4 s^{n-4} + \cdots$$

$$f_2(s) = a_1 s^{n-1} + a_3 s^{n-3} + a_5 s^{n-5} + \cdots$$

$$f_3(s) = b_1 s^{n-2} + b_3 s^{n-4} + \cdots$$
$$f_4(s) = c_1 s^{n-3} + c_3 s^{n-5} + \cdots \tag{23}$$
$$\cdots\cdots\cdots\cdots\cdots\cdots\cdots\cdots\cdots$$
$$f_{n-1}(s) = l_1 s^2 + a_n$$
$$f_n(s) = m_1 s$$
$$f_{n+1}(s) = a_n$$

The functions $f_3 \ldots f_{n+1}$ are called subsidiary functions. The variable s is customarily omitted, and the array of coefficients alone is called the Routh array. The elements of the third row, fourth row, etc, are obtained by a cross-multiplication process.

$$b_1 = \frac{a_1 a_2 - a_0 a_3}{a_1}, \qquad b_3 = \frac{a_1 a_4 - a_0 a_5}{a_1}$$
$$c_1 = \frac{b_1 a_3 - a_1 b_3}{b_1}, \qquad c_3 = \frac{b_1 a_5 - a_1 b_5}{b_1} \tag{24}$$

The process is repeated $n - 1$ times, giving $n + 1$ rows, as in the array, with the last row containing only a single element a_n. *For the system to be stable, it is necessary and sufficient that each term of the first column of* (23) *is positive, if $a_0 > 0$.* This corresponds to positiveness in the Hurwitz determinants.

Again consider a fourth order system

$$a_0 s^4 + a_1 s^3 + a_2 s^2 + a_3 s + a_4 = 0$$

Construct its Routh array:

$$
\begin{array}{ll}
f_1 = & a_0 \qquad\qquad\qquad\quad a_2 \qquad\qquad\qquad a_4 \\[4pt]
f_2 = & a_1 \qquad\qquad\qquad\quad a_3 \\[4pt]
f_3 = & b_1 = \dfrac{a_1 a_2 - a_0 a_3}{a_1}, \quad a_4 \\[10pt]
f_4 = & c_1 = \dfrac{b_1 a_3 - a_1 a_4}{b_1} \\[10pt]
f_5 = & a_4
\end{array}
$$

It is seen that

$$b_1 = \frac{\begin{vmatrix} a_1 & a_0 \\ a_3 & a_2 \end{vmatrix}}{a_1}$$

and

$$c_1 = \frac{b_1 a_3 - a_1 a_4}{b_1} = a_3 - \frac{a_1^2 a_4}{\begin{vmatrix} a_1 & a_0 \\ a_3 & a_2 \end{vmatrix}}$$

$$= \frac{\begin{vmatrix} a_1 & a_0 \\ a_3 & a_2 \end{vmatrix} a_3 - a_1^2 a_4}{\begin{vmatrix} a_1 & a_0 \\ a_3 & a_2 \end{vmatrix}}$$

$$= \frac{\begin{vmatrix} a_1 & a_0 & 0 \\ a_3 & a_2 & a_1 \\ 0 & 0 & a_3 \end{vmatrix}}{\begin{vmatrix} a_1 & a_0 \\ a_3 & a_2 \end{vmatrix}}$$

It is noted that the elements of the first row of the Routh array can be interpreted in terms of the Hurwitz determinants.

Hurwitz developed the criterion of stability in 1893; Routh established the array criterion in 1877. Evidently, Hurwitz did not know that the criterion had already been established by Routh.

Between these two dates, in the year 1892, another direct method was developed by a Russian named Liapunov. His direct method is more general and can be applied to nonlinear systems. The philosophy, the constructions, and the applications of Liapunov's direct method will be explained in detail in Chapter 8.

7. Example 1

Consider the characteristic equation of a system

$$s^4 + 5.3s^3 + 9s^2 + 9.3s + 9 = 0$$

The Routh array is as follows

1	9	9
5.3	9.3	
7.246	9	
2.718		
9		

The numbers of the first column are all positive. The system is stable. The Hurwitz criterion can also be used for the problem

$$\begin{vmatrix} 5.3 & 1 & 0 \\ 9.3 & 9 & 5.3 \\ 0 & 9 & 9.3 \end{vmatrix}$$

$$\Delta_1 = 5.3 > 0$$

$$\Delta_2 = \begin{vmatrix} 5.3 & 1 \\ 9.3 & 9 \end{vmatrix} > 0$$

and

$$\Delta_3 = \begin{vmatrix} 5.3 & 1 & 0 \\ 9.3 & 9 & 5.3 \\ 0 & 9 & 9.3 \end{vmatrix} > 0$$

8. Example 2

A system is shown in Fig. 3-19. The gain K is adjustable. Find the value of K which will put the system on the borderline of instability.

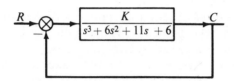

Fig. 3-19. A feedback system.

The characteristic equation of the over-all system function is

$$s^3 + 6s^2 + 11s + 6 + K = 0$$

Forming the Hurwitz determinants and meeting the requirement, we obtain

$$\begin{vmatrix} 6 & 1 \\ 6 + K & 11 \end{vmatrix} > 0$$

$$66 - 6 - K > 0$$

$$K < 60$$

9. Double Subscript Notation for the Routh Criterion

It is more convenient to adopt the double subscript notations to form the Routh array.

The original Routh array is rewritten as

$$
\begin{array}{cccc}
a_0 & a_2 & a_4 & \cdots \\
a_1 & a_3 & a_5 & \cdots \\
b_1 & b_2 & \cdot & \\
c_1 & c_2 & \cdot & \\
d_1 & d_2 & \cdot &
\end{array}
$$

Program 4. Using Routh's Criterion to Determine Stability

```
C   C   ROUTH HURWITZ CRITERION IN ARRAY FORM

C       CHARACTERISTIC EQUATION 3 TO 10 DEGREE
        DIMENSION A(11,6)
        READ,N
        GO TO (1,2,1,2,1,2,1,2,1,2),N
1       M=(N+1)/2
        READ,((A(J,K),K=1,M),J=1,2)
        I=N+1
        DO 3 J=3,I
3       A(J,M)=0.
        GO TO 5
2       M=(N+2)/2
        READ,(A(1,K),K=1,M)
        I=M-1
        READ,(A(2,K),K=1,I)
        I=N+1
        DO 4 J=2,I
4       A(J,M)=0.
5       I=M-1
        L=N+1
        DO 6 J=3,L
        DO 6 K=1,I
6       A(J,K)=A(J-2,K+1)-(A(J-2,1)*A(J-1,K+1))/A(J-1,1)
        DO 15 J=1,L
15      PUNCH,(A(J,K),K=1,M)
        DO 9 J=1,L
        IF (A(J,1))8,9,9
9       CONTINUE
        GO TO 10
8       PUNCH 50
        GO TO 14
10      DO 11 I=1,L
        IF(A(I,1))8,12,11
11      CONTINUE
        PUNCH 70
        GO TO 14
12      PUNCH 60
14      STOP
50      FORMAT(//18HSYSTEM IS UNSTABLE)
60      FORMAT(//21HSYSTEM IS OSCILLATORY)
70      FORMAT (//16HSYSTEM IS STABLE)
        END
```

It is changed to a new notation as follows

$$
\begin{array}{lll}
A_{11} & A_{12} & A_{13} \quad \cdots \\
A_{21} & A_{22} & A_{23} \quad \cdots \\
A_{31} & A_{32} \\
A_{41} & A_{42} \\
A_{51} & A_{52} \\
\;\vdots & \;\vdots \\
A_{n,1} \\
A_{n+1,1}
\end{array}
$$

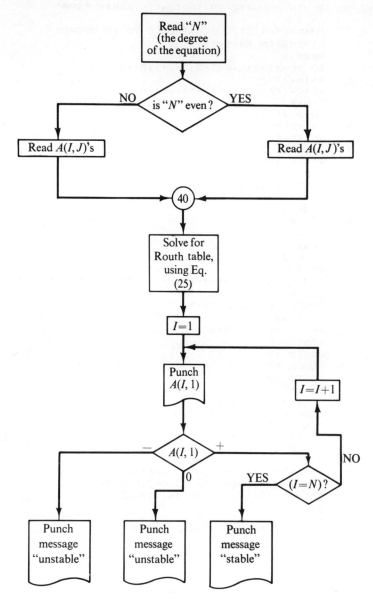

Fig. 3-20. Flow chart for Routh's criterion.

When n is even, a_n and a_{n-1} will appear at

$$a_n = A_{1, n/2+1} = A_{3, n/2} = \cdots = A_{n+1, 1}$$

$$a_{n-1} = A_{2, n/2}$$

and for n odd,

$$a_n = A_{2,(n+1)/2} = A_{4,(n-1)/2} = \cdots A_{n+1,1}$$

$$a_{n-1} = A_{1,(n+1)/2}$$

In terms of this new notation, the relations given for b_1, b_3, c_1, c_3, ... then correspond to the following single formula:

$$A_{jk} = A_{j-2,k+1} - \frac{A_{j-2,1}A_{j-1,k+1}}{A_{j-1,1}} \tag{25}$$

When a digital computer is used for evaluating the Routh array, (25) is particularly helpful. A program is written. It is for a characteristic equation up to 10 degrees. The corresponding flow chart is also shown in Fig. 3-20.

III.   The Root Locus Method of Evans

1. Introduction

From the direct method of Hurwitz, it is seen that the stability of a system depends on the coefficients of its characteristic equation. If any one of the coefficients varies, the roots of the equation change. Geometrically

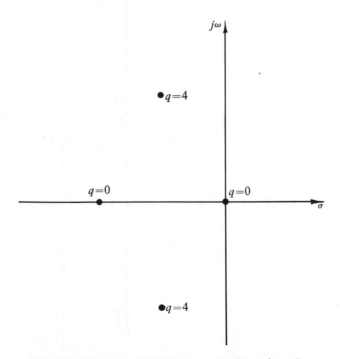

Fig. 3-21. The roots of $s^2 + 2s + q = 0$ with various q's.

speaking, this means that the original positions of the roots in the s-plane move to some new locations. To study the nature and the behavior of a system by investigating these corresponding movements of the roots of the characteristic equation in the s-plane (as a coefficient or a parameter varies) is a very powerful and comprehensive approach to the analysis of linear systems.

Consider a second order system such as

$$M_1(s) = \frac{1}{s^2 + 2s + q} \tag{1}$$

where q is a coefficient or a parameter. If q is changed from $q = 0$ to $q = 4$, the roots of the characteristic equation, originally lying on the real axis, move to two new positions on a vertical line in the s-plane, as shown in Fig. 3-21. If several values of q are given and the corresponding positions of the roots found, we obtain a locus as shown in Fig. 3-22.

Equation (1) can be physically interpreted as a mass-dashpot-spring

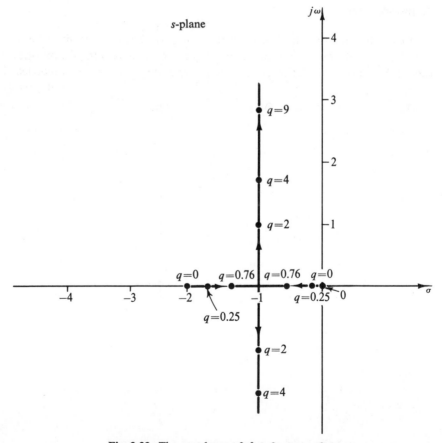

Fig. 3-22. The root locus of $s^2 + 2s + q = 0$.

system. The variations of q correspond to changes in the elastance of the spring. The effects of these changes of q can be found by analyzing the new positions of the roots in Fig. 3-22.

2. System with Feedback

One of the most important systems to the control engineer is the type of feedback system shown in Fig. 3-23. It is represented by

$$M(s) = \frac{KG(s)}{1 + KG(s)H(s)} \tag{2}$$

where $G(s)$ and $H(s)$ are defined as usual and the gain K can be adjusted.

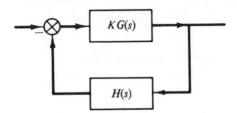

Fig. 3-23. A typical feedback system.

To vary K and to study the corresponding effects on the positions of the roots of

$$1 + KG(s)\,H(s) = 0 \tag{3}$$

is called the root locus method. In other words, Eq. (3) is an equation of complex variables, and the values of s which satisfy (3) constitute the root locus of the feedback system.

3. Basic Equations of Evans

To solve (3) is not very difficult. Since (3) is generally a higher order s polynomial, one needs only to know how to factor an algebraic higher order polynomial in order to solve (3). In the previous chapter, this was demonstrated with (1) Newton's method, (2) Lin's method, and others.

However, for this particular feedback system with K as a varying parameter, Evans developed a graphical solution which is easier and also presents a much clearer picture than the algebraic approach.

Evans' construction is based on two equations which are derived from (3). In other words,

$$1 + KG(s)\,H(s) = 0 \tag{3}$$

is converted into the following:

$$\begin{cases} |KGH| = 1 & (4) \\ \underline{/KGH} = (2k + 1)\,180° & (5) \end{cases}$$

where $k = 0, 1, 2, \ldots$.

Equations (4) and (5) imply that the roots of $1 + KGH = 0$ are those values of s which make the magnitude of KGH equal to 1 and the phase angle of KGH equal to $(2k + 1)\,180$ deg. Equations (4) and (5) can be considered as an alternate form of (3).

Example. A third order feedback system is shown in Fig. 3-24.

$$\frac{C}{R} = \frac{K}{(s + 1)(s + 3)(s + 6) + K} \tag{6}$$

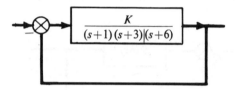

Fig. 3-24. An example illustrating the two basic equations.

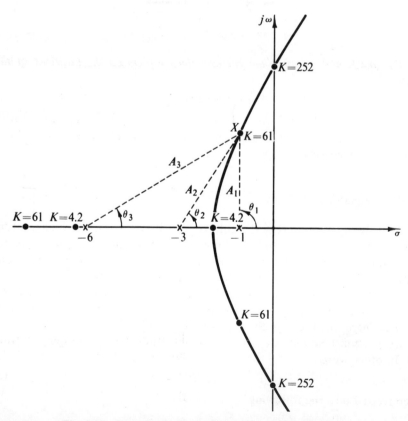

Fig. 3-25. An example illustrating the two basic equations.

One of the numerical methods can be used to factor out the characteristic equation for a given value of K.

Arbitrarily pick up a point X on the locus, at which $K = 61$ (Fig. 3-25); substitutions of the coordinates of X into (4) and (5) yield

$$|KGH| = \left| \frac{1}{A_1 A_2 A_3} \cdot 61 \right| = 1$$

and

$$\underline{/KGH} = \theta_1 + \theta_2 + \theta_3 = 180°$$

which mean that the value of X satisfies the two basic equations (4) and (5).

A logical question follows: Can one use the basic equations (4) and (5) to construct the locus without finding the roots numerically? The answer is affirmative. As a matter of fact, Evans developed several rules to construct the root locus of such a system.

4. Standard Expressions of the Open Loop Transfer Function

Seven rules form a procedure in constructing a root locus for a feedback system. They are either induced or extended from the two basic equations mentioned above, and they can be considered as an alternate form of the set of two basic equations.

Before deriving the rules in detail, we will first standardize the expression of KGH. The open loop transfer function of a feedback system can be expressed in any of the following three forms:

First: $\quad KGH = \dfrac{K(s^m + b_{m-1}s^{m-1} + \cdots + b_1 s + b_0)}{s^n + a_{n-1}s^{n-1} + \cdots + a_1 s + a_0}$ \hfill (7)

Second: $\quad KGH = \dfrac{k(T_1 s + 1)(T_2 s + 1) \ldots (T_m s + 1)}{(T_1 s + 1)(T_2 s + 1)(T_3 s + 1)(T_4 s + 1) \cdots (T_n s + 1)}$ \hfill (8)

Third: $\quad KGH = \dfrac{K(s + \beta_1)(s + \beta_2) \ldots (s + \beta_m)}{(s + \alpha_1)(s + \alpha_2) \ldots (s + \alpha_n)}$ \hfill (9)

Of course, (7), (8), and (9) are convertible. Evans' root locus technique, however, is based on (9). If the open loop transfer function of a system is given by (7) or (8), it should be converted into (9) before using Evans' technique. Thus, (9) is called the standard open loop transfer function for the root locus method.

The expression (9) has several advantages, one being that it has a proper graphical interpretation. Consider a typical factor:

$$(s + \beta_1)$$

It is easily interpreted as a vector in the s-plane, as shown in Fig. 3-26. A vector is written in polar form,

$$(s + \beta_1) = A_1 \underline{/\theta_1}$$

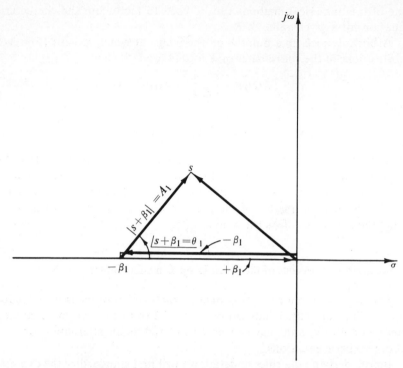

Fig. 3-26. Constructing a typical vector $(s + \beta_1)$.

where A_1 and θ_1 are the magnitude and phase angle of the vector $(s + \beta_1)$, respectively. Then (9) is converted into the following:

$$KGH = \frac{KB_1\underline{|\phi_1}\, B_2\underline{|\phi_2} \ldots B_m\underline{|\phi_m}}{A_1\underline{|\theta_1}\, A_2\underline{|\theta_2} \ldots A_n\underline{|\theta_n}}$$

$$= \frac{KB_1 B_2 \ldots B_m}{A_1 A_2 A_3 \ldots A_n}\underline{|(\phi_1 + \phi_2 + \cdots + \phi_m) - (\theta_1 + \theta_2 + \cdots + \theta_n)}$$

The two basic equations, (4) and (5), are then changed to the following forms:

$$\left| \frac{KB_1 B_2 \cdots B_m}{A_1 A_2 A_3 \cdots A_n} \right| = 1 \tag{4a}$$

$$(\phi_1 + \phi_2 + \cdots + \phi_m) - (\theta_1 + \theta_2 + \cdots + \theta_n) = (2k + 1)180° \tag{4b}$$

(4a) and (4b) are called the polar form of Evans' basic equations.

5. Auxiliaries for Drawing a Root Locus

In drawing the root locus for a feedback system by means of Evans' method, one has to find some auxiliary points and construct several auxiliary lines, which form a basis for sketching the locus rapidly.

Evans named the auxiliary points and auxiliary lines. A typical example is used here to define those names as well as to explain their meanings.

A given system is shown in Fig. 3-24. By assigning successive values to K and marking the corresponding roots on the phase plane, we obtain a plot as in Fig. 3-27. Certain regularities can be observed.

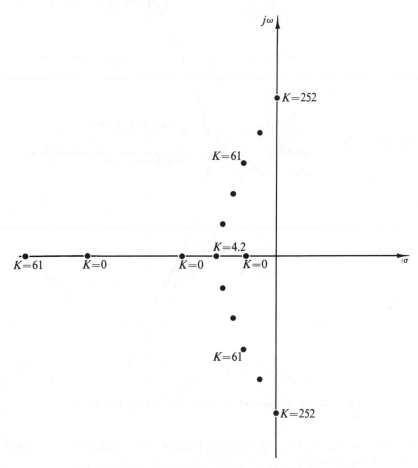

Fig. 3-27. Roots distribution of the example system with different gains indicated.

First, we can join points to make several branches such as the heavy lines in Fig. 3-28. Some portions of these usually can be approximated by straight lines. We then define the following:

1. The branches which lie on the real axis are "real-root branches."
2. The branches which extend to the complex region are "complex-root branches."

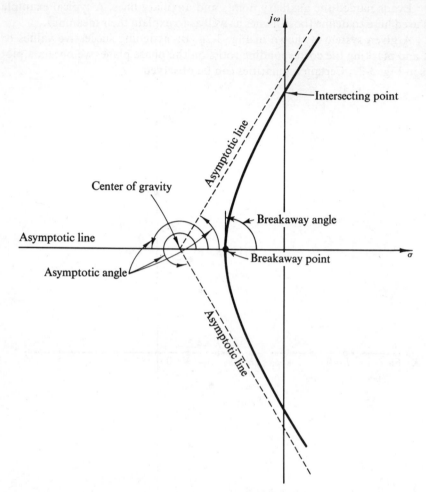

Fig. 3-28. Illustrations of the terms used in the root locus method.

3. The common point of a real-root branch and a complex-root branch is a breakaway point.

4. The points at which a branch and the imaginary axis intersect are intersecting points.

5. The straight lines which can approximate the branches are asymptotic lines.

6. The intersecting point of asymptotic lines and the *real axis* is the center of gravity.

7. The angles between asymptotic lines and the real axis are angles of asymptotic lines.

8. The tangent line at a starting point of a branch that makes an angle with the horizontal is called the departure angle.

All these auxiliary lines, points, and angles are illustrated in Fig. 3-28. Evans' graphical method is to find these auxiliaries first and then construct a root locus based upon them.

To help explain Evan's rules in the following sections, a typical feedback system with the following open loop transfer function is used:

$$KGH = \frac{K(s+4)}{s(s+3)(s+5)(s^2+2s+2)} \tag{10}$$

6. Rule 1. Determining the Starting and Ending Points

The pole-zero configuration of the open loop transfer function is drawn. The branches of root-locus start at each pole of KGH for $K = 0$ and terminate on the zeros of KGH for $K = \infty$.

Figure 3-29 shows the locations of the poles and zeros and some of the branches.

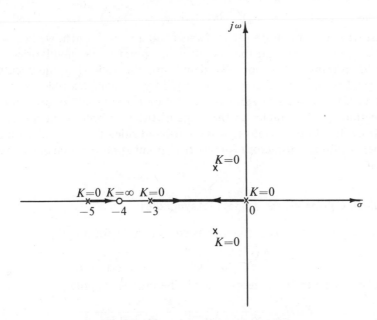

Fig. 3-29. Constructing real-root branches.

The proof of this rule is demonstrated by using the characteristic equation

$$s(s+3)(s+5)(s^2+2s+2) + K(s+4) = 0 \tag{11}$$

For $K = 0$, the second part of the equation is equal to zero, and the roots of the characteristic equation become

$$s(s + 3)(s + 5)(s^2 + 2s + 2) = 0 \qquad (11a)$$

In other words, the locus of $1 + KGH$ starts on the poles of KGH.

Similarly, the ends of the branches are defined. When $K = \infty$

$$s + 4 = 0 \qquad (11b)$$

which means that the branch of the locus of $1 + KGH$ ends at the zeros of KGH.

7. Rule 2. Finding Real-root Branches

Branches of the root locus on the real axis exist only if an odd number of singularities (poles plus zeros) are found to the right of the branch.

Refer to Fig. 3-29. It is seen that between 0 and -3 on the real axis a branch exists and also between -4 and -5, but there is no root on the real axis between -3 and -4, because two (an even number) of the singularities appear to the right of this segment.

In order to prove this rule, (4b) is recalled.

$$(\phi_1 + \phi_2 + \cdots + \phi_m) - (\theta_1 + \theta_2 + \cdots + \theta_n) = (2k + 1)180° \qquad (4b)$$

To test whether a point on the real axis is on a locus branch, we take a trial point on the real axis and then connect this point to all singularities of the open loop transfer function. By definition, the following statements are true: (1) The sum of the angles subtended by the complex poles or zeros is 360 deg. (2) Any poles or zeros to the left of a trial point do not contribute to the sum. (3) The angles to the right of the trial point contribute ±180 deg. It is, therefore, obvious that the left-hand side of (4b) has to be an odd number if (4b) is satisfactory (or the trial point is on the locus). The rule is proven.

8. Rule 3. Calculating the Angles of Asymptotes

If an open loop transfer function is written in the form of (7),

$$KGH = \frac{K(s^m + b_{m-1}s^{m-1} + \cdots + b_1 s + b_0)}{s^n + a_{n-1}s^{n-1} + \cdots + a_1 s + a_0} \qquad (7)$$

where m is the number of zeros and n is the number of poles,

$$KGH = \frac{K}{\dfrac{s^n + a_{n-1}s^{n-1} + \cdots + a_1 s + a_0}{s^m + b_{m-1}s^{m-1} + \cdots + b_1 s + b_0}}$$

$$= \frac{K}{s^{n-m} + (a_{n-1} - b_{m-1})s^{n-m-1} + \cdots + \dfrac{Q(s)}{s^m + b_{m-1}s^{m-1} + \cdots + b_1 s + b_0}} \qquad (12)$$

where $Q(s)$ is a polynomial of degree less than m. The characteristic equation of the system is

$$s^{n-m} + (a_{n-1} - b_{m-1})s^{n-m-1} + \cdots + \frac{Q(s)}{s^m + b_{m-1}s^{m-1}} + K = 0 \qquad (13)$$

When K becomes very large, only the first and last terms are significant, or

$$s^{n-m} + K \cong 0 \qquad (14)$$

or

$$s^{n-m} \cong -K$$

Two equivalent equations can be written:

$$\text{Mag of } s \cong \sqrt[n-m]{K} \qquad (14a)$$

$$\text{Arg of } s \cong \frac{180°(2k+1)}{n-m} \qquad (14b)$$

(14b) gives the angles of the asymptotic lines. When K is large, (14a) and (14b) almost coincide with the exact solutions. We have, therefore, established the following rule: The angles of the asymptotic lines are determined by

$$\frac{180°(2k+1)}{n-m} \qquad (15)$$

where $k = 0, 1, 2, \ldots$, etc., n is the number of poles, and m is the number of zeros of the open loop transfer function.

9. Rule 4. Locating the Center of Gravity

When the magnitude of s is large, Eq. (13) is an $(n - m)$ order polynomial and the sum of the roots is equal to $(a_{n-1} - b_{m-1})$, which is a constant and is independent of gain. In other words, the sum of the roots remains fixed for any value of gain. Hence, the asymptotic lines radiate from this point, which is called the center of gravity.

$$\text{C.G.} = \frac{a_{n-1} - b_{m-1}}{n-m} = \frac{\Sigma \text{ poles} - \Sigma \text{ zeros}}{n-m} \qquad (16)$$

The term "center of gravity" is by way of analogy, a_{n-1} being considered as the moment of poles and b_{m-1} as the moment of zeros with respect to the imaginary axis, and each pole or zero being considered as a positive or negative unit mass, respectively.

For our example, the angles of the asymptotic lines are calculated from (15)

$$\frac{(2k+1)180°}{5-1} = 45°, 135°, 225°, 315° \qquad k = 0, 1, 2, 3$$

and the center of gravity is obtained from substituting the data into (16).

$$\text{C.G.} = \frac{(-1 + j1 - 1 - j1 - 3 - 5) - (-4)}{5 - 1} = -1.5$$

The asymptotic lines are shown in Fig. 3-30.

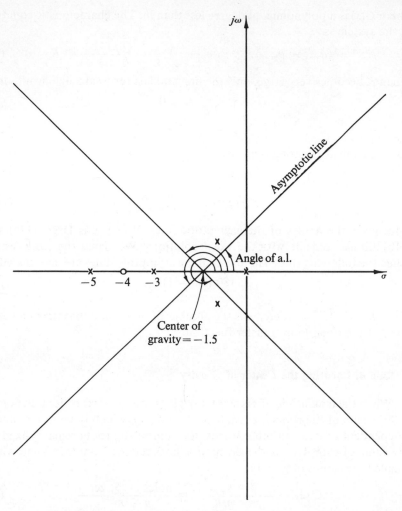

Fig. 3-30. Locating the center of gravity and drawing asymptotic lines.

10. Rule 5. Evaluating the Breakaway Points

By definition, at breakaway points several branches of the locus coalesce; such a point represents the case when the characteristic equation has several roots, or

$$1 + KGH = (s + b)^q A(s) \qquad (17)$$

where $A(s)$ does not contain the $(s + b)$ factor and q is an integer.

The derivative of (17) with respect to s is

$$q(s + b)^{q-1}A(s) + (s + b)^q \frac{d}{ds}A(s) = (s + b)^{q-1}\left[qA(s) + (s + b)\frac{d}{ds}A(s)\right] \quad (18)$$

At the breakaway point, $-b$, (18) should be zero, because $(s + b) = 0$ is a factor of (18) and $A(s)$ cannot be zero by definition.

Hence, the first method for finding a breakaway point is to find the value of s such that

$$\frac{d}{ds}(KGH) = 0 \quad (19)$$

However, Equation (19) is still a very high order polynomial, and to solve it is not very easy. Therefore, we develop the following graphical method. Rewriting the characteristic equation, we have

$$1 + KGH = 0$$

Take the logarithm:

$$\log K + \sum_i \log (s - z_i) - \sum_j \log (s - p_j) = 0$$

where z_i, p_i are zeros and poles of the GH function, respectively.

Taking the derivative with respect to s, we obtain

$$\sum_i \frac{1}{s - z_i} - \sum_j \frac{1}{s - p_j} = 0 \quad (20)$$

Equation (20) is a general necessary condition which must be satisfied at all breakaway points, and it can also be written as

$$\sum \frac{1}{z_i - b} - \sum \frac{1}{p_j - b} = \sum \frac{1}{b - z_m} - \sum \frac{1}{b - p_n} \quad (21)$$

$$\begin{array}{cccc} \uparrow & \uparrow & \uparrow & \uparrow \\ \text{zero to} & \text{poles to} & \text{zeros to} & \text{poles to} \\ \text{the left of} & \text{the left of} & \text{the right of} & \text{the right of} \\ -b & -b & -b & -b \end{array}$$

If complex zeros and complex poles are involved in the open loop transfer function, the breakaway point $-b$ must satisfy the following necessary condition, which can be derived by reasoning similar to the above.

$$\sum_i \frac{1}{z_i - b} - \sum_j \frac{1}{p_j - b} + \sum_k \frac{2(\alpha_k - b)}{(\alpha_k - b)^2 + \beta_k^2} - \sum_l \frac{2(\alpha_l - b)}{(\alpha_l - b)^2 + \beta_l^2}$$
$$= \sum_n \frac{1}{b - z_m} - \sum_n \frac{1}{b - p_n} + \sum_q \frac{2(b - \alpha_q)}{(b - \alpha_q)^2 + \beta_q^2} - \sum_u \frac{2(b - \alpha_u)}{(b - \alpha_u)^2 + \beta_u^2} \quad (22)$$

where the α's and β's are real and imaginary parts of the complex roots. There are i zeros, j poles, k complex zeros, and l pairs of complex poles to the left of the trial point $-b$; there are m zeros, n poles, q pairs of complex zeros and u pairs of complex poles on the right of the trial point $-b$.

The procedure for finding the value of $-b$ can be illustrated by the following two examples.

Example 1. Find the breakaway point of a unit feedback system with an open loop transfer function

$$\frac{K}{(s + 1)(s + 2)(s + 3)} \tag{23}$$

From Rule 2, the breakaway point has to be a point on the real axis between -1 and -2. A trial point $\tilde{b} = -1.5$ is assumed. Substituting the trial value into (21) yields

$$\frac{1}{3 - 1.5} + \frac{1}{2 - 1.5} \overset{?}{=} \frac{1}{1.5 - 1} \tag{24}$$

$$\frac{1}{0.375} \neq \frac{1}{0.5} \tag{24a}$$

We observe that -1.5 is not a good try; however, the next choice of trial number is governed by the information (24a) offered. Take a difference between the denominators of both sides, i.e.,

$$\tilde{\tilde{b}} = \tilde{b} + (+0.5 - 0.375) = -1.5 + 0.125 = -1.375 \tag{25}$$

Substituting $\tilde{\tilde{b}} = -1.375$ into (21) again, we have

$$\frac{1}{3 - 1.375} + \frac{1}{2 - 1.375} \overset{?}{=} \frac{1}{1.375 - 1} \tag{26}$$

$$\frac{1}{0.451} \neq \frac{1}{0.375} \tag{26a}$$

(26a) is not satisfactory either. Again, the difference between the denominators becomes our third trial number, or

$$\tilde{\tilde{\tilde{b}}} = \tilde{\tilde{b}} + (-0.451 + 0.375) = -1.451 \tag{27}$$

We repeat this iterative procedure until we obtain a balanced equation or a satisfactory breakaway value.

Example 2. Now we return to our old example and give the graphical procedure for finding the breakaway points for a system with complex roots.

Reconsider the general necessary condition formula (22) for breakaway points. The typical term contributed by complex roots in (22) is

$$\frac{2(b - \alpha)}{(b - \alpha)^2 + \beta^2} \tag{28}$$

(28) can be written in the form

$$\frac{(b - \alpha)}{\dfrac{\sqrt{(b - \alpha)^2 + \beta^2}}{2} \cdot \sqrt{(b - \alpha)^2 + \beta^2}} \tag{29}$$

In Fig. 3-31, through the middle point of AB, draw a line perpendicular to AB which intersects the real axis at F; (29) becomes

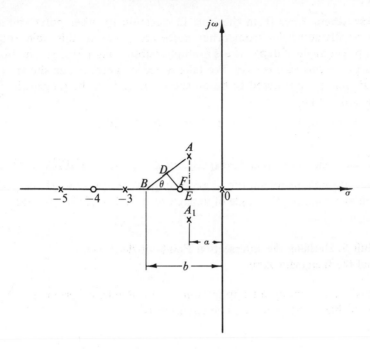

Fig. 3-31. Locating breakaway points.

$$\frac{\overline{BE}}{\overline{BD} \times \overline{AB}} = \frac{1}{\overline{BD}} \times \cos\theta = \frac{1}{\overline{BD}} \times \frac{\overline{BD}}{\overline{BF}} = \frac{1}{\overline{BF}} \qquad (30)$$

An equivalent point F is found. As far as finding the breakaway point is concerned, a pair of complex poles A, A' are replaceable by a real pole at F.

For this example, choose $b = -2.4$.

$$\frac{1}{5 - 2.4} + \frac{1}{3 - 2.4} - \frac{1}{4 - 2.4} \overset{?}{=} \frac{1}{2.4} + \frac{1}{1.05}$$

This condition is satisfactory, and the breakaway point is found to be at $(-2.4, 0)$ of the s-plane. If the first trial value of b were not satisfactory to the necessary condition, we would resort to the iterative procedure explained in the first example.

11. Rule 6. Obtaining the Angle of Departure

The angle of departure of a complex pole is the angle between the tangent line of the branch at the pole in question and the horizontal line through the pole.

In the example problem, θ_5 is a departure angle of the complex pole $(1 + j1)$. The angle of departure from a complex pole can be found by using Eq. (4b). In other words, the rule can be formulated as follows:

Connect several lines from the pole in question to other poles and zeros. Measure all the angles from other poles and zeros to this pole and sum them up. The angle of departure is found by subtracting 180 deg from this sum.

To prove this rule is easy. We take a nearby point P_1, as shown in Fig. 3-32. P_1 can be considered to be on the locus and on the tangent line also. From (4b), we have

$$\phi_1 - (\theta_1 + \theta_2 + \theta_5 + \theta_3 + \theta_4) = 180°(2k + 1) \qquad (31)$$

or

$$\theta_5 = \phi_1 + (-\theta_1) + (-\theta_2) + (-\theta_3) + (-\theta_4) - 180°(2k + 1)$$

For our example problem, the angle of departure at $(-1 + j1)$ is -67 deg. By similar reasoning, the angle of departure at $(-1 - j1)$ is $+67$ deg.

12. Rule 7. Deciding the Intersection Points of the Locus and the Imaginary Axis

This rule is simply an application of the Routh criterion. For the example problem, the characteristic equation is

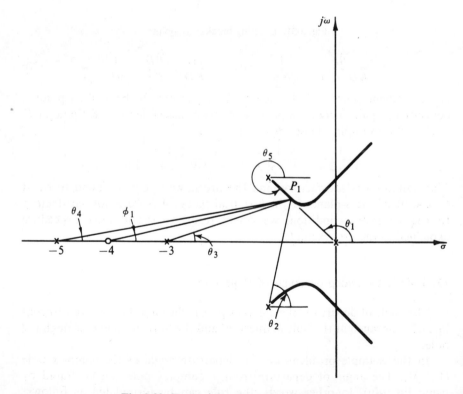

Fig. 3-32. Determining the angles of departure.

$$K(s + 4) + (s^2 + 8s^2 + 15s)(s^2 + 2s + 2)$$
$$= s^5 + 10s^4 + 33s^3 + 46s^2 + (30 + K)s + 4K \tag{32}$$

The corresponding Routh's array is

$$
\begin{array}{lll}
1 & 33 & 30 + K \\
10 & 46 & 4K \\
28.4 & 30 - 0.6K & \\
25.4 - 0.21K & 4K & \\
\dfrac{(30 + 0.6K)(35.4 - 0.211K) - 113.6}{35.4 - 0.211K} & & \\
4K & &
\end{array}
\tag{33}
$$

Solving

$$1.266K^2 - 98.72K + 1062 = 0 \tag{34}$$

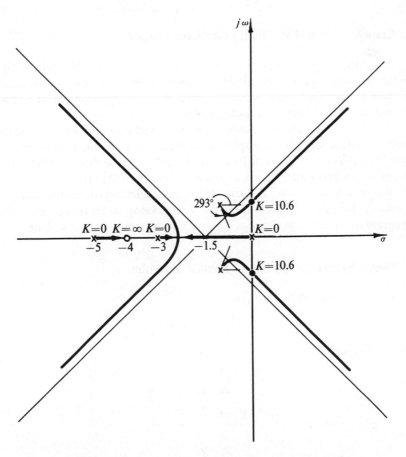

Fig. 3-33. The complete root locus plot of the example problem.

we obtain

$$K = 10.6$$

When the open loop gain K is 10.6, the locus intersects the imaginary axis.

If the plots shown in Fig. 3-29 through Fig. 3-32 are superimposed and combined into one, we obtain the complete root locus plot for the example problem, as shown in Fig. 3-33.

Evans not only derived the graphical rules mentioned above but also invented an instrument for drawing the root locus. The instrument is called a Spriule, which can be considered as a "rotate" (not slide) rule for calculating complex numbers.

IV.   The Root Locus Method of Routh

1. Graphical Method vs. Digital Computer Method

The root locus method was originated by Evans in 1950 before digital computers were put to extensive use in engineering analysis. Today, because of their widespread availability, we must look to the problem of adapting the graphics to a digital computer method.

Of course, this type of problem can be solved numerically by some root-finding routine such as by applying Newton's or Lin's method to the basic defining equation of the root locus. However, it can be done much better by developing a procedure which uses the algorithm of Routh.

We will first review Routh's algorithm, particularly emphasizing his double subscript notation; then we will develop a rearrangement and a program for setting out the root locus of a system on a complex plane.

2. Double Subscript Notation of Routh's Algorithm

Let us review the linear system

$$\sum_{i=0}^{n} a_i s^{n-i} = 0 \tag{1}$$

and its Routh's array:

$$
\begin{matrix}
a_0 & a_2 & a_4 \\
a_1 & a_3 & a_5 \\
b_1 & b_3 & b_5 \\
\vdots & \vdots & \\
g_1 & g_3 & \\
h_1 & &
\end{matrix}
\tag{2}
$$

where

$$b_1 = \frac{a_1 a_2 - a_0 b_3}{a_1}, \qquad b_3 = \frac{a_1 a_4 - a_0 a_5}{a_1} \tag{3}$$

$$c_1 = \frac{b_1 a_3 - a_1 b_3}{b_1}, \qquad c_3 = \frac{b_1 a_5 - a_1 b_5}{b_1} \tag{4}$$

etc.

Since it is more convenient to use the matrix notations for Routh's array, (2) is written as

$$
\begin{array}{ccc}
A_{11} & A_{12} & A_{13} \\
A_{21} & A_{22} & A_{23} \\
\vdots & \vdots & \\
A_{n-1,1} & A_{n-1,2} & \\
A_{n,1} & & \\
A_{n+1,1} & &
\end{array}
\tag{5}
$$

Using this notation, we rewrite the relations in (3) and (4) as

$$A_{j,k} = A_{j-2,k+1} - \frac{A_{j-2,1} A_{j-1,k+1}}{A_{j-1,1}} \tag{6}$$

Equation (6) will be repeatedly used in our method.

3. Condition for Oscillatory Mode

If (1) contains a pair of pure imaginary roots, it can be written as follows:

$$\sum_{i=0}^{n} a_i s^{n-i} = \sum_{i=0}^{n-2} (b_i s^{n-i-2})(s^2 + \omega_0^2) \tag{7}$$

in which the a_i's are known; the b_i's and ω_0^2 are to be determined.

It is easy to show that the a_i's and b_i's are related by

$$a_i = b_i + b_{i-2} \omega_0^2 \qquad i = 0, 1, 2, \ldots, n \tag{8}$$

where the nonexisting coefficients are considered to be zeros.

If we also write Routh's array for

$$\sum_{i=0}^{n-2} (b_i s^{n-i-2})$$

it would have $n - 1$ rows:

$$
\begin{array}{ccc}
B_{11} & B_{12} & B_{13} \\
B_{21} & B_{22} & B_{23} \\
\vdots & & \\
B_{n-2,1} & & \\
B_{n-1,1} & &
\end{array}
\tag{9}
$$

where for n even,

$$b_{n-2} = B_{1,n/2} = B_{3,(n/2)-1} = \cdots = B_{n-1,1}$$

$$b_{n-3} = B_{2,(n/2)-1}$$

and for n odd,

$$b_{n-2} = B_{2,(\overline{n-1}/2)} = B_{4,(\overline{n-3}/2)} = B_{n-1,1}$$

$$b_{n-3} = B_{1,(\overline{n-1}/2)}$$

and

$$B_{j,k} = B_{j-2,k+1} - \frac{B_{j-2,1}B_{j-1,k+1}}{B_{j-1,1}} \tag{10}$$

From (8) and (10) it follows that

$$A_{j,k} = B_{j,k} + B_{j,k-1}\omega_0^2 \tag{11}$$

and

$$\omega_0^2 = \frac{A_{n-1,2}}{A_{n-1,1}} \tag{11a}$$

Also, it can be seen that the necessary and sufficient condition for the relation in (7) to be true is that

$$A_{n,1} = 0 \tag{12}$$

Thus, using (11) and (12), we can find both the conditions for oscillatory modes and the values of the oscillating frequencies.

4. Determination of Pure Imaginary Roots

In the determination of the root loci of a feedback system, the characteristic equation can usually be written in the form

$$\prod_{i=1}^{n} (s - p_i) + K \prod_{j=1}^{m} (s - z_j) = 0 \tag{13}$$

where p_j's and z_j's are the poles and zeros of the open loop system. Expanding (13) and regrouping terms, we write

$$\sum_{i=0}^{n} a_i s^{n-i} = 0, \qquad a_0 \neq 0 \tag{14}$$

In (14), some of the coefficients will be simple functions of the unknown K. Consequently, the necessary and sufficient condition for the oscillatory mode becomes

$$A_{n,1}(K) = 0 \tag{15}$$

The solutions of Eq. (15) yield the values of K, and the substitution of K into Eq. (11a) automatically yields the value of ω_0.

The values of K and ω_0^2 obtained above determine a pair of pure imaginary roots.

5. Invariance Principle of Geometrical Configuration

The configuration of the root loci of (13) is completely defined by the relative positions of the poles and zeros and is independent of any linear coordinate transformation. This principle was developed by Bendrikov and Teodorchik. It is easily demonstrated by comparing root loci before and after shifting the imaginary axis. Thus the exact configuration can be determined by the family of equations which have the form

$$\prod_{i=1}^{n} [s - (p_i - \alpha)] + K \prod_{j=1}^{m} [s - (z_j - \alpha)] = 0 \tag{16}$$

where α is an arbitrary real quantity.

For any given $\alpha = \alpha_0$, Eq. (16) differs from (13) only in that the imaginary axis of the reference complex plane is shifted a distance α_0.

The procedure here is to combine the idea of the invariance of geometrical configuration with Routh's related formulas (4) and (5). By successively shifting the imaginary axis, we can draw a root locus.

The steps can be stated as follows:
1. Shifting the imaginary axis.
2. Forming Routh's array of the shifted equation (16).
3. Determining the critical K and ω_0 from the conditions shown in Eqs. (12) and (11).

The root locus is thus formed by scanning the s-plane.

To illustrate the steps, consider a system with the following open loop transfer function (Fig. 3-34):

$$\frac{K(s + 4)}{s(s^2 + 2s + 2)(s + 3)(s + 5)} \tag{17}$$

Typical computations are
1. Shift the ordinate by letting $s = s' - 4.5$.
2. Form Routh's array for s'.
3. Set $a_n(K) = 0$, and solve for K; then $K = 89.4$, which gives a root on the real axis. Set

$$A_{5,1}(K) = 0, \quad \text{i.e.,} \quad 0.241K^2 - 53.1K - 4748 = 0$$

$$K = 288.90$$

$$\omega_0^2 = 6.417, \qquad \omega_0 = \pm 2.533$$

Thus the roots on the imaginary axis of the s'-plane are determined.

Note that in this example a fifth order, complex-root-finding problem has been reduced to a second order, real-root-finding problem.

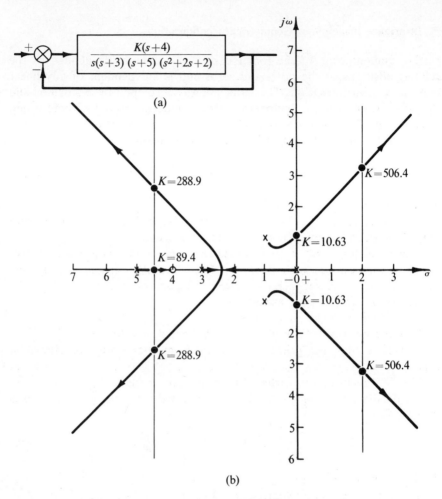

Fig. 3-34. Illustration of the root locus method of Routh:
(a) the system; (b) the root locus plot.

6. Qualitative Analysis

One of the advantages of Evans' graphical method is that it gives a qualitative picture of the dynamic system; that is, the direction of the movement of a root, the gain at which the system is critically damped, the dominant roots, and the region containing particular roots.

The Routh method also gives a qualitative picture. For example, the open loop transfer function of a system is given as follows:

$$\frac{C}{E} = \frac{K}{(s + 1)(s + 3)(s + 6)} \tag{18}$$

The root locus of the system is constructed as shown in Fig. 3-35. Four kinds of regions can be marked as regions A, B, C, and D.

Region A can be located from the Routh array by shifting the imaginary axis to a point in the region. For example, shifting the imaginary axis to the point -5, we have

$$(s' - 5 + 1)(s' - 5 + 3)(s' - 5 + 6) + K = 0 \qquad (19)$$

The corresponding Routh's array is

$$
\begin{array}{cc}
1 & 2 \\
-5 & (8 + K) \\
\left(2 + \dfrac{8 + K}{5}\right) &
\end{array}
\qquad (20)
$$

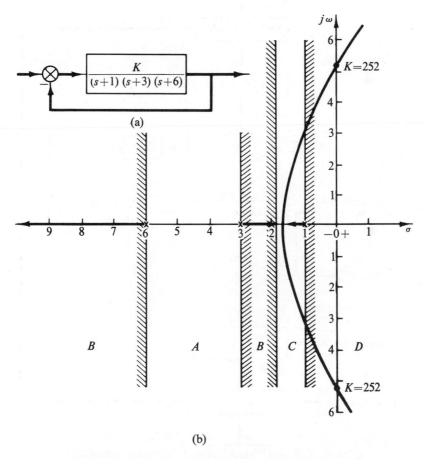

(a)

(b)

Fig. 3-35. Qualitative analysis in the root locus method of Routh: (a) the system; (b) the root locus plot and the various regions.

where $8 + K = a_n$. Clearly, we see that for any positive value of K, it is impossible to make $8 + K$ or a_n equal to zero. This means that there is no real root locus in this region.

By similar reasoning, region B can contain only real roots. Region C can have real and complex roots, whereas region D can contain only complex roots.

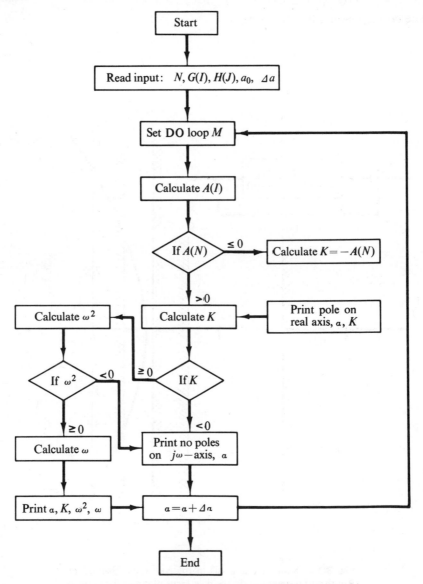

Fig. 3-36. The flow diagram of the example problem.

This demarcation enables us to understand the system better and also makes it easy to program on a digital computer. The flow diagram for such a program is shown in Fig. 3-36.

7. A Particular Case and a Difficult Case

For the feedback system shown in Fig. 3-37, after substituting $s = s' - 0.5$ into the characteristic equation, $s^2 + s + K = 0$, we obtain the shifted equation as follows:

$$s'^2 - 0.25 + K = 0 \qquad (21)$$

The general Routh's array in matrix form reads

$$A_{n-1,1} \quad A_{n-1,2}$$
$$A_{n,1}$$
$$A_{n+1,1}$$

and the corresponding array for the problem is then

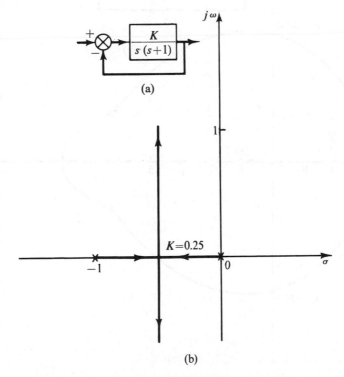

(a)

(b)

Fig. 3-37. A special case: (a) the system; (b) the root locus plot.

$$\begin{array}{cc} 1 & (-0.25 + K) \\ 0 & \\ (-0.25 + K) & \end{array} \tag{22}$$

The condition for the system to have an oscillatory mode is

$$A_{n,1} = 0$$

In this example, $0 = 0$, which means that the whole shifted imaginary axis $s = -0.5$ is a branch of the root locus. The corresponding ω_0 is then obtained from the equation

$$\omega_0^2 = \frac{A_{n-1,2}}{A_{n-1,1}}$$

Substitutions yield

$$\omega_0^2 = \frac{-0.25 + K}{1}$$

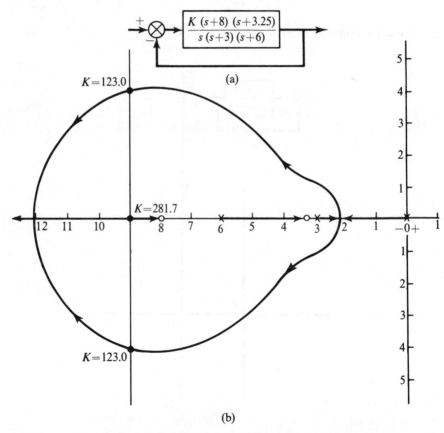

Fig. 3-38. A complicated case: (a) the system; (b) the root locus plot.

or

$$\omega_0 = \sqrt{K - 0.25} \tag{23}$$

Any value of K which is larger than 0.25 would be on the locus, and the positions of the various values of the parameter K are completely defined by (23).

From this particularly simple example, it is noted that this method offers much more information than Evan's method.

Let us try another example, one for which it is not easy to draw the root locus by Evans' technique.

A feedback system has the following open loop transfer function:

$$\frac{K(s + 8)(s + 3.25)}{s(s + 3)(s + 6)} \tag{24}$$

The root locus is drawn by this method as shown in Fig. 3-38. If Evans' method were used it would be very difficult to construct the *pear-shaped* branch of the locus. Practically, there is no rule available for constructing the pear-shaped locus described in Evans' classical work. The Routh method, however, shows the pear-shaped branch in a very detailed manner.

V. The Frequency Method of Nyquist

1. Definition of Frequency Response

For defining the frequency response of a system, without loss of generality, we choose a simple transfer function as follows:

$$M(s) = \frac{1}{s + 1} \tag{1}$$

If a sine wave is applied to the system, the corresponding response function, $C(s)$, should be

$$C(s) = \frac{1}{s + 1} \cdot \frac{\omega}{s^2 + \omega^2} \tag{2}$$

Inverse Laplace transformation yields

$$c(t) = \frac{\omega}{1 + \omega^2} e^{-t} + \frac{1}{\sqrt{1 + \omega^2}} \sin\left[\omega t - \left(\tan^{-1}\frac{\omega}{1}\right)\right] \tag{3}$$

The second term of the right-hand side is usually called the steady-state part of the response, and the first term is called the transient part. This is because the former has values as $t \to \infty$, whereas the latter dies out as $t \to \infty$.

If only the steady-state part is considered, we have

$$c(t)_{\text{steady state}} = \frac{1}{\sqrt{1+\omega^2}} \sin\left[\omega t - \left(\tan^{-1}\frac{\omega}{1}\right)\right] \tag{4}$$

Equation (4) is a sine wave, with an amplitude $A = 1/\sqrt{1+\omega^2}$ and a phase angle $\varphi = -\tan^{-1}\omega/1$. Both A and φ are functions of ω. Equation (4) can be written into many forms; for example, the following are some well-known expressions.

Trignometric form:
$$A \sin(\omega t + \varphi)$$
Polar form:
$$A/\underline{\varphi}, \text{ with the angular velocity } \omega$$
Rectangular form:
$$A(\cos\varphi + j\sin\varphi), \text{ or } U + jV, \text{ with } \omega$$
Exponential form:
$$Ae^{+j\varphi}, \text{ with } \omega$$
where
$$A = \frac{1}{\sqrt{1+\omega^2}}, \qquad \varphi = -\tan^{-1}\frac{\omega}{1}$$
and
$$U(\omega) = \text{Re }(Ae^{j\varphi}) = A\cos\varphi$$
$$V(\omega) = \text{Im}(Ae^{j\varphi}) = A\sin\varphi$$

Indeed, once the amplitude and the phase angle of a sine wave in a certain frequency ω are determined, the sine wave is uniquely defined. The form in which it is expressed is only a matter of convenience.

By considering ω as a variable, we determine a set of corresponding A and φ. In other words, once ω is given, the A's and φ's of a linear system are uniquely defined.

The procedure mentioned above can be considered as a definition of the frequency response of a system, as well as a procedure to obtain the frequency response of a system in practice. As a matter of fact, the procedure is a major method for identifying a system in the laboratory.

2. Transfer Function and Frequency Response

If a transfer function of a system is given, can we obtain the frequency response of the system from it?

We use our example system again:
$$M(s) = \frac{1}{s+1} \tag{1}$$
Replacement of s by $j\omega$ yields
$$M(j\omega) = \frac{1}{1+j\omega}$$

or

$$= \frac{1}{\sqrt{1 + \omega^2}} \underline{/-\tan^{-1} \frac{\omega}{1}} \tag{5}$$

Equation (5) is the polar form of the frequency response. The result is just what we expected. This method can be called a shortcut method for obtaining the frequency response; it is based on the Fourier transformation. We will treat the topics in detail in the chapter on the identification problem.

If the transfer function is expressed in a pole-zero form, the pole-zero configuration of the transfer function usually can be drawn in a complex plane. Then the frequency response can be obtained by a couple of measurements and some simple calculations.

For our example, the pole-zero configuration of the transfer function is shown in Fig. 3-39.

$$\overline{OQ} = -1 \quad \text{and} \quad \overline{QO} = +1$$

$$M(j\omega_1) = \frac{1}{1 + j\omega_1} = \frac{1}{\overline{QO} + j\omega_1}$$

$$= \frac{1}{\overline{QR}\underline{/\varphi}} = \frac{1}{|A(\omega_1)| \underline{/\varphi(\omega_1)}} \tag{5a}$$

Equation (5a) is the same as (5).

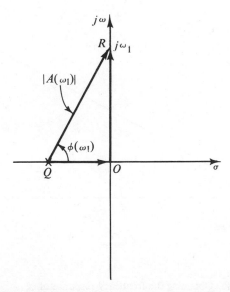

Fig. 3-39. Obtaining the frequency response of $1/(1 + s)$ by measurements.

3. Two Graphical Representations

As we have seen, three variables are involved in a frequency response, namely: (1) $A(\omega)$, (2) $\varphi(\omega)$, and (3) ω. Either consider ω as a parameter and draw $A(\omega)$ and $\varphi(\omega)$ on one graph or consider ω as an independent variable and draw two graphs: ω vs. A, ω vs. φ.

For example, a transfer function given as

$$T(s) = \frac{10}{s(1 + 0.02s)(1 + 0.5s)(1 + s)}$$

can also be written as

$$T(s) = \frac{1000}{s(s + 50)(s + 2)(s + 1)}$$

The frequency response is evaluated:

$$T(j\omega) = \frac{1000}{j\omega(j\omega + 50)(j\omega + 2)(j\omega + 1)}$$

The pole-zero configuration is shown in Fig. 3-40.

$$T(j\omega) = \frac{1000}{|A_1| \cdot |A_2| \cdot |A_3| \cdot |A_4| \underline{/\varphi_1 + \varphi_2 + \varphi_3 + \varphi_4}}$$

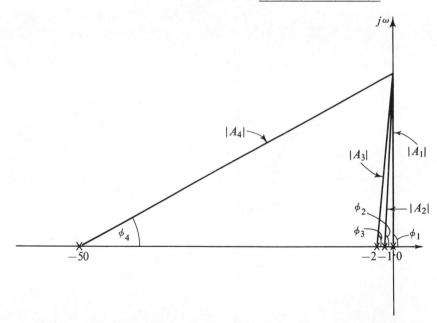

Fig. 3-40. Obtaining the frequency response of $\dfrac{1000}{s(s + 50)(s + 2)(s + 1)}$ by measurements.

Figure 3-41 shows the first graphical representation of the frequency response or the polar form of the frequency response. The ω's are indicated as parameters.

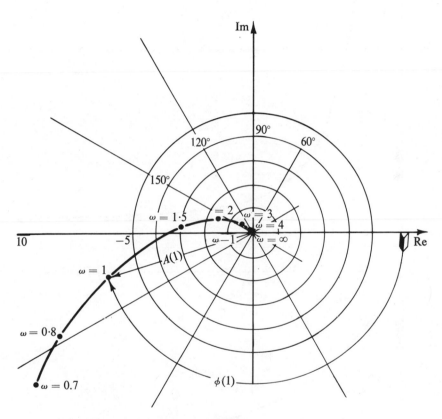

Fig. 3-41. The polar form of the frequency response of

$$\frac{10}{s(1 + 0.02s)(1 + 0.5s)(1 + s)}.$$

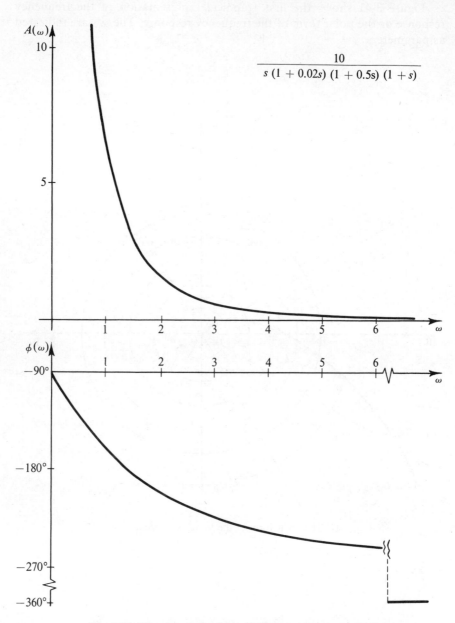

$$\frac{10}{s\,(1+0.02s)\,(1+0.5s)\,(1+s)}$$

Fig. 3-42. The rectangular form of the frequency response of $\dfrac{10}{s(1+0.02s)(1+0.5s)(1+s)}$: (a) $A(\omega)$ vs. ω; (b) $\varphi(\omega)$ vs. ω.

The second graphical representation is shown in Fig. 3-42, which has two subgraphs: ω is used as the independent variable, and $A(\omega)$ and $\varphi(\omega)$ are the ordinates of the two subgraphs, respectively.

4. Nyquist Criterion

Nyquist criterion is a criterion for determining the stability of a feedback system by the use of its open loop frequency response information.

Consider a typical feedback system, shown in Fig. 3-43. The over-all transfer function is shown as

$$\frac{C}{R} = \frac{KG}{1 + KGH} \qquad (6)$$

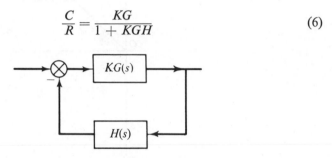

Fig. 3-43. A typical feedback system.

Let the transfer function KGH have the following properties:

1. KGH is expressed as a ratio of two real algebraic polynomials of s in their lowest terms, $KGH = A(s)/B(s)$, where $A(s)$ and $B(s)$ are two real algebraic polynomials of integral power and have no common factor involving s.

2. The degree of $A(s)$ is not higher than that of $B(s)$, so as s increases without limit, KGH remains finite or approaches 0.

3. KGH has no poles on the imaginary axis but may have a simple or multiple pole at the origin.

The system will be stable only if $1 + KGH$ has no zero in the right half plane. Now

$$1 + KGH = \frac{(A + B)}{B} \qquad (7)$$

The zeros of $1 + KGH$ are also those of $(A + B)$. In other words, a stable system requires that all zeros of $A(s) + B(s)$ be in the left half plane.

Consider the function $s - s_0$, where s_0 is a complex number and $s = j\omega$, ω being a real number varying continuously from $-\infty$ to ∞. In terms of the s-plane, s varies along the entire imaginary axis. Let

$$s - s_0 = \mathrm{R}e^{j\theta}$$

as shown in Fig. 3-44, in which s_0 has a positive real component or is in the right half plane. As ω varies from $-\infty$ to ∞, s describes 180 deg or π radians around s_0 in a clockwise direction, so that the change of phase $\Delta\theta$ equals $-\pi$.

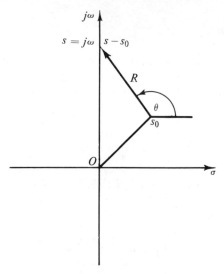

Fig. 3-44. Deriving the Nyquist criterion.

In a similar manner, if s_0 has a negative real component or is in the left half plane, the phase change $\Delta\theta$ equals π.

It should be noted that although the initial and final values of θ are the same in both cases, the change of θ depends upon the relative location of s_0 and upon the path. In the graphical plot of θ, the essential point is that it must be continuous, and this determines its change.

For a product such as $(s - s_1)(s - s_2)\ldots$ the phase angle θ equals $\theta_1 + \theta_2 + \cdots$; therefore, $\Delta\theta = \Delta\theta_1 + \Delta\theta_2 + \cdots$. The total phase change of the product is then equal to the sum of the phase changes of its individual factors.

A real polynomial has only real or complex conjugate pairs of roots. Let ω vary from 0 to ∞. For a real root, the phase change $\Delta\theta$ will be $-\pi/2$ or $\pi/2$ for a root in the right or the left half plane, respectively. For a complex conjugate pair of roots in the right half plane, the total phase change for ω varying from $-\infty$ to ∞ will be -2π. Because of the symmetrical location of the two roots, $\Delta\theta$ for ω varying from 0 to ∞ will be half of the above amount, or $-\pi$. Similarly, for a complex pair of roots in the left half plane, $\Delta\theta$ for ω varying from 0 to ∞ will be π radians. Thus the following statement may be made:

The net phase change of a polynomial for ω varying from 0 to ∞ is $-\pi/2$ for each root, real or complex, in the right half plane, and $\pi/2$ for each one in the left half plane.

Hereafter, the phase change $\Delta\theta$ will be for variation of ω from 0 to ∞, i.e., along the upper half of the imaginary axis for s.

If s is only to vary along the positive half of the imaginary axis, a factor $s = j\omega$ has a *constant* phase of $\pi/2$; hence, it will not contribute anything to the phase change.

The above results may be applied to the phase change of a polynomial for variation of ω from 0 to ∞. Let the degree of the polynomial be r, with m roots in the right half plane and an $n$ple root at the origin. Since it has r roots in all, the remaining ones, $r - m - n$, will be in the left half plane. The total phase change is, therefore,

$$\Delta\theta = (r - m - n)\frac{\pi}{2} - \frac{m\pi}{2} = (r - 2m - n)\frac{\pi}{2} \text{ radians}$$

Return now to the expression $1 + KGH = (A + B)/B$. Let the degree of B be r. Since A is not a higher degree than B, $A + B$ will still have a degree of r. If the system is stable, all the roots of $A + B$ will be in the left half plane. Thus the contribution to the plot of $1 + KGH$ for ω varying from 0 to ∞ will be $r\pi/2$. On the other hand, B may have m roots in the right half plane and an $n$ple root at the origin; the phase change for B will be $(r - 2m - n)\pi/2$, as given above. The net phase change for the quotient will be the difference of the two,

$$\Delta\theta = \frac{r\pi}{2} - \frac{(r - 2m - n)\pi}{2} = \frac{(2m + n)\pi}{2}$$

The preceding discussion leads to the following modified form of the Nyquist stability criterion:

Let the open loop transfer function KGH of a feedback system satisfy the following conditions:

1. KGH is expressed as a ratio of two real algebraic polynomials of s in their lowest terms, $KGH = A/B$.

2. The degree of A is not higher than that of B, so the open loop transfer function remains finite (or zero) at very high frequencies.

3. KGH has m poles in the right half plane, an $n$ple at the origin, but none along the j axis.

If the transfer function KGH is plotted for s varying from 0 to $j\infty$ along the imaginary axis, and the curve of KGH thus plotted encircles the point $(-1, 0)$ by an angle of $(n + 2m)\pi/2$ in the counterclockwise direction, the system is stable.

5. Illustrative Examples

Example 1. A unity feedback system has the following open loop transfer function:

$$\frac{C}{E} = \frac{4}{s(1 + 0.02s)(1 + 0.5s)(1 + s)}$$

Determine the stability of the feedback system.

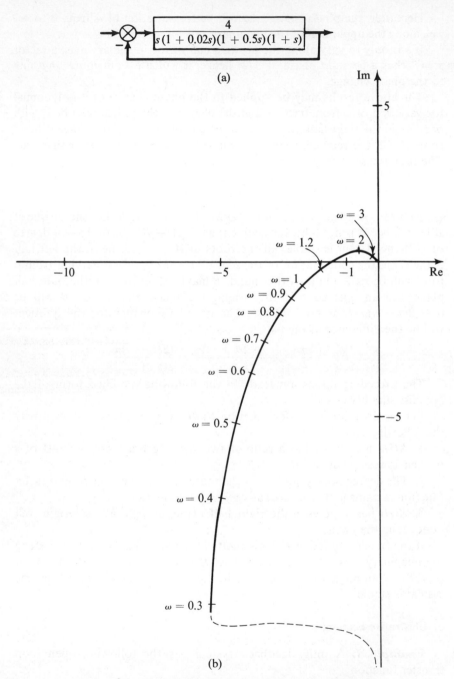

Fig. 3-45. The Nyquist plot of a feedback system: (a) the system; (b) the Nyquist plot, which is the frequency response polar plot of the open loop transfer function.

The open loop frequency response is plotted as shown in Fig. 3-45. Here $m = 0$ (which means there is no pole on the right half plane); $n = 1$ (there is a single pole at the origin). Therefore, the criterion for stability is

$$\Delta\theta = (n + 2m)\frac{\pi}{2} = \frac{\pi}{2} \text{ or } 90°$$

Inspection of Fig. 3-45 indicates that $\Delta\theta = 3 \times (-90°) = -270$ deg (minus sign means clockwise); the system is unstable.

Example 2. A feedback system has the following functions:

$$KG = \frac{10}{s^2(1 + 0.7s)}$$

$$H = \frac{s + 1}{0.2s + 1}$$

The open loop transfer function is

$$KGH = \frac{10(s + 1)}{s^2(1 + 0.7s)(0.2s + 1)}$$

The open loop frequency response is shown in Fig. 3-46. We have $m = 0$ and $n = 2$, so the criterion for stability is

$$\Delta\theta = (n + 2m)\frac{\pi}{2} = 180°$$

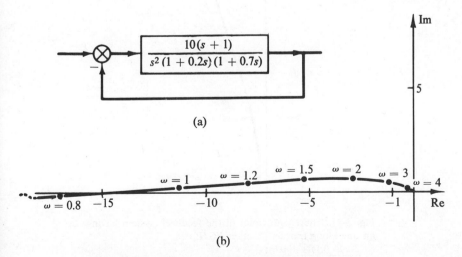

(a)

(b)

Fig. 3-46. The Nyquist plot of another feedback system: (a) the system; (b) the Nyquist plot.

By inspecting Fig. 3-46, we see that $\Delta\theta = -180$ deg, or 180 deg in the clockwise direction. So the system is unstable.

Example 3. A feedback system is with the following elements:

$$KG = \frac{2(s1 + 0.1)(s + 0.6)(s^2 + s + 1)}{s^3(s - 0.2)(s + 1)}$$

$$H = 1$$

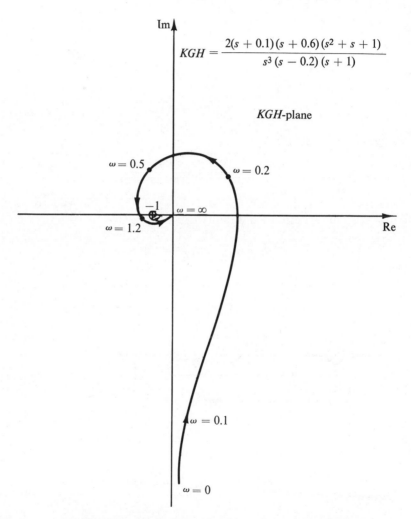

Fig. 3-47. The Nyquist plot of the feedback system having an open loop transfer function KGH.

$$= \frac{2(s + 0.1)(s + 0.6)(s^2 + s + 1)}{s^3(s - 0.2)(s + 1)}.$$

A sketch of the open loop frequency response is shown as Fig. 3-47 (the magnitudes are not in scale). We have $m = 1$, $n = 3$, so the criterion for stability is

$$\Delta\theta = (n + 2m)\frac{\pi}{2} = 450°$$

By inspecting Fig. 3-47, we see that $\Delta\theta = 450$ deg, so the system is stable.

VI.   Performance

1. Figures of Merit

If a system is stable, the next question is: How stable is it? Engineers have established several figures of merit to aid in analysis and to serve as goals in design. They have been developed as the result of experience as well as from specifications set by people requiring the design.

Investigations of the relationships among the various figures of merit and the parameters of the system are called performance studies.

In linear system analysis, we usually look at a system in three domains: frequency, time, and complex. Accordingly, there are three corresponding groups of figures of merit. The best-known of these are

1. In the frequency domain:
   a. Gain margin
   b. Phase margin
   c. M peak
2. In the time domain:
   a. Steady-state values
   b. Rise time and overshoot
   c. Settling time
3. In the complex domain:
   a. Damping ratio
   b. Damping factor

These quantities tell us much about the nature of a system. If they are not satisfactory the performance of the system will be questionable, so we must change the system to improve it.

2. In the Frequency Domain

In the previous section we discussed the stability of a feedback system and established the Nyquist criterion. We will herein consider a certain class of system which is very common, that in which the open loop transfer function has no roots in the right half plane ($m = 0$) and has two, one, or no root at the origin ($n = 0, 1, 2$). For such systems the Nyquist criterion can be simplified as follows.

If the transfer function KGH is plotted for s varying from 0 to $j\infty$ along the imaginary axis, and the curve GH encircles the point $(-1, 0)$, by an angle of $n(\pi/2)$ in the counterclockwise direction, the system is stable.

For example, a feedback system is shown in Fig. 3-48. If the open loop transfer function plot is as shown in Fig. 3-49(a), the system is stable; if it is like Fig. 3-49(b), the system is unstable. It can be seen that the relative

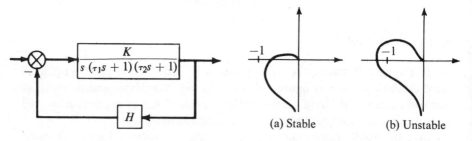

(a) Stable   (b) Unstable

Fig. 3-48. A feedback system (K can be varied).

Fig. 3-49. Nyquist plots: (a) stable system; (b) unstable system.

positions of the locus and the $(-1, 0)$ point are significant. Not only do we know that if the locus passes through the $(-1, 0)$ point the system would be oscillatory and that if the locus does not encircle the $(-1, 0)$ point the system is stable, but we might also surmise that the degree of stability would be related to the distance between the point and the locus.

Two new terms arise in considering this distance from the locus to the $(-1, 0)$ point. Figure 3-50 illustrates the two figures of merit called the phase margin and the gain margin.

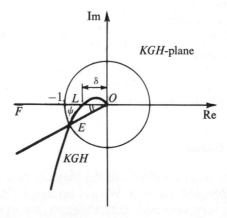

Fig. 3-50. Illustration of "gain margin" and "phase margin."

(a) PHASE MARGIN

Describe a unit circle through $(-1, 0)$ with the origin as center. Draw line OE. The angle EOF is called the phase margin of the feedback system whose open loop transfer function is KGH.

(b) GAIN MARGIN

If the locus intersects the negative axis at L, the distance LO is called δ. The inverse, $1/\delta$, is the gain margin of the system.

Example 1. For a given unity feedback system, the open loop transfer function is

$$KG(s) = \frac{20}{s\left(\frac{s}{10} + 1\right)\left(\frac{s}{30} + 1\right)} \tag{1}$$

The Nyquist plot of (1) is drawn as in Fig. 3-51. It is found that the locus intersects the unit circle when ω_c is 14.0. This is called the crossover frequency. The phase margin is measured and turns out to be 14.5 deg. The distance δ is approximately 0.5, making the gain margin equal to 2.

(c) M PEAK

The maximum magnitude of the frequency response of a system over-all transfer function is called the M peak. This is usually found by direct measurement on the plot of the frequency response of the over-all transfer function.

(d) BANDWIDTH

The bandwidth is that frequency at which the closed loop frequency response falls to one-half of its mid-band value.

Example 2. Consider a unity feedback system with the following open loop transfer function:

$$KGH = \frac{10}{s(s+1)(1.2 \cdot 10^{-4}s^2 + 3 \cdot 10^{-3}s + 1)}$$

The over-all transfer function of the system is formed:

$$\frac{C}{R} = \frac{KG}{1 + KGH} = \frac{10}{s(s+1)(1.2 \cdot 10^{-4}s^2 + 3 \cdot 10^{-3}s + 1) + 10}$$

The frequency response of the over-all transfer function can be obtained by substituting s for $j\omega$.

$$\frac{C}{R}(j\omega) = \frac{10}{j\omega(j\omega + 1)(1.2 \cdot 10^{-4}(j\omega)^2 + 3 \cdot 10^{-3}(j\omega) + 1) + 10}$$

A frequency response plot for the over-all transfer function is shown in Fig. 3-52. M_p or M_{max} is directly read at the point which is the greatest distance from the center. For this problem, $M_p = 3.2$ at $\omega = 3.1$.

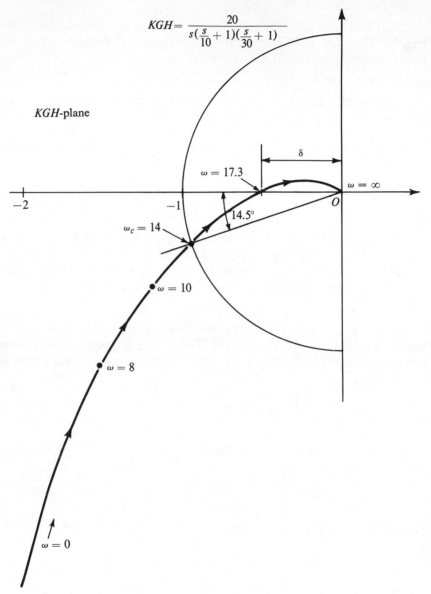

$$KGH = \frac{20}{s(\frac{s}{10}+1)(\frac{s}{30}+1)}$$

KGH-plane

$\omega = 17.3$

$\omega = \infty$

$\omega_c = 14$

$14.5°$

$\omega = 10$

$\omega = 8$

$\omega = 0$

δ

-2 -1 O

Fig. 3-51. The Nyquist plot of $\dfrac{20}{s\left(\frac{s}{10}+1\right)\left(\frac{s}{30}+1\right)}$

Describe a circle with radius of 0.707 and with the origin as the center. The circle intersects the locus at the point where $\omega = 4.8$ rad/sec, which is the bandwidth.

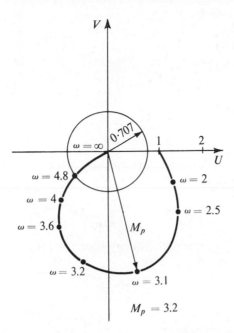

Fig. 3-52. Obtaining the maximum magnitude and the band-width of the system with the over-all transfer function

$$\frac{10}{s(s+1)(1.2 \times 10^{-4}s^2 + 3 \times 10^{-3}s + 1) + 10}.$$

3. In the Time Domain

One of the main reasons for using feedback in systems is to obtain a response that is a faithful reproduction of the stimulus. In other words, in the initial period the response should rise as quickly as possible when a step input is applied; when steady state is attained, the error (difference) between the input and the response should be as small as possible.

The figures of merit in the time domain are of great interest to engineers. After they have been found, the performance of the system can be evaluated directly and easily.

From among the many considerations in the time domain we will first define the terms concerning the steady-state values:

(a) STEADY-STATE ERRORS

Steady-state measurements are usually the first made in a performance study. In the case of feedback control systems it is desirable to find the steady-state errors which result from the application of several different inputs, such as:

1. Unit impulse function.
2. Unit step function.
3. Unit ramp function.

Consider the feedback system in Fig. 3-53. The error is defined as

$$e(t) = r(t) - c(t) \tag{2}$$

or

$$E = R - C \tag{2a}$$

By definition,

$$E \cdot KG(s) = C \tag{3}$$

Elimination of C gives

$$E = \frac{1}{1 + KG(s)} R \tag{4}$$

If R is a unit step function, the steady-state error, $e(t)_{\text{s.s.}}$, is also a constant. The error is shown as in Fig. 3-54.

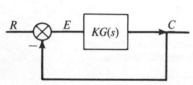

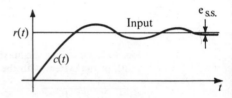

Fig. 3-53. A unity feedback system.

Fig. 3-54. Illustration of the steady-state error of a system with a unit step input.

As another example, consider a unity feedback system whose open loop transfer function is as follows:

$$KG = \frac{20\left(\frac{s}{1.1} + 1\right)\left(\frac{s}{1.5} + 1\right)}{s\left(\frac{s}{10} + 1\right)\left(\frac{s}{30} + 1\right)\left(\frac{s}{0.1} + 1\right)\left(\frac{s}{4.5} + 1\right)} \tag{5}$$

A unit ramp function is applied. The input and the output are shown in Fig. 3-55. The difference between them, measured in the middle of the two curves, is the steady-state error: $e_{\text{s.s.}} = 0.0459$.

Some shortcut methods have been developed for calculating $e_{\text{s.s.}}$ directly, such as the final value theorem of Laplace transformation theory. The technique is straightforward.

(b) ERROR CONSTANTS

Based on unity feedback systems, the error constants are defined as follows:

position error const $= K_p = \dfrac{1}{e_{\text{s.s.}}}$ with $u(t)$ as input

velocity error const $= K_v = \dfrac{1}{e_{\text{s.s.}}}$ with t as input

acceleration error const $= K_a = \dfrac{1}{e_{\text{s.s.}}}$ with $t^2/2$ as input

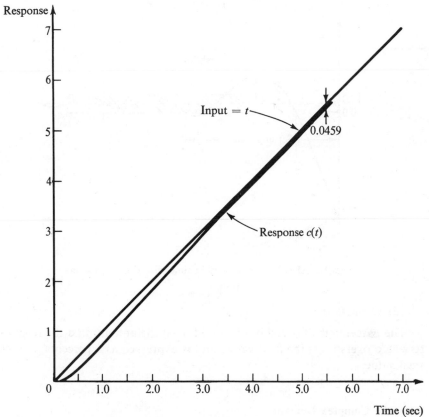

Fig. 3-55. Illustration of the steady-state error of a system with a unit ramp input.

Error constants are usually required by buyers of control systems as part of the specifications. They show the accuracy of the system in steady state and also indicate which types of input the system can follow.

For example, if a unity feedback system has (5) for its open loop transfer function, the velocity constant for a ramp input to the system is

$$K_v = \frac{1}{e_{\text{s.s.}}} = \frac{1}{0.0459} = 21.8 \tag{6}$$

(c) RISE TIME AND SETTLING TIME

The rise time is the time required for the response to a unit step function to rise from 10 to 90 per cent of its final value. Settling time is the time required for the response to a unit step function to reach a specified percentage of its final value. These terms are illustrated in Fig. 3-56.

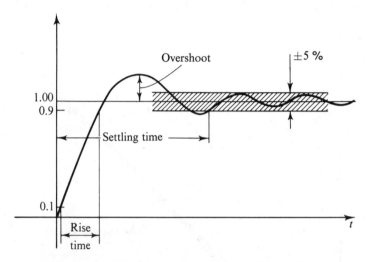

Fig. 3-56. Definitions of settling time, rise time, and overshoot.

(d) OVERSHOOT

The percentage of overshoot is found from the amount that the response to a step overshoots the final value, and is expressed as a percentage of the final value.

4. In the Complex Domain

In the pole-zero configuration of a closed loop transfer function there is often a pair of poles near the imaginary axis. These are called the dominant poles, and behavior of the system is largely dependent on them. In the case of a second order system, they describe its response exactly.

The important study of the dominant poles has led to some new terms:

1. $\zeta =$ damping ratio.
2. $\omega_n =$ undamped natural frequency.
3. $\alpha =$ damping factor.

Suppose the two poles are $(-\alpha \pm j\beta)$. We have, by definition, the characteristic equation

$$(s + \alpha + j\beta)(s + \alpha - j\beta) = s^2 + 2\alpha s + \alpha^2 + \beta^2 = s^2 + 2\zeta\omega_n s + \omega_n^2$$

Therefore,

$$\omega_n = \sqrt{\alpha^2 + \beta^2}$$

and

$$\alpha = \zeta\omega_n$$

Figure 3-57 shows that $\cos\theta = \zeta\omega_n/\omega_n = \zeta$. This could be considered an alternate definition of damping ratio.

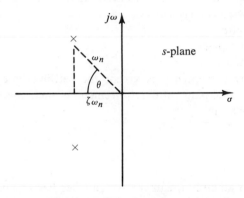

Fig. 3-57. Definitions of damping ratio, undamped natural frequency, and damping factor.

Example. A feedback system has the following open loop transfer function:

$$KG = \frac{1638(s^2 + 2.6s + 1.65)}{s(s^2 + 40s + 300)(s^2 + 4.65 + 0.45)}$$

from which the over-all or closed loop transfer function is found.

$$\frac{C}{R} = \frac{KG}{1 + KG}$$

$$= \frac{1638(s^2 + 2.6s + 1.65)}{(s + 32.417)(s + 0.855 + j0.672)(s + 0.855 - j0.672)(s + 5.24 + j6.566)(s + 5.24 - j6.566)}$$

The dominant poles of this system are

$$s + 0.855 \pm j0.672$$

from which we find θ first.

$$\tan\theta = \frac{0.672}{0.855} = 0.786$$

$$\theta = 38.2°$$

and

$$\zeta = \cos 38.2° = 0.785$$

which is the damping ratio of the dominant poles of the system.

5. Relationship Among the Figures of Merit

The various figures of merit are not all independent. They fall into three groups: overshoot, damping ratio, and phase margin; settling time and crossover frequency; rise time, bandwidth, and frequency at M peak. The relationships among them cannot be found explicitly except for an individual case.

A set of specifications usually consists of one of the terms from each group. The following is a typical set.
1. Damping ratio.
2. Crossover frequency.
3. Error constant.
When the crossover frequency is given, the settling time is not needed. Similarly, when a damping ratio requirement is met, there is no need to check the phase margin.

6. Summary

In this chapter stability for linear systems has been defined and studied. The Cauchy theorem was developed and used as the criterion for the subsequent methods of analysis. These were

1. *The direct method.* We can predict the stability of a system by analyzing the coefficients of the differential equations which represent it without actually solving the equations. Two approaches, Hurwitz' determinant and Routh's algorithmic forms, were discussed. The latter is particularly suitable for programming on a digital computer.

2. *The root-locus method.* By varying a parameter and observing the effect on stability, a general picture of the behavior of a system can be obtained. Evans' original technique is a graphic approach; Routh's algorithm is digital computer oriented.

3. *The frequency method.* Necessary information for stability study can be obtained directly from experimental data when the Nyquist criterion is applied. This makes it the most popular of all methods. The derivation and presentation in this chapter is different from Nyquist's original. We dealt with positive frequency response only and made a derivation based on Cauchy's theorem. It is believed to be simpler and easier this way.

PROBLEMS

3-1. Determine the stability of each of the feedback systems shown in Fig. P3-1 by using Routh-Hurwitz' criterion.

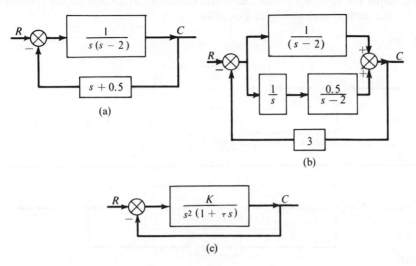

Fig. P3-1

3-2. Determine the root locus for each of the systems shown in Fig. P3-2. Then let K equal several arbitrary values and find the corresponding roots to verify that they are on the locus you have drawn.

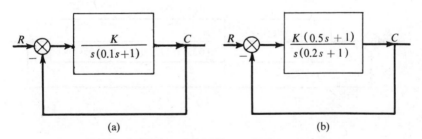

Fig. P3-2

3-3. Find the breakaway points of the root locus for each system in Fig. P3-3.

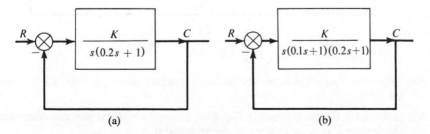

Fig. P3-3

3-4. Draw the asymptotic lines and locate the center of gravity of the root locus for each system shown in Fig. P3-4.

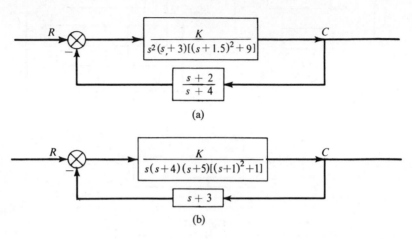

(a)

(b)

Fig. P3-4

3-5. Draw the root locus for each system shown in Fig. P3-5.

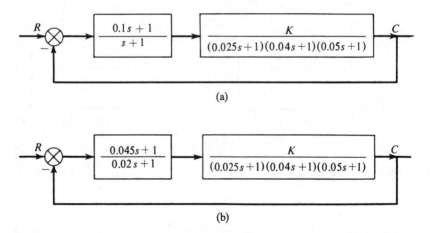

(a)

(b)

Fig. P3-5

3-6. Draw the Nyquist plots for the feedback systems shown in Figs. P3-1(a) and P3-1(b).

3-7. A feedback system is shown in Fig. P3-7. Draw the Nyquist plot and determine the stability, if $K = 1$. What is the stability if $K = 2$?

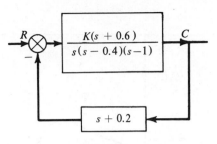

Fig. P3-7

3-8. The function $1 + KGH$ is called the return difference of a feedback system with an open loop transfer function KGH. Based on the return difference, the Nyquist criterion can be restated as follows: Draw the phase plot of $1 + KGH(j\omega)$ against ω, if and only if $\theta(\infty) - \theta(0) = (n + 2m)\pi/2$, the system is stable where $\theta(\infty)$ and $\theta(0)$ are the phase angles of $1 + KGH(j\omega)$ at $\omega = \infty$ and $\omega = 0$, respectively. By using the return difference version of Nyquist's criterion, determine the stability of each of the systems shown.

(a) $\quad KGH = \dfrac{2(s + 0.1)(s + 0.6)(s^2 + s + 1)}{s^3(s - 0.2)(s + 1)}$

The phase plot of $1 + KGH$ is shown in Fig. P3-8(a).

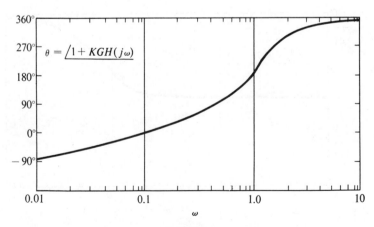

Fig. P3-8(a)

(b) $\quad KGH = \dfrac{(s + 0.1)(s + 0.6)(s^2 + s + 1)}{s^3(s - 0.2)(s + 1)}$

The phase plot of $1 + KGH$ is shown in Fig. P3-8(b).

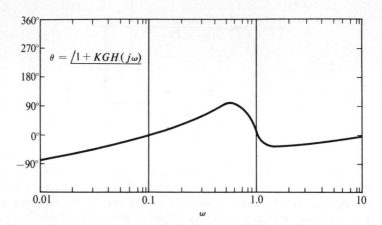

Fig. P3-8(b)

(c) $KGH = \dfrac{3(s + 0.5)}{s(s - 2)}$

The phase plot of $1 + KGH$ is shown in Fig. P3-8(c).

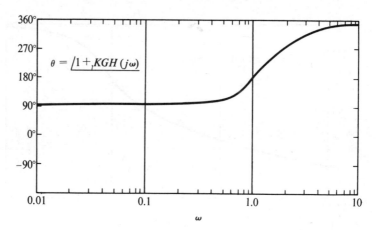

Fig. P3-8(c)

(d) $KGH = \dfrac{s + 0.5}{s(s - 2)}$

The phase plot of $1 + KGH$ is shown in Fig. P3-8(d).

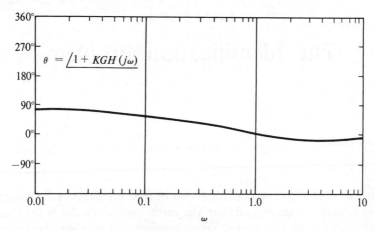

$$\theta = \underline{/1 + KGH\,(j\omega)}$$

Fig. P3-8(d)

3-9. Sometimes the system shown in Fig. P3-9(b) can be considered as a simplified model of (a). Compare the rise times, settling times, overshoots, damping ratios and natural frequencies of the two systems.

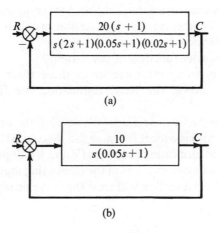

(a)

(b)

Fig. P3-9

CHAPTER 4

The Identification Problem

The nature of a system is usually found by measurements in the frequency, time, or complex domain, and then the mathematical model is constructed by analysis. In some instances the measurements are easy, but the analysis is difficult; in others the opposite is the case. Therefore, it is necessary to be able to convert the information from one domain to the other. This conversion is a part of the identification problem. In most cases, the easy and accurate way for measuring the output of a system is to obtain its frequency response. In other words, sinusoidal waves at various frequencies can be applied to the system and the steady-state responses measured. If the system is linear, with time invariant elements, these outputs will be sinusoidal waves of some sort. Then comes the conversion problem:

1. What is the transient response of the system?
2. What is the transfer function of the system?

Once the corresponding information in these three (time, frequency, and s) domains is known, the basic foundation for further analysis is established.

In this chapter the first two sections will develop the analyses of periodic and aperiodic waves as basic tools for later use. Section III will show the use of Bode diagrams in identification; sections IV and V will present the Wiener-Lee theorem, and Bush's decomposition for identifying highly oscillatory systems, respectively. The last section will treat the curve fitting method of Levy.

I. Analysis of a Periodic Signal: Fourier Series

1. Graphical Method for Determining Coefficients

Any periodic signal can be represented as a Fourier series. The problem that engineers usually encounter is how to determine the Fourier coefficients when only experimental data or numerical values of the periodic signal are given.

Consider the general form of the Fourier series:

$$f(x) = \tfrac{1}{2}a_0 + a_1 \cos x + a_2 \cos 2x + \cdots + a_n \cos nx$$
$$+ \, b_1 \sin x + b_2 \sin 2x + \cdots + b_n \sin nx \tag{1}$$

where

$$a_n = \frac{1}{\pi} \int_0^{2\pi} f(x) \cos nx \, dx \tag{2}$$

$$b_n = \frac{1}{\pi} \int_0^{2\pi} f(x) \sin nx \, dx \tag{3}$$

Rewriting (2) and (3) into summation forms

$$a_n = \frac{1}{m} \sum_{k=1}^{k=2m} f_k \cos nk \frac{360°}{2m} \tag{2a}$$

$$b_n = \frac{1}{m} \sum_{k=1}^{k=2m} f_k \sin nk \frac{360°}{2m} \tag{3a}$$

where the notations m and k are shown in Fig. 4-1.

A graphical method for determining the coefficients can be established from the following consideration: Combining (2a) and (3a) gives

$$a_n + jb_n = \frac{1}{m} \sum_{k=1}^{k=2m} f_k \left(\cos nk \frac{360°}{2m} + j \, \sin nk \frac{360°}{2m} \right) \tag{4}$$

$$= \frac{1}{m} \sum_{k=1}^{k=2m} f_k e^{jnk\,360°/2m} \tag{5}$$

But the preceding equation may be written as

$$a_n + jb_n = \frac{1}{m} \sum_{k=1}^{k=2m} f_k \left/ nk \frac{360°}{2m} \right. \tag{5a}$$

Equation (5a) is easily recognized as a vector summation.

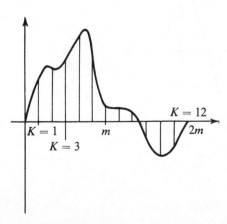

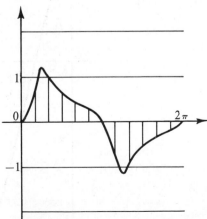

Fig. 4-1. A periodic but nonsinusoidal wave form.

Fig. 4-2. A periodic wave to be analyzed.

In Fig. 4-2, m is 6 and $f_1, f_2, f_3, \ldots$ can be measured directly. For various values of k, corresponding vectors,

$$f_1 \underline{/n360°/(2 \times 6)}, \quad f_2 \underline{/n \times 2 \times 360°/(2 \times 6)}, \quad f_3 \underline{/n \times 3 \times 360°/(2 \times 6)}, \ldots$$

are obtained. When these vectors are plotted, as shown in Fig. 4-3, $(a_n + jb_n)$ is seen to be the resultant of the sum of the vectors divided by m.

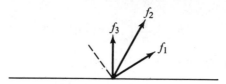

Fig. 4-3. Vectors taken from Fig. 4-2.

2. Example 1

Find the coefficient of the fundamental of the triangular wave shown in Fig. 4-4 by the graphical method.

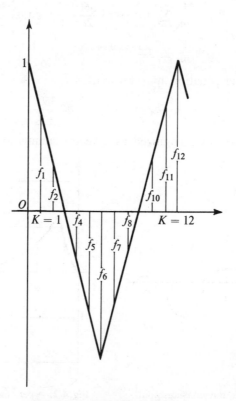

Fig. 4-4. A triangular wave form.

From Fig. 4-4 we obtain the ordinates and angular displacements as listed in Table 1.

TABLE 1

k	f	Vectors
1	2/3	2/3 $\underline{/30°}$
2	1/3	1/3 $\underline{/60°}$
3	0	0
4	−1/3	−1/3 $\underline{/120°}$
5	−2/3	−2/3 $\underline{/150°}$
6	−1	−1 $\underline{/180°}$
7	−2/3	−2/3 $\underline{/210°}$
8	−1/3	−1/3 $\underline{/240°}$
9	0	0
10	1/3	1/3 $\underline{/300°}$
11	2/3	2/3 $\underline{/330°}$
12	1	1 $\underline{/360°}$

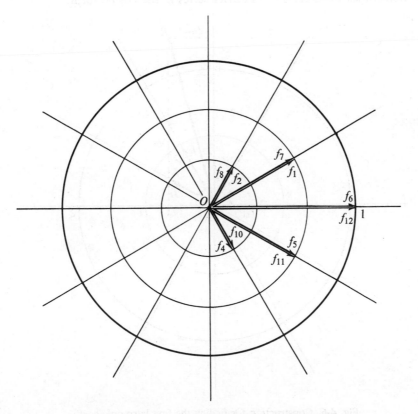

Fig. 4-5. Construction for finding a_1 coefficient.

Draw a unit circle. For the unit, use that of the vertical scale of the original $f(x)$ curve in Fig. 4-1. Then divide the circle into $2m$ angular segments through the origin.

Draw f_k vectors on the polar coordinate plane shown as in Fig. 4-5. The resultant of these vectors is 4.97. We then have

$$a_1 = \frac{1}{m} \sum_{k=1}^{2m} f_k \left| k \frac{360°}{2m} \right. = \frac{1}{6}(4.97) = 0.829$$

The exact value of $a_1 = 8/\pi^2 = 0.813$.

From Fig. 4-5, we can see that $b_1 = 0$, because there is no vertical component in the resultant vector.

In order to obtain a_2 and b_2 in this example, the angle between two consecutive vectors should be 60 deg, 120 deg, etc.

3. Example 2

A wave form is shown in Fig. 4-2. Find its Fourier coefficients. Sometimes we express the Fourier series in the cosine form.

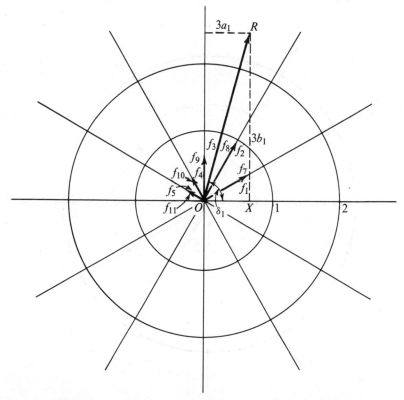

Fig. 4-6. Construction for finding the first harmonic coefficients of the wave shown in Fig. 4-2.

$$f(x) = A_0 + A_1 \cos(x - \delta_1) + A_2 \cos(2x - \delta_2) + \cdots + \cdots$$
$$+ A_n \cos(nx - \delta_n) + \cdots \tag{6}$$

The derivation of (6) is self-explanatory. A_n and δ_n are defined as

$$A_n = \sqrt{a_n^2 + b_n^2}$$

$$\delta_n = \tan^{-1} \frac{b_n}{a_n}$$

Find A_1, δ_1, a_1, b_1 for the wave form in Fig. 4-2. Draw $f_1, f_2, \ldots, f_n$ on polar coordinate paper as shown in Fig. 4-6. OR is the resultant vector equal to $3A_1$, and the angle XOR is equal to δ_1. Figure 4-7 shows that the second harmonic of the wave is zero. Figure 4-8 is a construction for finding the third harmonic.

Directly measuring the lengths and the angles of the resultant lines in Figs. 4-6 and 4-8 yields Table 2. Thus the Fourier series expansion can be written as follows:

$$f(x) = 0.85 \cos(x - 74.5°) + 0.23 \cos(3x - 145°) + \cdots$$

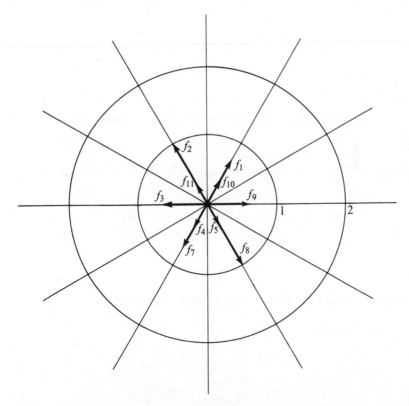

Fig. 4-7. Construction for finding the second harmonic coefficients.

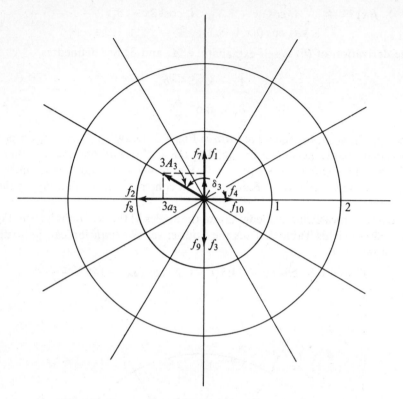

Fig. 4-8. Construction for finding the third harmonic coefficients.

or

$$f(x) = 0.22 \cos x + 0.2 \cos 3x + \cdots + 0.8 \sin x - 0.125 \sin 3x + \cdots$$

TABLE 2

$A_1 = 0.85,$	$\delta_1 = \quad 74.5°$
$a_1 = 0.22,$	$b_1 = \quad 0.8$
$A_3 = 0.23,$	$\delta_3 = \quad 145°$
$a_3 = 0.2,$	$b_3 = -0.125$

4. Digital Method for Determining Fourier Coefficients

Recall Eqs. (2a) and (3a).

$$a_n = \frac{1}{m} \sum_{k=1}^{k=2m} f_k \cos nk \frac{360°}{2m} \qquad (2a)$$

$$b_n = \frac{1}{m} \sum_{k=1}^{k=2m} f_k \sin nk \frac{360°}{2m} \qquad (3a)$$

These summation forms for the coefficients lend themselves readily to digital methods of computation. They call for a multiplication and division to be repeated over and over again, each time with some changed factors, usually by incrementing a factor by one. Figure 4-9 is the flow diagram of the consecutive

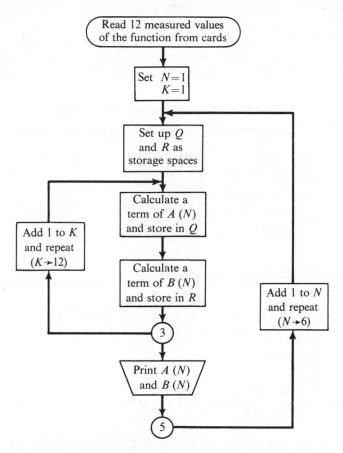

Fig. 4-9. Flow chart for the calculation of the first six coefficients of a Fourier series.

steps necessary to find a and b for the first six terms in the Fourier series. Program 5 shows Fortran statements that must be punched on cards, one line per card, to produce the object deck that will execute this program. Note that the core of the program is

$$T(K) = (F(K) * COS ((D*C*3.14159)/6.0))/6.0$$

$$U(K) = (F(K) * SIN ((D*C*3.14159)/6.0))/6.0$$

T and U are calculated 12 times with 12 successive values of F(K). These twelve quantities are added respectively and stored as Q and R. The Fortran statement

$$Q = Q + T(K)$$

means that Q_{new} is to be Q_{old} plus the present value of $T(K)$.

When all twelve values have been accumulated in Q and R, they are placed in A(1) and B(1), respectively, and these quantities are printed out on the typewriter.

Then N and D are changed from 1 to 2 and the core calculations are repeated another twelve times, ending with the printing of A(2) and B(2). The process is again repeated with N and D changed to 3; then 4 and so on up to 6.

Program 5. CALCULATING FOURIER COEFFICIENTS

```
C   C   CALCULATING FOURIER COEFFICIENTS
        DIMENSION F(12),A(6),T(12),U(12),B(12)
        READ,(F(K),K=1,12)
        PRINT2
   2 FORMAT(10HINPUT DATA)
        PRINT,(F(K),K=1,12)
        PRINT7
   7 FORMAT(//16HTHE COEFFICIENTS)
        DO 5 N=1,6
        Q=0.0
        R=0.0
        DO 3 K=1,12
        C=K
        D=N
        T(K)=(F(K)*COS((D*C*3.14159)/6.0))/6.0
        U(K)=(F(K)*SIN((D*C*3.14159)/6.0))/6.0
        Q=Q+T(K)
   3 R=R+U(K)
        A(N)=Q
        B(N)=R
        PRINT4,N,A(N),N,B(N)
   4 FORMAT(2H(AI2,3H)   E10.3,5H   (BI2,3H)   E10.3)
   5 CONTINUE
        END
```

The measured (input) data are twelve sample values of F(K) taken at 30 deg intervals across the cycle of the wave. These are read in at the beginning of the program and assigned names, F(1), F(2), ..., etc., according to the order in which they are fed into the machine.

This program could be used to calculate more than six Fourier coefficients with slight modifications. For L coefficients:

To the beginning, add two cards:

      (1)   READ6,L

      (2)   6FORMAT(I4)

Change   DO 5 N = 1,6   *to*   DO 5 N = 1,L

                (*add a card*)   M = 2*L

Change   DO 3 K = 1,12   *to*   DO 3 K = 1,M

Change   T(K) = (F(K)*COS((D*C*3.14159)/6.0))/6.0

to   T(K) = (F(K)*COS((D*C*3.14159)/M))/M

The number of coefficients, L, can be very large—any number that does not cause the program to exceed the memory capacity of the machine.

The result shown here is what was typed out in computing Example 1. Note that A(2), B(2), and the other even harmonics are expressed as very small numbers in the precise calculations of the computer. Effectively they are zero.

```
INPUT DATA

   6.6660E-01   3.3330E-01   0.0000        -3.3330E-01

  -6.6660E-01  -1.0000        -6.6660E-01  -3.3330E-01

   0.0000       3.3330E-01

    6.6660E-01   1.0000
```

```
THE COEFFICIENTS

(A 1)    0.829        (B 1)   -0.261E-05

(A 2)   -0.161E-05    (B 2)   -0.938E-06

(A 3)    0.111        (B 3)   -0.218E-05

(A 4)   -0.102E-05    (B 4)   -0.168E-05

(A 5)    0.596E-01    (B 5)   -0.303E-05

(A 6)   -0.100E-07    (B 6)   -0.267E-05

         END OF PROGRAM AT S. 0005 + 00 L.
```

II.   Analysis of an Aperiodic Signal: Fourier Integral

1. Fourier Integral

Formulas for expressing a Fourier series and its coefficients can be written in the following form:

$$f(x) = \sum_{-\infty}^{\infty} F_n e^{jnx} \tag{1}$$

$$F(n) = \frac{1}{2\pi} \int_0^{2\pi} f(x) e^{-jnx} \, dx \tag{2}$$

where $e^{jnx} = \cos nx + j \sin nx$.

Equations (1) and (2) are sometimes called the exponential form.
Let $x = \omega t$. When $t = 0$, $x = 0$ and when $t = T$, $x = 2\pi$.

$$f(\omega t) = \sum_{-\infty}^{\infty} F_n e^{jn\omega t} \tag{1a}$$

We have then

$$F(n) = \frac{1}{2\pi} \int_0^T f(t) e^{-jn\omega t} \, d\omega t \tag{2a}$$

or

$$f(t) = \sum_{-\infty}^{\infty} F_n e^{jn\omega t}$$

$$\tag{2b}$$

$$F(n) = \frac{1}{T} \int_{-T/2}^{T/2} f(t) e^{-jn\omega t} \, dt$$

If the period is increased to a very large value, then the following changes
in notation are appropriate.

$$n\omega \longrightarrow \omega$$

$$\omega \longrightarrow \Delta\omega$$

$$T \longrightarrow \frac{2\pi}{\omega}$$

From (1a) and (2a),

$$f(t) = \sum_{\omega=-\infty}^{\infty} F_\omega e^{j\omega t} \tag{3}$$

$$F(\omega) = \frac{\Delta\omega}{2\pi} \int_{-T/2}^{T/2} f(t) e^{-j\omega t} \, dt \tag{4}$$

and substituting (4) into (3), we have

$$f(t) = \frac{\Delta\omega}{2\pi} \left[\sum_{\omega=-\infty}^{\infty} \int_{-T/2}^{T/2} f(t) e^{-j\omega t} \, dt \, e^{j\omega t} \right] \tag{5}$$

If we increase the period to infinity, $f(t)$ becomes an aperiodic function.
$\Delta\omega \rightarrow d\omega$, and $\sum \rightarrow \int$.

$$f(t) = \frac{1}{2\pi} \int_{-\infty}^{\infty} \left[\int_{-\infty}^{\infty} f(t) e^{-j\omega t} \, dt \right] e^{j\omega t} \, d\omega \tag{6}$$

Equation (6) is the Fourier integral of $f(t)$.

2. Fourier Transform

A split form of the Fourier integral can be written as follows:

$$f(t) = \frac{1}{2\pi} \int_{-\infty}^{\infty} g(\omega) e^{j\omega t} \, d\omega \tag{7}$$

$$g(\omega) = \int_{-\infty}^{\infty} f(t) e^{-j\omega t} \, dt \tag{8}$$

which are called a Fourier transform pair. Their relationship can be written
symbolically as

$$f(t) = \mathscr{F}^{-1}[g(\omega)] \tag{7a}$$

$$g(\omega) = \mathscr{F}[f(t)] \tag{8a}$$

$\mathscr{F}^{-1}$ is said to "take inverse Fourier transform of," and $\mathscr{F}$ is said to "perform the Fourier transform on."

3. Time-frequency Correlation

Rewriting (7) gives

$$f(t) = \frac{1}{2\pi} \int_{-\infty}^{\infty} g(\omega) e^{j\omega t}\, d\omega \tag{7}$$

and rewriting (8) gives

$$g(\omega) = \int_{-\infty}^{\infty} f(t) e^{-j\omega t}\, dt \tag{8}$$

This pair expresses exactly the relation between time and frequency domains. Separate the time function into its even and odd parts by writing

$$f(t) = f_e(t) + f_o(t) \tag{9}$$

and let the $g(\omega)$ function be replaced by the form

$$g(\omega) = g_1(\omega) + jg_2(\omega) \tag{10}$$

Substitution into (8) gives

$$g_1(\omega) + jg_2(\omega) = \int_{-\infty}^{\infty} f_e(t) \cos \omega t\, dt - j \int_{-\infty}^{\infty} f_e(t) \sin \omega t\, dt$$
$$+ \int_{-\infty}^{\infty} f_o(t) \cos \omega t\, dt - j \int_{-\infty}^{\infty} f_o(t) \sin \omega t\, dt \tag{11}$$

The integrands of the middle two terms of the right-hand side of (11) are odd functions of t; these integrals have zero value. Therefore

$$g_1(\omega) = \int_{-\infty}^{\infty} f_e(t) \cos \omega t\, dt \tag{12}$$

$$g_2(\omega) = \int_{-\infty}^{\infty} f_o(t) \sin \omega t\, dt \tag{13}$$

The inverses of (12) and (13) can be obtained by manipulation analogous to that just carried out.

$$f_e(t) = \frac{1}{2\pi} \int_{-\infty}^{\infty} g_1(\omega) \cos \omega t\, d\omega \tag{14}$$

$$f_o(t) = \frac{1}{2\pi} \int_{-\infty}^{\infty} g_2(\omega) \sin \omega t\, d\omega \tag{15}$$

The integrand of (14) is an even function. It can also be written as

$$f_e(t) = \frac{1}{\pi} \int_{0}^{\infty} g_1(\omega) \cos t\omega\, d\omega$$

Then $f(t)$ is as follows:

$$f(t) = \frac{2}{\pi} \int_{0}^{\infty} g_1(\omega) \cos t\omega\, d\omega \tag{16}$$

Equations (8) and (16) are important formulas for converting an aperiodic function in the time domain into the corresponding function in the frequency domain, and vice versa. This would be done by simply substituting in (8)

$$g(\omega) = \int_{-\infty}^{\infty} f(t) e^{-j\omega t} \, dt \tag{8a}$$

provided $f(t)$ is given analytically.

However, if the $f(t)$ function is a graph or a list of experimental data, Formula (8a) cannot be directly applied, and some computational techniques need to be developed.

Take a segment of the given transient response curve $t_1 t_2 A_1 A_2$ in Fig. 4-10, which can be approximated by a trapezoid. In an analytical expression

$$f(t)_s = \left[\left(\frac{A_2 - A_1}{t_2 - t_1}\right)t + A_1 - \left(\frac{A_2 - A_1}{t_2 - t_1}\right)t_1\right][u(t - t_1) - u(t - t_2)] \tag{17}$$

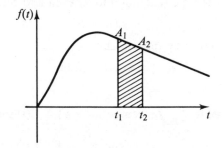

Fig. 4-10. A trapezoid to approximate a segment of the given transient curve.

The first factor is the equation of the line segment $A_1 A_2$, and the second factor is a gate function between t_1 and t_2.

In general for the segment between t_n and t_{n+1}, we can write:

$$f(t) = (m_n t + b_n)[u(t - t_n) - u(t - t_{n+1})] \tag{17a}$$

where

$$m_n = \frac{A_{n+1} - A_n}{t_{n+1} - t_n} \tag{18}$$

$$b_n = A_n - \left(\frac{A_{n+1} - A_n}{t_{n+1} - t_n}\right) t_n \tag{19}$$

Substituting (17a) into (8a) we have

$$g_n(\omega) = m_n \int_{t_n}^{\infty} t e^{-j\omega t} \, dt + b_n \int_{t_n}^{\infty} e^{-j\omega t} \, dt - m_n \int_{t_{n+1}}^{\infty} t e^{-j\omega t} \, dt - b_n \int_{t_{n+1}}^{\infty} e^{-j\omega t} \, dt$$

$$= -\frac{m_n}{\omega^2}(e^{-j\omega t_n} - e^{-j\omega t_{n+1}}) + \frac{1}{j\omega}[(m_n t_n + b_n)e^{-j\omega t_n} - (m_n t_{n+1} + b_n)e^{-j\omega t_{n+1}}] \tag{20}$$

Combining (18), (19), and (20) gives

$$g_n(\omega) = -\frac{1}{\omega^2}\left(\frac{A_{n+1} - A_n}{t_{n+1} - t_n}\right)(e^{-j\omega t_n} - e^{-j\omega t_{n+1}}) + \frac{1}{j\omega}(A_n e^{-j\omega t_n} - A_{n+1}e^{-j\omega t_{n+1}})$$

The summation of these straight line approximations of the time function gives the entire corresponding function in the frequency domain:

$$g(\omega) = \sum g_n(\omega)_s \qquad (\text{if } t_0 = 0; \; A_{k+1} \to 0)$$

$$= \frac{A_0}{j\omega} + \sum_{n=0}^{k}\left\{\left(\frac{A_{n+1} - A_n}{t_{n+1} - t_n}\right)\frac{1}{(-\omega^2)}(\cos \omega t_n - j \sin \omega t_n)\right.$$

$$\left. -\left[\frac{(A_{n+1} - A_n)}{t_{n+1} - t_n}\right]\left(\frac{1}{-\omega^2}\right)(\cos \omega t_{n+1} - j \sin \omega t_{n+1})\right\} \qquad (21)$$

A digital computer program to accomplish this is shown (see Program 6).

Program 6. EVALUATING FREQUENCY RESPONSE FROM TRANSIENT

```
C   C   EVALUATING FREQUENCY RESPONSE FROM TRANSIENT
        READ,V,Z,AO
        L=Z-2.
        K=Z-1.
        D=V/(Z-1.)
        DIMENSION A(100),Z(100)
        READ,(A(I),I=1,K)
        W=0.0
        ACCEPT,WMAX
        DO 1 I=1,24
        W=W+(WMAX/25.)
        DIV=-D*W**2
        T=0.0
        SUMRE=(A(1)-AO)/(DIV)
        SUMXI=-AO/W
        DO 2 J=1,L
        T=T+D
        J1=J+1
        JM1=J-1
        IF(I-1)9,9,8
   9    IF(J-1)3,3,7
   3    Z(1)=A(J1)-2.*A(J)+AO
        GO TO 8
   7    Z(J)=A(J1)-2.*A(J)+A(JM1)
   8    W1=-W*T
        RE=COS(W1)*Z(J)
        XI=SIN(W1)*Z(J)
        SUMRE=SUMRE+RE/(+DIV)
   2    SUMXI=SUMXI+XI/(DIV)
        XM=SQRT(SUMRE**2+SUMXI**2)
        IF(SUMRE)5,6,6
   5    PHD=(ATAN(SUMXI/SUMRE))*180./3.141592+180.
        GO TO 10
   6    PHD=(ATAN(SUMXI/SUMRE))*180./3.141592
  10    PUNCH11,W,XM,PHD
  11    FORMAT(3E15.8)
   1    CONTINUE
        END
```

Example. An impulse response which is generated from the following system function

$$\frac{1 + 0.5s}{1 + 0.8s + 0.02s^2}$$

is shown in Fig. 4-11. The frequency response of the system is desired.

In Fig. 4-11, 200 samples in the first given two seconds are taken as the input data. The frequency response is then obtained by using the digital computer program 6. Figure 4-12 shows the high accuracy, compared with the actual response.

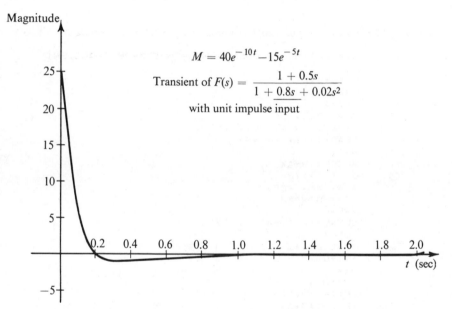

$$M = 40e^{-10t} - 15e^{-5t}$$

$$\text{Transient of } F(s) = \frac{1 + 0.5s}{1 + 0.8s + 0.02s^2}$$

with unit impulse input

Fig. 4-11. The transient response of $\dfrac{1 + 0.5s}{1 + 0.8s + 0.02s^2}$.

COMPARISON OF DATA

ω rad/sec	Approx. 200 values in 2 sec		Actual response	
	M (nepers)	Angle (deg)	M (nepers)	Angle (deg)
1.	1.096	9.636	1.091	9.54
5.	1.705	−3.39	1.703	−3.36
10.	1.614	−29.76	1.612	−29.74
20.	1.091	−55.12	1.090	−55.11
50.	0.489	−75.27	0.488	−75.27
90.	0.2761	−81.76	0.276	−81.75

III. Bode's Decomposition

1. Decomposing a Transfer Function

The transfer function for a time invariant system is usually expressed by a ratio of two polynomials of s.

$$M(s) = \frac{b_m s^m + b_{m-1} s^{m-1} + \cdots + b_1 s + b_0}{a_n s^n + a_{n-1} s^{n-1} + \cdots + a_1 s + a_0} \tag{1}$$

It is well known that (1) can be decomposed into a partial fraction as follows:

$$M(s) = \frac{k_1}{s + \alpha_1} + \frac{k_2}{s + \alpha_2} + \frac{k_3}{s + \alpha_3} + \cdots + \frac{k_n}{s + \alpha_n} \tag{2}$$

where $\alpha_1, \alpha_2, \ldots, \alpha_n$ are poles of (1), and $k_1, k_2, \ldots, k_n$ are residues. This form is sometimes called Heaviside's decomposition form, and it is very useful when performing the inverse Laplace transformation.

The second form of decomposition is to factor both the numerator and the denominator:

$$M(s) = \frac{K(s + \beta_1)(s + \beta_2) \cdots (s + \beta_m)}{(s + \alpha_1)(s + \alpha_2) \cdots (s + \alpha_n)} \tag{3}$$

where $\beta_1, \beta_2, \ldots, \beta_n$ are zeros of (1) and K is determined by b_m/a_n.

Equation (3) is called the pole-zero form or Bode's decomposition form. The third form is Wiener's:

$$M(s) = A_0 + A_1\left(\frac{1-s}{1+s}\right) + A_2\left(\frac{1-s}{1+s}\right)^2 + \cdots + A_n\left(\frac{1-s}{1+s}\right)^n + \cdots \tag{4}$$

where $A_0, A_1, \ldots, A_n$ can be determined by Fourier theory. The three forms described above indicate that decomposing a transfer function is not unique.

There is another method, known as Bush's form, which is as follows:

$$M(s) = \frac{(\ldots((b_m s + b_{m-1})s + b_{m-2})s + \cdots)s + b_0}{(\ldots((a_n s + a_{n-1})s + a_{n-2})s + \cdots)s + a_0} \tag{5}$$

(1), (2), (3), (4), and (5) are equivalent, and each form can be converted into any of the remaining four forms. Although (1) is usually to be considered as the fundamental form, (2) (3), (4), and (5) have certain advantages and usage for certain aspects. In this section, we will concentrate our effort on (3), or Bode's form.

2. Bode's Basic Equations

Bode's form has two advantages in analyzing a system:

1. It facilitates drawing a frequency response curve, if a transfer function is given.

2. It helps with complex curve fitting, if a frequency response is given. We will demonstrate the first advantage. Rewrite (3):

$$M(s) = \frac{K(s + \beta_1)(s + \beta_2)(s + \beta_3) \cdots (s + \beta_m)}{(s + \alpha_1)(s + \alpha_2)(s + \alpha_3) \cdots (s + \alpha_n)} \tag{3}$$

which can be rearranged as the following standard form:

$$M(s) = \frac{Q\left(\dfrac{s}{\beta_1} + 1\right)\left(\dfrac{s}{\beta_2} + 1\right) \cdots \left(\dfrac{s}{\beta_m} + 1\right)}{\left(\dfrac{s}{\alpha_1} + 1\right)\left(\dfrac{s}{\alpha_2} + 1\right) \cdots \left(\dfrac{s}{\alpha_n} + 1\right)} \tag{6}$$

where

$$Q = \frac{K \cdot \beta_1 \cdot \beta_2 \cdot \beta_3 \cdots \beta_m}{\alpha_1 \cdot \alpha_2 \cdot \alpha_3 \cdots \alpha_n}$$

The frequency response of the system can be evaluated by simply substituting s by $j\omega$ into (6); or

$$M(j\omega) = \frac{Q\left(\dfrac{j\omega}{\beta_1} + 1\right)\left(\dfrac{j\omega}{\beta_2} + 1\right) \cdots \left(\dfrac{j\omega}{\beta_m} + 1\right)}{\left(\dfrac{j\omega}{\alpha_1} + 1\right)\left(\dfrac{j\omega}{\alpha_2} + 1\right) \cdots \left(\dfrac{j\omega}{\alpha_n} + 1\right)} \tag{7}$$

Equation (7) can be interpreted as two equations: first, the magnitude response equation,

$$\begin{aligned} \log |M(\omega)| = \log Q &+ \log\left|\frac{j\omega}{\beta_1} + 1\right| + \log\left|\frac{j\omega}{\beta_2} + 1\right| \cdots + \log\left|\frac{j\omega}{\beta_m} + 1\right| \\ &- \log\left|\frac{j\omega}{\alpha_1} + 1\right| - \log\left|\frac{j\omega}{\alpha_2} - 1\right| \cdots - \log\left|\frac{j\omega}{\alpha_n} + 1\right| \end{aligned} \tag{8}$$

and secondly, the phase response equation,

$$\begin{aligned} \phi(\omega) = \tan^{-1}\frac{\omega}{\beta_1} &+ \tan^{-1}\frac{\omega}{\beta_2} + \cdots + \tan^{-1}\frac{\omega}{\beta_m} \\ &- \tan^{-1}\frac{\omega}{\alpha_1} - \tan^{-1}\frac{\omega}{\alpha_2} - \cdots - \tan^{-1}\frac{\omega}{\alpha_n} \end{aligned} \tag{9}$$

where

$$|M(\omega)| \, \underline{/\phi(\omega)} = M(j\omega) \tag{10}$$

Bode's technique is based on these two equations.

4. Magnitude vs. Frequency Plot

Equations (8) and (9) are called *frequency response* curves: (8) is for "magnitude vs. frequency," whereas (9) is for "phase vs. frequency."

A typical term of (8) reads

$$\log \left| \frac{j\omega}{\beta_1} + 1 \right| \tag{11}$$

Of course, (11) can be plotted by making the following table first:

ω	$\log \left\| \dfrac{j\omega}{\beta_1} + 1 \right\|$
0	$\log 1$
1	$\log \sqrt{\left(\dfrac{1}{\beta_1}\right)^2 + 1^2}$
2	$\log \sqrt{\left(\dfrac{2}{\beta_1}\right)^2 + 1^2}$
...	....
100	$\log \sqrt{\left(\dfrac{100}{\beta_1}\right)^2 + 1^2}$

and then draw the magnitude vs. frequency curve as shown in Fig. 4-12. There is a shortcut method, however, for plotting (11) by reasoning in the following way: Consider $\omega \ll \beta_1$,

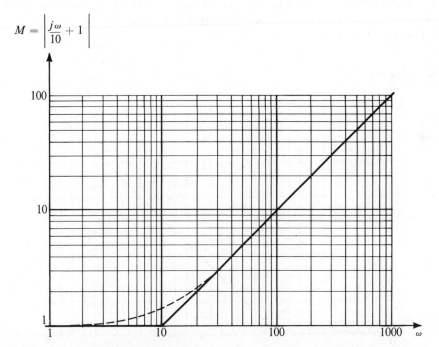

Fig. 4-12. The magnitude of the frequency response of $\left(\frac{s}{10} + 1\right)$: the exact curve and its asymptotes.

$$\log\left|\frac{j\omega}{\beta_1} + 1\right| \cong 0 \tag{11a}$$

because $j\omega/\beta_1$ is much less than 1 and can be omitted.

On the other hand, if $\omega \gg \beta_1$, 1 of (11) can be omitted, and (11) becomes

$$\log\left|\frac{j\omega}{\beta_1} + 1\right| = \log\left|\frac{j\omega}{\beta_1}\right| = \log|j\omega| - \log|\beta_1| = \log|\omega| - \log|\beta_1|$$

Differentiation with respect to $(\log \omega)$ yields

$$\frac{d\log\left|\frac{j\omega}{\beta_1} + 1\right|}{d\log|\omega|} \cong 1 \tag{11b}$$

If the magnitude vs. ω curve is plotted on a log-log paper, (11b) means that when $\omega \gg \beta_1$, the slope of the curve has $+1$ slope.

The magnitude vs. frequency plot of (11) is shown as in Fig. 4-12. It is noted that the background of Fig. 4-12 is a log-log paper, whereas that of Fig. 4-13 is regular linear scale paper. If the two planes are superimposed together, they present the same information, In the x-y plane, the scales are $x = \log\omega$, $y = 20\log|1 + (j\omega/\beta_1)|$, whereas in the M-ω plane, the scales are logarithmic. When we say that a curve has $+1$ slope, it means that the slope in the x-y plane is $+1$; however, this $+1$ slope can be directly read from the M-ω plane.

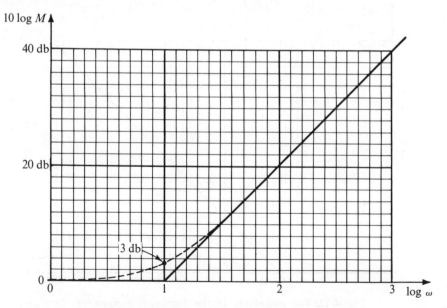

Fig. 4-13. The magnitude of the frequency response of $\left(\frac{s}{10} + 1\right)$ in the log plane.

The dotted line in Fig. 4-12 is the actual curve; the solid heavy line is called the asymptotic approximation of the actual curve. It is much easier to recognize β_1 on a log-log paper and then draw two straight lines than to plot one point after another. This kind of asymptotic approximation is called Bode's diagram in recognition of his contribution in feedback theory.

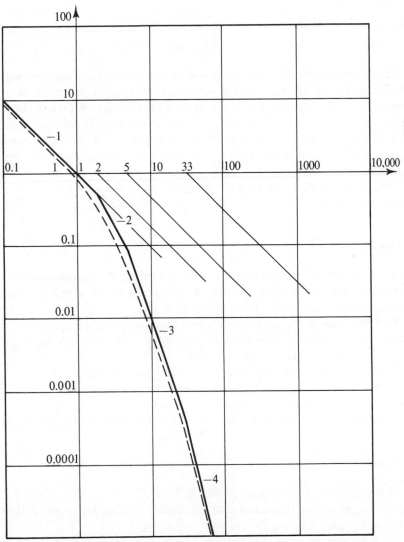

Fig. 4-14. The Bode diagram of $\dfrac{1}{s(1 + 0.03s)(1 + 0.2s)(1 + 0.5s)}$.

5. Example

Plot the frequency response of the following system:

$$M(s) = \frac{1}{s(1 + 0.03s)(1 + 0.2s)(1 + 0.5s)} \tag{12}$$

Substitution of s by $j\omega$ gives

$$M(j\omega) = \frac{1}{(j\omega)(1 + 0.03j\omega)(1 + 0.2j\omega)(1 + 0.5j\omega)} \tag{12a}$$

The standard magnitude curve is generated from

$$\log|M(\omega)| = -\log|\omega| - \log\left|1 + \frac{j\omega}{33}\right| - \log\left|1 + \frac{j\omega}{5}\right| - \log\left|1 + \frac{j\omega}{2}\right| \tag{13}$$

Several particular frequencies should be recognized on the graph paper; they are 2, 5, and 33. From these particular frequency corners, three "-1 slope" straight lines are drawn. The first term of (13) is a straight line through the origin (1,1) of the M-ω plane or $(0, 0)$ of x-y plane. These four component asymptotics are then combined, as shown in Fig. 4-14. The piecewise linear approximation is fairly good, compared with the frequency response as shown by the dotted line.

There is no simpler way to draw a phase curve than by plotting it one point by one point.

6. Identification

If the frequency response of a system is experimentally obtained and it is desired to find its transfer function, we can perform the reverse process of the procedures mentioned in the last section.

1. Approximate the magnitude curve by a series of piecewise straight lines or asymptotes. The asymptotes are chosen such that they have "$+1$" or "-1" or any "integer number" slope.

2. Decompose the asymptotes into several components.

3. Recognize the corner frequencies and then the corresponding transfer function in the pole-zero form can be written.

The process can be easily understood by the following illustrative example.

7. Example

Given experimental data as shown in Table 3.

The magnitude vs. frequency data are plotted on a log-log paper, as shown in Fig. 4-15.

Then we approximate the curve by several straight lines (Fig. 4-16). The corner frequencies are recognized as 0.7, 2.7, 19, 30, etc. The next step is to

TABLE 3

ω	Mag.	Ang.
0.5	33.086	$-117.5°$
1.0	12.207	$-129.61°$
2.0	4.079	$-134.01°$
3.0	2.226	$-133.54°$
4.0	1.494	$-133.58°$
5.0	1.113	$-134.61°$
6.0	0.880	$-136.41°$
7.0	0.721	$-138.74°$
8.0	0.607	$-141.41°$
9.0	0.519	$-144.30°$
10.0	0.449	$-147.30°$
15.0	0.246	$-162.39°$
20.0	0.149	$-177.07°$
25.0	0.097	$-187.13°$

decompose the asymptotes into several simple broken lines, namely, (1), (2), (3), (4), (5), and (6) of Fig. 4-17

Curve (1) of Fig. 4-17 is a constant, and curve (2) is recognized as $1/s$, etc. The following component transfer functions are established:

$$M_1 = 20$$

$$M_2 = \frac{1}{s}$$

$$M_3 = \frac{1}{\dfrac{s}{0.7} + 1}$$

$$M_4 = \frac{s}{2.7} + 1 \tag{14}$$

$$M_5 = \frac{1}{\dfrac{s}{19} + 1}$$

$$M_6 = \frac{1}{\dfrac{s}{30} + 1}$$

The combination of these components is as follows:

$$M = \prod_i M_i$$

$$= \frac{20\left(\dfrac{s}{2.7} + 1\right)}{s\left(\dfrac{s}{0.7} + 1\right)\left(\dfrac{s}{19} + 1\right)\left(\dfrac{s}{30} + 1\right)} \tag{15}$$

which is the transfer function of the system in question.

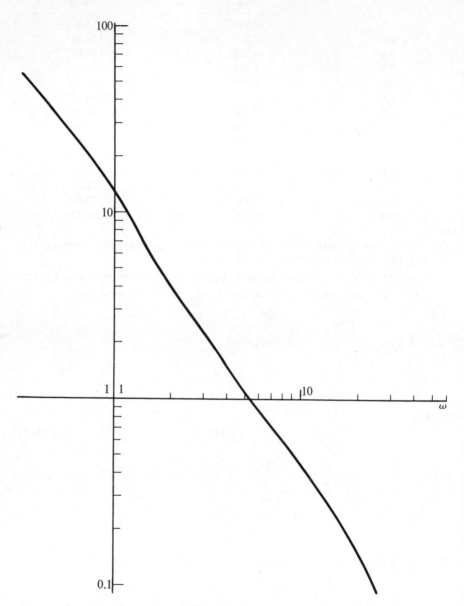

Fig. 4-15. A given magnitude curve of the frequency response of a system.

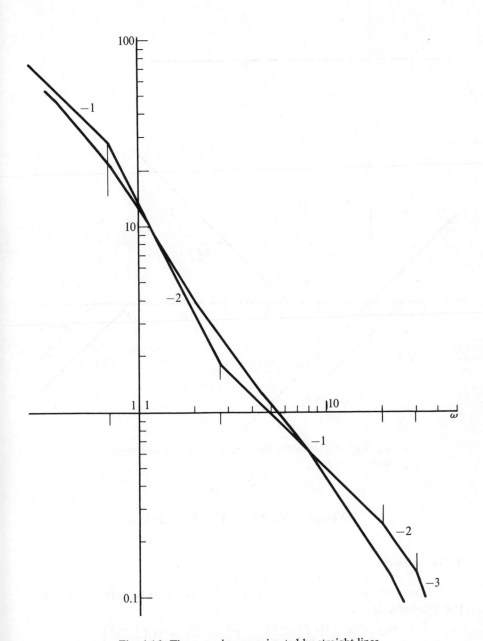

Fig. 4-16. The curve is approximated by straight lines.

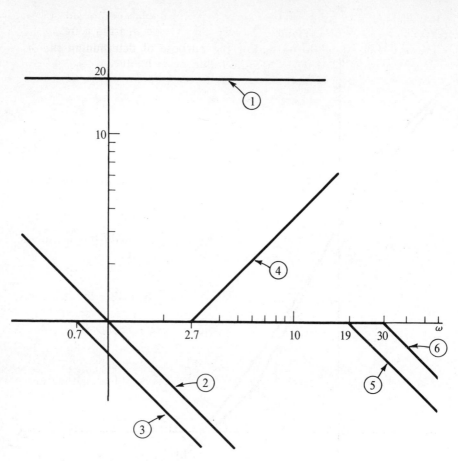

Fig. 4-17. Figure 4-16 is then decomposed into six components.

IV. Wiener-Lee's Method of Decomposition

1. The Theory

Wiener and Lee propose that a transfer function can be decomposed into the following form:

$$F(s) = A_0 + A_1\left(\frac{1-s}{1+s}\right) + A_2\left(\frac{1-s}{1+s}\right)^2 + A_3\left(\frac{1-s}{1+s}\right)^3$$

$$+ \cdots + A_k\left(\frac{1-s}{1+s}\right)^k + \cdots \tag{1}$$

where $F(s)$ could be a rational fraction of two polynomials in s, or simply the frequency response of a system. However, to evaluate A_k from a frequency response is not straightforward. For the purpose of determining the A_k values, Wiener and Lee developed the following technique.

With the use of a bilinear transformation

$$s = \frac{1 - e^{-j\phi}}{1 + e^{-j\phi}}\bigg|_{s \to j\omega} \tag{2}$$

or

$$e^{-j\phi} = \frac{1 - s}{1 + s}\bigg|_{s \to j\omega} \tag{3}$$

$F(j\omega)$ can be transformed from the ω domain to the ϕ domain. Substitution of (3) into (1) yields

$$\hat{F}(\phi) = A_0 + A_1 e^{-j\phi} + A_2 e^{-j2\phi} + A_3 e^{-j3\phi} + \cdots \tag{4}$$

where $\hat{F}(\phi)$ is a complex function in general and is obtained from measured data. It can be separated into its real and imaginary parts:

$$\hat{F}(\phi) = P(\phi) + jQ(\phi) \tag{5}$$

The right-hand side of (4) can also be separated. Equating the reals, we have

$$P(\phi) = A_0 + A_1 \cos \phi + A_2 \cos 2\phi + A_3 \cos 3\phi + \cdots$$

Once $P(\phi)$ is obtained numerically, the A_k's can be determined by a Fourier series expansion, as shown in Sec. I of this chapter.

This is the core of the Wiener-Lee method. It is a way of finding the transfer function of a system when the frequency response is known. The techniques necessary to apply it will now be developed.

2. The Bilinear Transformation

First, it is necessary to express the real part of the given frequency response, Re $\{F(j\omega)\}$, in terms of the new independent variable ϕ by using

$$e^{-j\phi} = \frac{1 - s}{1 + s}\bigg|_{s \to j\omega} \tag{3}$$

The transformation bridge can be written in another form:

$$\underline{/-\phi} = \frac{1 - j\omega}{1 + j\omega}$$

Rationalize the right side.

$$\underline{/-\phi} = \frac{1 - \omega^2 - 2j\omega}{1 + \omega^2}$$

Separate into real and imaginary parts.

$$\cos \phi - j \sin \phi = \frac{1 - \omega^2}{1 + \omega^2} - \frac{2j\omega}{1 + \omega^2} = 1\underline{/\tan^{-1} \frac{-2\omega}{1 - \omega^2}} \tag{6}$$

Rewrite into polar form.

$$\tan \phi = \frac{2\omega}{1 - \omega^2}$$

or

$$\tan \left(\frac{\phi}{2} + \frac{\phi}{2} \right) = \frac{2\omega}{1 - \omega^2}$$

Finally,

$$\omega = \tan \frac{\phi}{2} \qquad (6a)$$

or

$$\phi = 2 \tan^{-1} \omega \qquad (6b)$$

It is apparent that the interval $(-\infty, \infty)$ for the variable ω corresponds to the interval $(-\pi, \pi)$ for the variable ϕ. This conversion from one variable to the other can be done with a circular chart, as shown in Fig. 4-18.

The right side of Eq. (6) indicates a magnitude of 1 and an angle ϕ [Fig. 4-18(b)]. The same point in the ω plane corresponds to $\omega = \tan (\phi/2)$ [Fig. 4-18(a)].

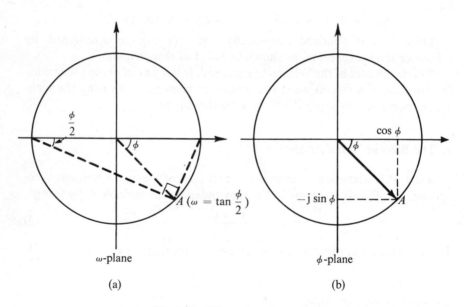

Fig. 4-18. Bilinear transformation.

Example 1. A set of data which was generated from the real part of $F(j\omega) = 1/(j\omega + 1)$ is shown in Table 4. These data are plotted on the ω plane to produce the curve in Fig. 4-19(a). Then the independent variable is changed, as shown in Fig. 4-19(b). By interpolation the latter can be cut into uniform intervals. The values of the variable can then be measured at these intervals to produce Table 5.

The $P(\phi)$ curve in Fig. 4-19(b) is a function with a period of 2π.

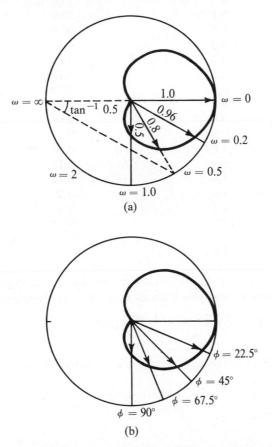

(a)

(b)

Fig. 4-19. Converting the data in the ω domain into the ϕ domain.

3. Fourier Series Coefficients

From the list in Table 5 the Fourier coefficients $A_0, A_1, A_2, \ldots, A_n$ can easily be found by applying Eq. (2a) in Sec. I of this chapter. It was

$$A_k = \frac{1}{m} \sum_{k=1}^{2m} P_k \cos nk \frac{360}{2m} \tag{7}$$

In our example problem, $m = 18$ and the P_k's are listed in Table 5:

$$A_0 = 0.5, \quad A_1 = 0.5, \quad A_2 = 0, \quad A_3 = 0, \quad \ldots$$

TABLE 4

ω	Re $\{F(j\omega)\}$
0.0	1.00
0.1	0.99
0.2	0.96
0.3	0.92
0.4	0.86
0.5	0.80
0.6	0.74
0.7	0.67
0.8	0.61
0.9	0.55
1.0	0.50
1.2	0.41
1.4	0.34
1.6	0.28
1.8	0.24
2.0	0.20
2.5	0.14
3.0	0.10

Substituting these into (1), we obtain Wiener-Lee's decomposition form of the transfer function of a system whose frequency response is given in Table 4:

$$F(s) = 0.5 + 0.5\left(\frac{1-s}{1+s}\right) + 0\left(\frac{1-s}{1+s}\right)^2 + \cdots$$

Example 2. The frequency response generated from

$$G(s) = \frac{1}{s+2}$$

is given as follows:

$$G(j\omega) = \frac{1}{j\omega + 2} = \sqrt{\omega^2 + 2^2}\left/-\tan^{-1}\frac{\omega}{2}\right. \tag{8}$$

By substituting in (6a) and, interpolating the data, we make a new set in terms of ϕ, as shown in Table 6.

Then substitute into (7) to obtain the A_k's, as shown in Table 7.

Thus the Wiener-Lee form for this system is

$$F(s) = 0.33332120 + 0.22222173\left(\frac{1-s}{1+s}\right) - 0.07407394\left(\frac{1-s}{1+s}\right)^2 + \cdots \tag{9}$$

4. Laguerre Polynomials

Why describe a system by the Wiener-Lee decomposition form? The ultimate objective is to get the transient response of the system, and this form

TABLE 5	
ϕ	$P(\phi)$
0	1.00
10	0.99
20	0.97
30	0.93
40	0.88
50	0.82
60	0.75
70	0.67
80	0.59
90	0.50
100	0.41
110	0.33
120	0.25
130	0.18
140	0.12
150	0.07
160	0.03
170	0.01
180	0.00

TABLE 6	
ϕ	$P(\phi)$
0	0.4999
10	0.4990
20	0.4961
30	0.4912
40	0.4840
50	0.4742
60	0.4615
70	0.4454
80	0.4242
90	0.4000
100	0.3690
110	0.3311
120	0.2857
130	0.2326
140	0.1732
150	0.1116
160	0.0553
170	0.0149
180	0.0000

TABLE 7

A_0	0.33332120
A_1	0.222221730
A_2	−0.07407397
A_3	0.02469134
A_4	−0.00823044
A_5	0.00274347
A_6	−0.00091449
A_7	0.00030482
A_8	−0.00010162
A_9	0.00003387
A_{10}	−0.00001130

is useful as a means to that end only if there is some way to find the inverse Laplace transform of it.

As a matter of fact, there is an elegant way to perform the inverse transformation. But in order to understand how it is done, it will first be necessary to examine a special function known as Laguerre's function.

Laguerre polynomials, $L_n(t)$, are defined by

$$L_n(t) = \frac{e^t}{n!} \frac{d^n}{dt^n}(t^n e^{-t}) \tag{10}$$

Differentiating this expression leads to the recursive formula

$$\frac{d}{dt}L_n(t) = \frac{d}{dt}L_{n-1}(t) - L_{n-1}(t) \tag{11}$$

The explicit forms of these polynomials are

$$L_0(t) = 1$$
$$L_1(t) = -t + 1$$
$$L_2(t) = \frac{t^2}{2} - 2t + 1$$
$$L_3(t) = -\frac{1}{3!}t^3 + \frac{3}{2}t^2 - 3t + 1 \tag{12}$$
$$L_4(t) = \frac{1}{4!}t^4 - \frac{2}{3}t^3 + 3t^2 - 4t + 1$$

and the derivatives are

$$L_0'(t) = 0$$
$$L_1'(t) = -1$$
$$L_2'(t) = t - 2$$
$$L_3'(t) = -\frac{1}{2}t^2 + 3t - 3 \tag{13}$$
$$L_4'(t) = \frac{1}{3!}t^3 - 2t^2 + 6t - 4$$

5. A Laplace Transform Pair

Before the inverse Laplace transform of (1) being explained, a Laplace transform pair related to a Laguerre function is developed:

$$\mathscr{L}\{e^{-at}L_n[(a - b)t]\} = \frac{(s + b)^n}{(s + a)^{n+1}} \tag{14}$$

From the basic definition we obtain the transforms of (12) as follows:

$$\mathscr{L}L_0(t) = \int_0^\infty 1e^{-st}\, dt = \frac{1}{s}$$

$$\mathscr{L}L_1(t) = \int_0^\infty (-t + 1)e^{-st}\, dt = -\frac{1}{s^2} + \frac{1}{s}$$

$$\mathscr{L}L_2(t) = \int_0^\infty \left(\frac{t^2}{2} - 2t + 1\right)e^{-st}\, dt = \frac{1}{s^3} - \frac{2}{s^2} + \frac{1}{s} \tag{15}$$

. . . .

$$\mathscr{L}L_n(t) = \frac{1}{s}\left(\frac{s - 1}{s}\right)^n$$

Using the scaling theorem from Laplace theory, we have

$$\mathscr{L}L_n(qt) = \frac{\dfrac{1}{q}}{\dfrac{s}{q}}\left(\dfrac{\dfrac{s}{q} - 1}{\dfrac{s}{q}}\right)^n$$

where q is a real number. Simplification yields

$$\mathscr{L}L_n(qt) = \frac{1}{s}\left(\frac{s-q}{s}\right)^n$$

Applying the translation theorem from Laplace theory gives

$$\mathscr{L}e^{-at}L_n(qt) = \frac{1}{s+a}\left(\frac{s+a-q}{s+a}\right)^n$$

Let $a - q = b$ or $q = a - b$:

$$\mathscr{L}\{e^{-at}L_n[(a-b)t]\} = \frac{(s+b)^n}{(s+a)^{n+1}} \tag{16}$$

6. Inverse Laplace Transform

Rewriting (1), we have

$$F(s) = A_0 + A_1\left(\frac{1-s}{1+s}\right) + A_2\left(\frac{1-s}{1+s}\right)^2 + A_3\left(\frac{1-s}{1+s}\right)^3 + \cdots \tag{1}$$

Dividing the right-hand side by $(1 + s)$ and multiplying it by the same factor gives

$$F(s) = (s+1)\left[\frac{A_0}{s+1} + (-1)^1 A_1\frac{s-1}{(s+1)^2} + (-1)^2 A_2\frac{(s-1)^2}{(s+1)^3} + \cdots\right]$$

Applying the transform pair (16), we have

$$F(s) = (s+1)\mathscr{L}[A_0 e^{-t} + (-1)^1 A_1 e^{-t}L_1(2t)$$
$$+ (-1)^2 A_2 e^{-t}L_2(2t) + (-1)^3 A_3 e^{-t}L_3(2t) + \cdots]$$
$$= (s+1)\mathscr{L}e^{-t}[\sum_{k=0}^{n}(-1)^k A_k L_k(2t)]$$

Inverse Laplace transform yields

$$f(t) = \frac{d}{dt}\left[e^{-t}\sum_{k=0}^{k=n}(-1)^k A_k L_k(2t)\right]$$
$$+ \delta(t)\sum(-1)^k A_k L_k(0) + e^{-t}\left[\sum_{k=0}^{k=n}(-1)^k A_k L_k(2t)\right]$$
$$= e^{-t}\left[\sum_{k=0}^{k=n}(-1)^k A_k L'(2t)\right] + \left[\sum_{k=0}^{k=n}(-1)^k A_k L_k(2t)\right]\cancel{(-1)(e^{-t})}$$
$$+ \delta(t)\sum(-1)^k A_k L_k(0) + e^{-t}\cancel{\left[\sum_{k=0}^{k=n}(-1)^k A_k L_k(2t)\right]} \tag{17}$$

For the common case where the numerator of the transfer function is of lower order than the denominator, the third term of (17) is equal to zero.

Therefore,

$$f(t) = e^{-t}[\sum_{k=0}^{n}(-1)^k A_k L'_k(2t)] \tag{17a}$$

The inverse Laplace transformation of (1) is (17). This formula is even more useful than the one for the exponential form which we used in Chapter 2.

7. Comparison of Two Methods

The regular inverse Laplace transformation, discussed in Chapter 2, involves the following steps:

1. Factoring the denominator of the transfer function

$$F(s) = \frac{s^m + b_{n-1}s^{m-1} + \cdots + b_0}{s^n + a_{n-1}s^{n-1} + a_{n-2}s^{n-2} + \cdots + a_0}$$

into $\prod_{k=1}^{n} (s - s_k)$.

2. Using Heaviside partial fraction expansion to find the values of K_k.

3. Substituting into the inverse formula:

$$f(t) = \sum_k K_k e^{s_k t} \tag{18}$$

where s_k is the argument of the exponential function.

The method for obtaining the inverse Laplace transform in terms of the Laguerre functions involves:

1. Plotting $G(j\omega)$ on a circular chart in order to change the independent variable by a bilinear transformation.

2. Using Fourier series expansion to find the coefficients A_k, where A_k is defined as

$$G(s) = A_0 + A_1\left(\frac{1-s}{1+s}\right) + A_2\left(\frac{1-s}{1+s}\right)^2 + A_3\left(\frac{1-s}{1+s}\right)^3 + \cdots$$

3. Substituting A_k into the following formula:

$$g(t) = e^{-t}[\sum_{k=0}^{n} (-1)^k A_k L_k'(2t)] \tag{19}$$

where $L_k'(2t)$ is the derivative of the Laguerre polynomial with respect to t. The argument is $2t$.

8. Example

Perform the inverse Laplace transform of (9) as follows:

$$\mathcal{L}^{-1}\{F(s)\} = \mathcal{L}^{-1}\left[0.33332120 + 0.22222173\left(\frac{1-s}{1+s}\right)\right.$$
$$\left. - 0.07407394\left(\frac{1-s}{1+s}\right)^2 + \cdots\right] \tag{8}$$

For the sake of clarity, only the first five terms will be taken. That means we let $n = 4$.

$$f(t) = e^{-t}[(-1)^0 A_0 L_0'(2t) + (-1)^1 A_1 L_1'(2t) + (-1)^2 A_2 L_2'(2t)$$
$$+ (-1)^3 A_3 L_3'(2t) + (-1)^4 A_4 L_4'(2t) + \cdots]$$
$$= e^{-t}\{0 + (-1)^1 (0.22222173)(-1)(2)$$
$$+ (-1)^2 (-0.07407394)(2t - 2)(2)$$
$$+ (-1)^3 (0.02469134)[-\tfrac{1}{2}(2t)^2 + 3(2t) - 3](2)$$
$$+ (-1)^4 (-0.00823044)[\tfrac{1}{6}(2t)^3 - 2(2t)^2 + 6(2t) - 4](2)\}$$

Rearrangement gives

$$f(t) = e^{-t}(0.95473078 - 0.7901224t + 0.2304524t^2 - 0.0219478t^3)$$

The exact answer for $\mathscr{L}^{-1}F(s) = \mathscr{L}^{-1}1/(s+2)$ should be a power series expansion of e^{-2t}:

$$f(t) = e^{-t}\left(1 - t + \frac{t^2}{2} - \frac{t^3}{6} + \cdots\right)$$

It can be seen that, even when only five A_k's are considered, the result is fairly good. When more coefficients are taken into consideration, the approximation is more accurate.

V. Bush's Decomposition

1. Open Loop and Closed Loop Transfer Functions

A typical feedback system, shown in Fig. 4-20, can be combined into a single block or a closed loop transfer function by the use of the following formula:

$$\frac{G(s)}{1 + G(s)} = M(s) \tag{1}$$

On the other hand, a system is easily decomposed into a feedback system by applying the inverse formula:

$$\frac{M(s)}{1 - M(s)} = G(s) \tag{2}$$

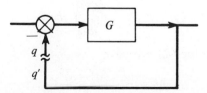

Fig. 4-20. A feedback system; the loop is opened at q, q'.

Combining a feedback system into a single transfer function or decomposing a transfer function into a feedback system are techniques used very often in analysis and synthesis of control systems, particularly in the identification problem.

Before explaining how one can identify a system by repeated use of (1) and (2), we have to illustrate the corresponding conversion formula in the frequency domain as well as the related charts.

2. Open Loop and Closed Loop Frequency Responses

In practice, the frequency response of an open loop transfer function $G(j\omega)$ is obtained from measurements and the closed loop frequency response $M(j\omega)$ is calculated. This is done by opening the loop at qq', applying a sinusoidal wave on q, and measuring the response on q'. The over-all or closed loop frequency response can be obtained by using the formula

$$M(j\omega) = \frac{G(j\omega)}{1 + G(j\omega)} \tag{1a}$$

or

$$|M(\omega)|\underline{/\alpha(\omega)} = \frac{|G(\omega)|\underline{/\phi(\omega)}}{1 + |G(\omega)|\underline{/\phi(\omega)}} \tag{1b}$$

where $|M(\omega)|\underline{/\alpha(\omega)}$ and $|G(\omega)|\underline{/\phi(\omega)}$ are the magnitude and phase angle of the closed loop and the open loop frequency responses, respectively.

3. Graphical Conversion Technique I: Hall's Chart

The conversion formula (1b) can be presented graphically. Rewrite the open loop frequency response into the rectangular form:

$$G(j\omega) = |G(\omega)|\underline{/\phi(\omega)} = a(\omega) + jb(\omega) \tag{3}$$

where $a(\omega)$ and $b(\omega)$ are called the real part and imaginary part responses of $G(s)$, respectively.

The function $M(\omega)$ is then given by

$$|M(\omega)| = \frac{|a(\omega) + jb(\omega)|}{|1 + a(\omega) + jb(\omega)|} \tag{4}$$

For clarity, omitting the (ω), we have

$$M = \frac{\sqrt{a^2 + b^2}}{\sqrt{(1 + a)^2 + b^2}}$$

Rearrangement gives

$$a^2(1 - M^2) - 2M^2a - M^2 + (1 - M^2)b^2 = 0 \tag{5}$$

or finally,

$$\left(a - \frac{M^2}{1 - M^2}\right)^2 + b^2 = \left(\frac{M}{1 - M^2}\right)^2 \tag{6}$$

Equation (6) is recognized as the equation for a circle with a radius

$$r = \left|\frac{M}{1 - M^2}\right| \tag{7a}$$

and a center

$$\text{Center:} \left(\frac{M^2}{1 - M^2}, 0\right) \tag{7b}$$

Based on (7a) and (7b) we compile the following table:

M	Radius: $\left\lvert\dfrac{M}{1-M^2}\right\rvert$	Center: $\left(\dfrac{M^2}{1-M^2}, 0\right)$
0.2	0.208	+0.0416, 0
0.5	0.666	+0.333, 0
0.9	4.73	+4.26, 0
1.0	∞	∞, 0
1.1	5.238	−5.762, 0
1.5	1.2	−1.8, 0
2.0	0.666	−1.333, 0
5.0	0.208	−1.041, 0

A family of constant M contours is drawn in Fig. 4-21. By the aid of this chart, if a and b are located, the corresponding M can be directly read from the contour.

If a similar process is used, we can obtain a family of constant α contours. Starting from the formula

$$/\alpha = \frac{\left/\tan^{-1}\dfrac{b}{a}\right.}{\left/\tan^{-1}\dfrac{b}{1+a}\right.} \tag{8}$$

or

$$\alpha = \tan^{-1}\frac{b}{a} - \tan^{-1}\frac{b}{1+a} \tag{9}$$

and taking the tangent on both sides, we have

$$\tan \alpha = \tan\left(\tan^{-1}\frac{b}{a} - \tan^{-1}\frac{b}{1+a}\right)$$

This yields

$$\tan \alpha = \frac{b}{a^2 + a + b^2}$$

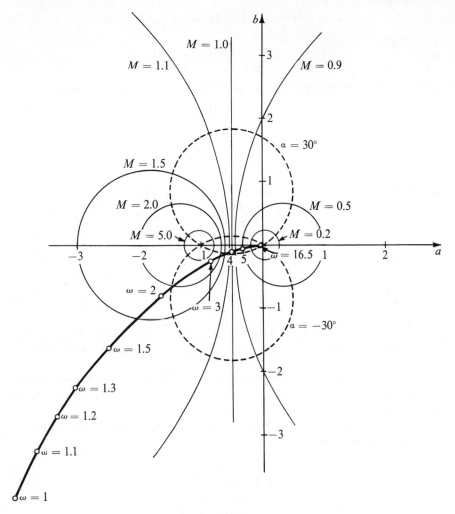

Fig. 4-21. Hall's chart: converting open loop response $a(\omega) + jb(\omega)$ into the corresponding closed loop response $M\underline{/\alpha}$, and vice versa.

Finally, we obtain

$$\left(a + \frac{1}{2}\right)^2 + \left(b - \frac{1}{2\tan\alpha}\right)^2 = \left[\sqrt{\frac{1}{4} + \left(\frac{1}{2\tan\alpha}\right)^2}\right]^2 \tag{10}$$

This equation is recognized as a family of circles also. In Fig. 4-21 the dotted circles are the constant α contours. The two families of circles are usually called Hall's chart.

The closed loop frequency response of a system can be obtained from the contour readings after the open loop frequency response on the G plane or $a + jb$ plane has been drawn.

4. Graphical Conversion Technique II: Chen-Shen's Chart

Instead of using polar coordinates M, α, Chen and Shen develop a chart by using $M \underline{/\alpha} = U + jV$. Equation (1a) and the following two forms are equivalent:

$$\frac{G(j\omega)}{1 + G(j\omega)} = M\underline{/\alpha} \tag{1c}$$

or

$$\frac{a + jb}{1 + a + jb} = U + jV \tag{1d}$$

Equation (1d) can be resolved into two equations involving the real and the imaginary parts respectively; i.e.,

$$U = \frac{a^2 + b^2 + a}{(1 + a)^2 + b^2} \tag{11}$$

$$V = \frac{b}{(1 + a)^2 + b^2} \tag{12}$$

Elimination of b from Eqs. (11) and (12) gives

$$\left[U - \frac{2a + 1}{2(1 + a)}\right]^2 + V^2 = \left[\frac{1}{2(1 + a)}\right]^2 \tag{13}$$

and elimination of a yields

$$(U - 1)^2 + \left(V - \frac{1}{2b}\right)^2 = \left(\frac{1}{2b}\right)^2 \tag{14}$$

Equation (13) represents a family of constant a circles having centers on the U axis at $(2a + 1)/[2(1 + a)]$ and radii of $1/[2(1 + a)]$. Equation (14) represents constant b circles having centers on the $U = 1$ line at $1/2b$ and radii of $1/2b$. These two families of circles are orthogonal, and they intersect at the common point $U = 1$, $V = 0$.

Figure 4-22 is the conversion chart which shows the constant a contours and the constant b contours as indicated.

The open loop frequency response of a system can be obtained from the contour readings by drawing the closed loop frequency response on the M plane or $U + jV$ plane.

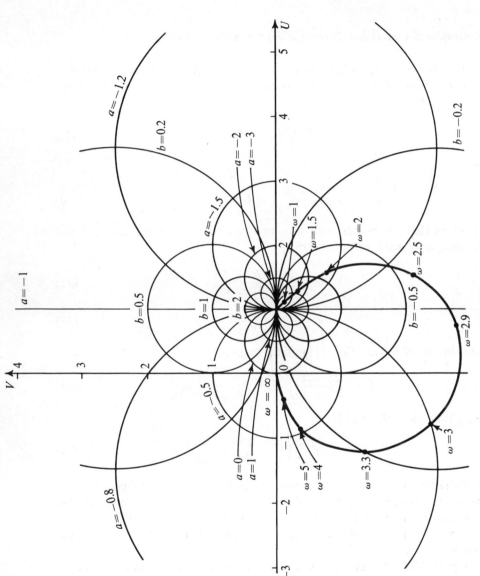

Fig. 4-22. Chen-Shen chart: converting an open loop response $a(\omega) + jb(\omega)$ into the corresponding closed loop response $U(\omega) + jV(\omega)$, and vice versa.

5. Graphical Conversion Technique III: Nichols' Chart

If the magnitude M is expressed by a log scale and α is in degrees, the frequency response plot of a system usually shows a better distribution. Of course, one can present the conversion chart in the log and degree scale.

Consider

$$G(j\omega) = |G(\omega)| e^{j\phi(\omega)} \tag{15}$$

and

$$M(j\omega) = |M(j\omega)| e^{j\alpha(\omega)} \tag{16}$$

Taking the logarithm of both sides of (15) and (16), we have

$$\log G(j\omega) = \log|G(\omega)| + j\phi(\omega) \tag{15a}$$

$$\log M(j\omega) = \log|M(\omega)| + j\alpha(\omega) \tag{16a}$$

If $\log|G(\omega)|$ is considered as an axis, then $\phi(\omega)$ is another axis. The right-hand side can be presented as a rectangular plane.

The M and α contours can be derived in a similar way.

$$M(j\omega) = \frac{G(j\omega)}{1 + G(j\omega)}$$

Rewrite

$$M(j\omega) = \frac{1}{1 + \dfrac{1}{G(j\omega)}}$$

or

$$Me^{j\alpha} = \frac{1}{1 + \dfrac{1}{|G|}e^{-j\phi}}$$

Taking the logarithm on both sides, we have

$$\ln M + j\alpha = -\ln\left(1 + \frac{1}{|G|}e^{-j\phi}\right)$$

$$= -\ln\left(1 + \frac{\cos\phi - j\sin\phi}{|G|}\right)$$

$$= -\ln\left[\left(1 + \frac{\cos\phi}{|G|}\right) - j\left(\frac{\sin\phi}{|G|}\right)\right]$$

$$= -\ln\left[\sqrt{\left(1 + \frac{\cos\phi}{|G|}\right)^2 + \left(\frac{\sin\phi}{|G|}\right)^2} \Big/ \tan^{-1}\frac{-\sin\phi}{|G| + \cos\phi}\right]$$

from which two equations can be written:

$$M = \left[\sqrt{\left(1 + \frac{\cos\phi}{|G|}\right)^2 + \left(\frac{\sin\phi}{|G|}\right)^2} \right]^{-1} \tag{17}$$

$$\alpha = \tan^{-1} \frac{\sin\phi}{|G| + \cos\phi} \tag{18}$$

Two families of contours, based on (17) and (18), can be drawn as shown in Fig. 4-23.

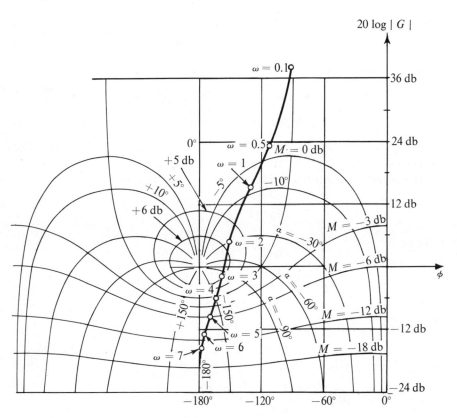

Fig. 4-23. Nichols' chart: converting an open loop frequency response $20 \log (G) + j\phi$ into the corresponding closed loop response $20 \log (M) + j\alpha$, and vice versa.

6. An Example

Find the closed loop frequency response of a unity feedback system which has an open loop frequency response as shown in Figs. 4-24 and 4-25. (The response is generated from $8/s(1 + s)(10^{-4}s^2 + 3.5 \times 10^{-3}s + 1)$.)

The answer can be obtained by drawing the open loop frequency response either on the a, b plane of Hall's chart (Fig. 4-21), or on the G, ϕ plane of

Nichols' chart (Fig. 4-23) and then taking the readings from the M, α contours.

Two loci are plotted on the Hall chart and Nichols' chart, respectively. Figures 4-26 and 4-27 present the answer in the regular form.

The answer can also be obtained by plotting the open loop frequency response on the a, b contours of Chen-Shen's chart (Fig. 4-22).

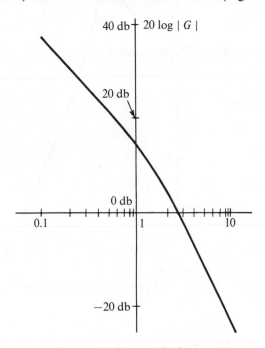

Fig. 4-24. The magnitude part of a frequency response.

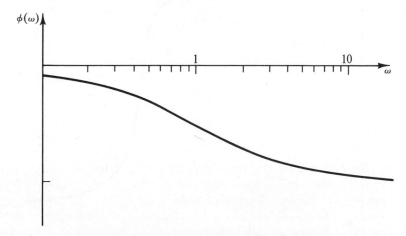

Fig. 4-25. The phase part of the frequency response.

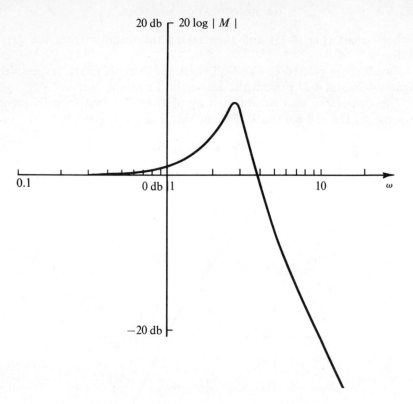

Fig. 4-26. The magnitude part of the frequency response.

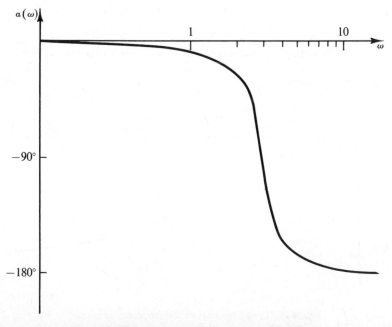

Fig. 4-27. The phase part of the frequency response.

212

7. A Pair of Generalized Conversion Formulas

The conversion charts that have been discussed are applicable for unity feedback systems; however, if the gain of the feedback path is not unity, the charts cannot be directly applied.

With the tremendous improvement and availability of digital computers, it is only logical that they be used to perform the conversion, especially when the feedback gain is not unity.

Two basic formulas are needed in converting a generalized feedback system that has a feedback gain H.

It is obvious that the following formula holds:

$$M(s) = \frac{G}{1 + GH} = \frac{1}{H + \dfrac{1}{G}} \tag{19}$$

The frequency responses $M(j\omega)$ and $G(j\omega)$ are related by

$$|M|\underline{/\alpha} = \frac{1}{H + \dfrac{1}{\overline{|G|\underline{/\phi}}}} \tag{20}$$

The conversion formulas are easily derived as follows:

$$U = \frac{a(1 + aH) + b^2 H}{(1 + aH)^2 + (bH)^2} \tag{21}$$

$$V = \frac{b}{(1 + aH)^2 + (bH)^2} \tag{22}$$

where a, b, U, and V are defined as before.

Formulas (21) and (22) are used for converting an open loop frequency response into a closed loop frequency response. On the other hand, if the closed loop frequency response is given, the corresponding open loop frequency response is then determined by

$$a = \frac{U(1 - HU) - H^2 V^2}{(1 - HU)^2 + (HV)^2} \tag{23}$$

$$b = \frac{V}{(1 - HU)^2 + (HV)^2} \tag{24}$$

8. Bush's Decomposition

Consider a typical transfer function

$$M(s) = \frac{1}{a_2 + a_1 s + s^2} \tag{25}$$

Bush regroups it as follows:

$$M(s) = \frac{1}{a_2 + (a_1 + s)s} \tag{26}$$

The expression can be interpreted as a feedback system

$$M(s) = \frac{1}{H_1 + \dfrac{1}{G_1}} \tag{27}$$

where

$$H_1 = a_2 \quad \text{and} \quad \frac{1}{G_1} = s(s + a_1)$$

Again, G_1 can be decomposed as another feedback system. Finally,

$$M(s) = \frac{1}{a_2 + \dfrac{1}{\dfrac{1}{s} \cdot \dfrac{1}{a_1 + \dfrac{1}{\dfrac{1}{s}}}}} \tag{28}$$

This is a continuous fraction expansion of (25) and the corresponding block diagram is particularly clear, as shown in Fig. 4-28.

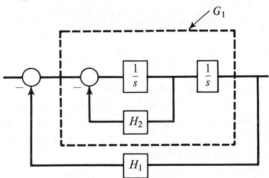

Fig. 4-28. Block diagram interpretation of continuous fraction expansion of a transfer function.

If $M(s)$ can be decomposed as $G_1(s)$ and H, is it possible to decompose $M(j\omega)$ into $G_1(j\omega)$ and H also? If this can be done, the identification problem becomes a problem of how to decompose a frequency response into a generalized feedback open loop frequency response.

The generalized conversion formulas (21) and (22) definitely will play the main role. One difficulty still remains: how to determine the values of H_i.

9. Determination of H

Consider the following two frequency responses: Fig. 4-29 is generated from

$$G_1 = \frac{s + \lambda}{s(s + \delta)}$$

and Fig. 4-29(b) is generated from

$$G_2 = \frac{(s + \gamma)}{(s + \alpha)(s + \beta)}$$

The two Bode diagrams have a basic difference: $|G_1(j\omega)|$ has an initial slope -1 in the low frequency region, whereas $|G_2(j\omega)|$ has an initial slope of 0 in that region. This phenomenon helps us determine the value of H in decomposing a system.

Now we return to our original problem. The solid lines of Fig. 4-30(a) is our system, and Fig. 4-30(b) is an equivalent feedback system of (a). If

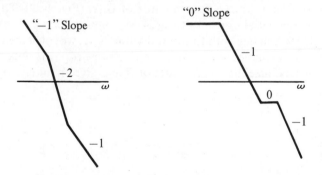

Fig. 4-29. Bode diagrams: (a) with "-1" initial slope; (b) with "0" initial slope.

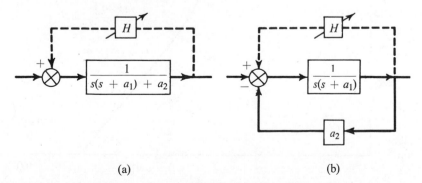

(a)   (b)

Fig. 4-30. A positive feedback is applied: (a) H is adjusted to a_2; (b) the constant term becomes zero.

a positive feedback gain a_2 is added to both systems, it is easy to see that Fig. (b) becomes $1/[s(s + a_1)]$.

From the frequency response viewpoint, the difference between the original system and (b) is evident, because the former has a 0 initial slope whereas (b) has a "-1" slope. If and only if $H = a_2$, the whole system of (b) has "-1" slope. If H is greater than or less than a_2, the frequency response of (b) still has 0 initial slope. This nature helps us to determine $a_2 = H$ accurately.

This decomposing technique is repeated until we find a straight line in the Bode diagram. The coefficients a_i have been determined.

10. First Example of Identification

Given a frequency response curve $M_1(j\omega)$, as shown in Fig. 4-31, find the transfer function of the system.

First, apply a positive feedback gain H to $M_1(j\omega)$ such that the resulting frequency response is $G_1(\omega)$, which has a -1 slope. This means that there is a single s as a factor in the denominator of $G_1(s)$ (Fig. 4-32). To be more specific, from the initial portion of Fig. 4-31, one sees that as $\omega \to 0$, $M \to -9.5$ db or $1/2.9$. Starting with this initial value of H, we calculate G_1 by (23) and (24) for various values of H near 2.9 until a slope of -1 is obtained. Only one value will maximize the initial values of G and give an initial slope of -1.

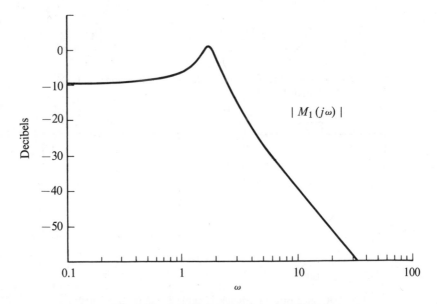

Fig. 4-31. A given frequency response $M_1(\omega)$.

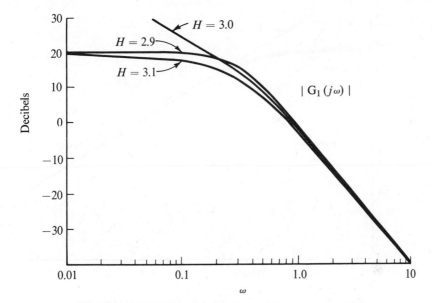

Fig. 4-32. After several trials, it is found that only when $H = 3.0$ the open loop response $G_1(\omega)$ has an initial slope of -1.

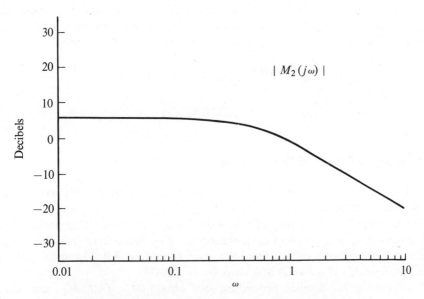

Fig. 4-33. Multiplying $G_1(\omega)$ by $j\omega$, we obtain $M_2(\omega)$.

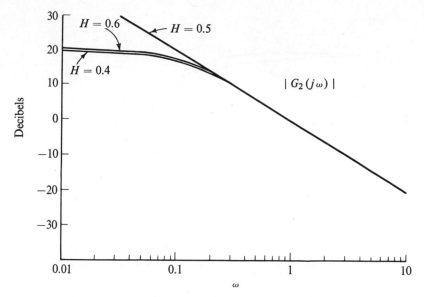

Fig. 4-34. Only when $H = 0.5$ does the second open loop response have -1 slope.

Note, in Fig. 4-32, that $H = 3.0$, while other values $H = 2.9$ and 3.1 still give the 0 initial slope.

If G_1 is multiplied by $j\omega$, then M_2 (Fig. 4-33), a new response, is obtained, and the process may be repeated until a straight line is obtained. From Fig. 4-34, the best value of H_2 is easily recognized; $H_2 = 0.5$. By synthesis,

$$M = \frac{1}{(s + 0.5)s + 3}$$

$$= \frac{1}{s^2 + 0.5s + 3}$$

11. Second Example of Identification

The frequency response of a more complicated system is shown in Fig. 4-35. Its transfer function is required.

From the initial reading of Fig. 4-35, an approximate value around 3.0 is assumed. Using (23) and (24), we obtain Fig. 4-36, from which we see that the exact value of the feedback gain should be 3.0. Using this value, we can find M_{II} (Fig. 4-37) and then G_{II} (Fig. 4-38).

Following the similar procedure, we obtain M_{III}, G_{III}, M_{IV}, and G_{IV}, as shown in Figs. 4-39, 4-40, 4-41, and 4-42, respectively.

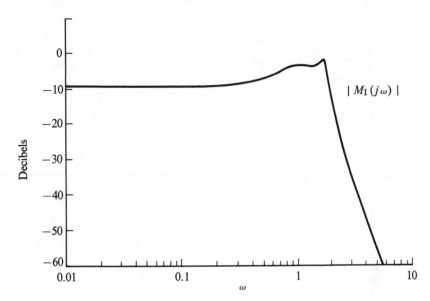

Fig. 4-35. Another given response.

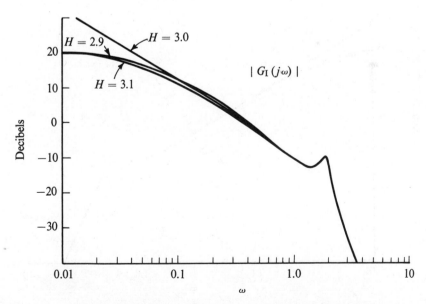

Fig. 4-36. Adjusting H values such that the initial slope is
with "-1" value.

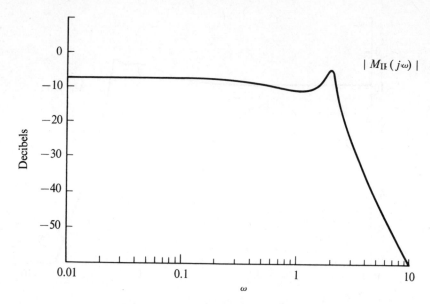

Fig. 4-37. M_{II} is obtained by multiplying $j\omega$ to G_{I}.

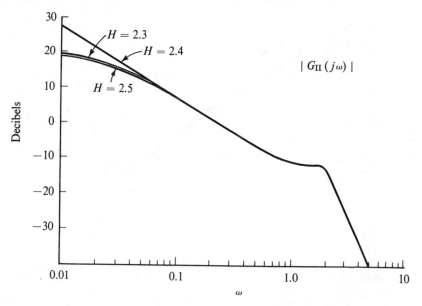

Fig. 4-38. G_{II} is found by adjusting H.

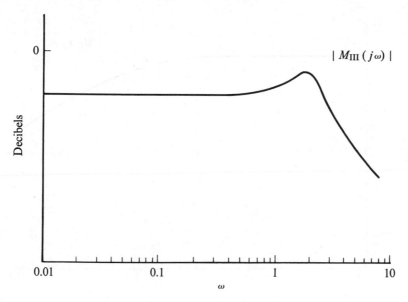

Fig. 4-39. M_{III} is obtained.

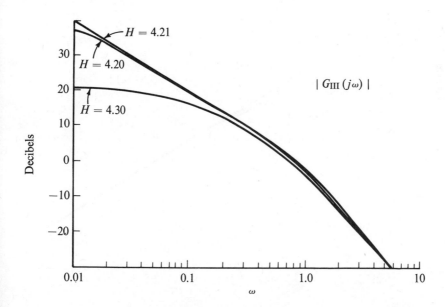

Fig. 4-40. G_{III} is found.

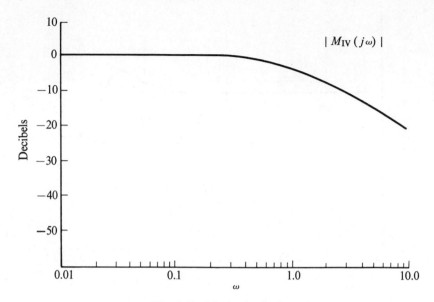

Fig. 4-41. M_{IV} is obtained.

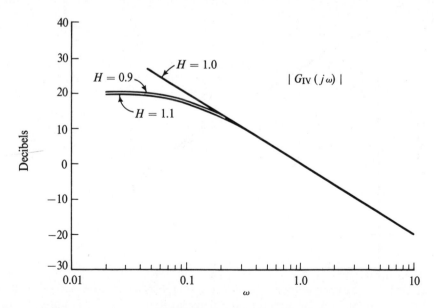

Fig. 4-42. G_{IV} is found.

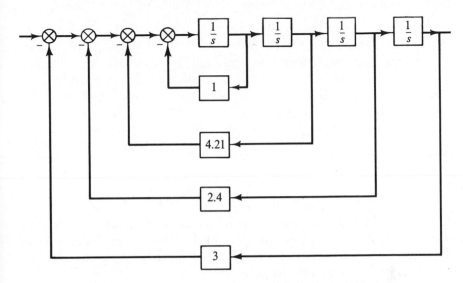

Fig. 4-43. Finally the over-all transfer function is established.

Finally, by synthesis, we have

$$M_I = \frac{1}{\{[(s+1)s + 4.21]s + 2.4\}s + 3}$$

$$= \frac{1}{s^4 + s^3 + 4.21s^2 + 2.45 + 3}$$

Figure 4-43 is a block diagram of the process of synthesis for the second example.

VI.   Levy's Curve Fitting Technique

1. Methods Based on Decomposition

The identification techniques so far developed are based on either Bode's decomposition, Wiener's decomposition, or Bush's decomposition. The general procedure for these methods can be summarized as follows: (1) decomposing the frequency response in question into many elementary components, (2) identifying each component and expressing it by a corresponding element in the s domain, and (3) combining the elements so obtained into the following fundamental transfer function form:

$$M(s) = \frac{b_0 + b_1 s + b_2 s^2 + \cdots}{a_0 + a_1 s + a_2 s^2 + \cdots} \tag{1}$$

Dividing an object and distinguishing its elements are common procedures for identification. Methods based on decomposition mentioned above are merely applications of this procedure to a mathematical model. A question naturally arises: can we determine Eq. (1) by directly fitting the given frequency response curve? In other words, is there any direct complex curve-fitting method for constructing the mathematical model for a system?

2. Levy's Basic Theory

Most notable among the complex curve-fitting methods is the one developed by Levy. Levy's method is to fit the frequency response curve of a linear system into expression (1). He first separates the numerator and denominator into real and imaginary parts:

$$M(j\omega) = \frac{(b_0 - b_2\omega^2 + b_4\omega^4 - \cdots) + j\omega(b_1 - b_3\omega^2 + b_5\omega^4 - \cdots)}{(a_0 - a_2\omega^2 + a_4\omega^4 - \cdots) + j\omega(a_1 - a_3\omega^2 + a_5\omega^4 - \cdots)} \tag{1a}$$

$$= \frac{\alpha + j\omega\beta}{\sigma + j\omega\tau} \tag{1b}$$

$$= \frac{N(\omega)}{D(\omega)} \tag{1c}$$

Let us further suppose the existence of a function $F(j\omega)$ that represents exactly the data points of the experimental frequency response curve; i.e., $F(j\omega)$ is assumed to coincide exactly with the values indicated by the experimental curve. $F(j\omega)$ will also have real and imaginary parts:

$$F(j\omega) = R(\omega) + jI(\omega) \tag{2}$$

At any specific value of frequency, ω_κ, the error in fitting is then

$$\epsilon(\omega_\kappa) = F(j\omega_\kappa) - M(j\omega_\kappa) \tag{3}$$

$$= F(j\omega_\kappa) - \frac{N(\omega_\kappa)}{D(\omega_\kappa)} \tag{3a}$$

Thus the problem is to minimize this error at each sampling point on the curve.

3. Levy's Special Technique

One might suppose the minimization problem could be done rather simply by summing up $|\epsilon(\omega_\kappa)|^2$ over the sampling frequencies and setting the partial derivatives of this summation with respect to each of the coefficients equal to zero. This would correspond to a least-square fit and would result in a set of linear simultaneous algebraic equations which might be solved for the desired coefficients of $M(j\omega)$. However, the problem becomes quite difficult when solved in this manner. Levy modifies the least-squares philosophy and does the following instead:

Equation (3a) is multiplied by $D(\omega_\kappa)$:

$$D(\omega_\kappa)\epsilon(\omega_\kappa) = D(\omega_\kappa)F(j\omega_\kappa) - N(\omega_\kappa) \tag{4}$$

Again, (4) may be separated into real and imaginary parts:

$$D(\omega_\kappa)\epsilon(\omega_\kappa) = A(\omega_\kappa) + jB(\omega_\kappa) \tag{5}$$

The magnitude of this expression is

$$| D(\omega_\kappa)\epsilon(\omega_\kappa)| = \sqrt{A(\omega_\kappa)^2 + B(\omega_\kappa)^2} \tag{6}$$

By squaring Eq. (6), a weighted error function which is summed up over the sampling frequencies can be defined; hence,

$$E = \sum_{\kappa=0}^{m} | D(\omega_\kappa)\epsilon(\omega_\kappa)|^2 = \sum_{\kappa=0}^{m} [A(\omega_\kappa)^2 + B(\omega_\kappa)^2] \tag{7}$$

and since from (1), (2), (3), and (4)

$$\begin{aligned}
A(\omega_\kappa) &= \text{real part of } [D(\omega_\kappa)F(j\omega_\kappa) - N(\omega_\kappa)] \\
&= \text{Re}\,\{\sigma_\kappa + j\omega_\kappa\tau_\kappa)[R(\omega_\kappa) + jI(\omega_\kappa)] - (\alpha_\kappa - j\omega_\kappa\beta_\kappa)\} \\
&= \sigma_\kappa R_\kappa - \omega_\kappa\tau_\kappa - \alpha_\kappa \\
B(\omega_\kappa) &= \text{imaginary part of } [D(\omega_\kappa)F(j\omega_\kappa) - N(\omega_\kappa)] \\
&= \text{Im}\,\{(\sigma_\kappa + j\omega_\kappa\tau_\kappa)[R(\omega_\kappa) + jI(\omega_\kappa)] - (\alpha_\kappa + j\omega_\kappa\beta_\kappa)\} \\
&= \omega_\kappa\tau_\kappa R_\kappa + \sigma_\kappa I_\kappa - \omega_\kappa\beta_\kappa
\end{aligned} \tag{8}$$

we have

$$E = \sum_{\kappa=0}^{m} [(\sigma_\kappa R_\kappa - \omega_\kappa\tau_\kappa I_\kappa - \alpha_\kappa)^2 + (\omega_\kappa\tau_\kappa R_\kappa + \sigma_\kappa I_\kappa - \omega_\kappa\beta_\kappa)^2] \tag{9}$$

Recalling (1b), we have

$$\begin{aligned}
\alpha &= b_0 - b_2\omega^2 + b_4\omega^4 - \cdots \\
\beta &= b_1 - b_3\omega^2 + b_5\omega^4 - \cdots \\
\sigma &= a_0 - a_2\omega^2 + a_4\omega^4 - \cdots \\
\tau &= a_1 - a_3\omega^2 + a_5\omega^4 - \cdots
\end{aligned}$$

Equation (9) can be differentiated with respect to each of the coefficients and the results can be set equal to zero; therefore,

$$\frac{\partial E}{\partial b_0} = \sum_{\kappa=0}^{m} 2(\sigma_\kappa R_\kappa - \omega_\kappa\tau_\kappa I_\kappa - \alpha_\kappa)(-1) = 0$$

$$\frac{\partial E}{\partial b_1} = \sum_{\kappa=0}^{m} 2(\omega_\kappa\tau_\kappa R_\kappa + \sigma_\kappa I_\kappa - \beta_\kappa)(-\omega_\kappa) = 0$$

$$\frac{\partial E}{\partial b_2} = \sum_{\kappa=0}^{m} 2(\sigma_\kappa R_\kappa - \omega_\kappa \tau_\kappa I_\kappa - \alpha_\kappa)(+\omega_\kappa^2) = 0$$

$$\frac{\partial E}{\partial b_3} = \sum_{\kappa=0}^{m} 2(\omega_\kappa \tau_\kappa R_\kappa + \sigma_\kappa I_\kappa - \omega_\kappa \beta_\kappa)(+\omega_\kappa^3) = 0$$

$$\vdots \qquad\qquad \vdots$$

(10)

$$\frac{\partial E}{\partial a_1} = \sum_{\kappa=0}^{m} 2(\sigma_\kappa R_\kappa - \omega_\kappa \tau_\kappa I_\kappa - \alpha_\kappa)(-\omega_\kappa I_\kappa)$$
$$+ 2(\omega_\kappa \tau_\kappa R_\kappa + \sigma_\kappa I_\kappa - \omega_\kappa \beta_\kappa)(\omega_\kappa \beta_\kappa) = 0$$

$$\frac{\partial E}{\partial a_2} = \sum_{k=0}^{m} 2(\sigma_\kappa R_\kappa - \omega_\kappa \tau_\kappa I_\kappa - \alpha_\kappa)(-\omega_\kappa^2 R_\kappa)$$
$$+ 2(\omega_\kappa \tau_\kappa R_\kappa + \sigma_\kappa I_\kappa - \omega_\kappa \beta_\kappa)(-\omega_\kappa^2 I_\kappa) = 0$$

$$\frac{\partial E}{\partial a_3} = \sum_{\kappa=0}^{m} 2(\sigma_\kappa R_\kappa - \omega_\kappa \tau_\kappa I_\kappa - a_\kappa)(\omega_\kappa^3 I_\kappa)$$
$$+ 2(\omega_\kappa \tau_\kappa R_\kappa + \sigma_\kappa I_\kappa - \omega_\kappa \beta_\kappa)(-\omega_\kappa^3 R_\kappa) = 0$$

$$\vdots \qquad\qquad \vdots$$

Apply the following linear transformations to the terms involving the unknown coefficients.

$$\alpha_\kappa = b_0 - \alpha_\kappa'$$
$$\beta_\kappa = b_1 - \beta_\kappa'$$
$$\sigma_\kappa = a_0 - \sigma_\kappa'$$
$$\tau_\kappa = a_1 - \tau_\kappa'$$

(11)

where

$$a_0 = 1$$
$$\alpha_\kappa' = b_2 \omega_\kappa^2 - b_4 \omega_\kappa^4 + b_6 \omega_\kappa^6 - \cdots$$
$$\beta_\kappa' = b_3 \omega_\kappa^2 - b_5 \omega_\kappa^4 + b_7 \omega_\kappa^6 - \cdots$$
$$\sigma_\kappa' = a_2 \omega_\kappa^2 - a_4 \omega_\kappa^4 + a_6 \omega_\kappa^6 - \cdots$$
$$\tau_\kappa' = a_3 \omega_\kappa^2 - a_5 \omega_\kappa^4 + a_7 \omega_\kappa^6 - \cdots$$

Substituting into Eq. (10), we obtain

$$\sum_{\kappa=0}^{m} -2R_\kappa(1 - \sigma_\kappa') + 2\omega_\kappa I_\kappa(a_1 - \tau_\kappa') + 2(b_0 - \alpha_\kappa') = 0$$

$$\sum_{\kappa=0}^{m} -2\omega_\kappa^2 R_\kappa(a_1 - \tau_\kappa') - 2\omega_\kappa I_\kappa(1 - \sigma_\kappa') + 2\omega_\kappa^2(b_1 - \beta_\kappa') = 0$$

$$\sum_{\kappa=0}^{m} 2\omega_\kappa^2 R_\kappa(1 - \sigma_\kappa') - 2\omega_\kappa^3 I_\kappa(a_1 - \tau_\kappa') - 2\omega_\kappa^2(b_0 - \alpha_\kappa') = 0$$

$$\sum_{\kappa=0}^{m} 2\omega_\kappa^4 R_\kappa(a_1 - \tau_\kappa') + 2\omega_\kappa^3 I_\kappa(1 - \sigma_\kappa') - 2\omega_\kappa^4(b_1 - \beta_\kappa') = 0$$

$$\begin{matrix} \cdot & & \cdot \\ \cdot & & \cdot \\ \cdot & & \cdot \end{matrix} \qquad\qquad (12)$$

$$\sum_{\kappa=0}^{m} -2\omega_\kappa I_\kappa R_\kappa(1 - \sigma_\kappa') + 2\omega_\kappa^2 I_\kappa^2(a_1 - \tau_\kappa') + 2\omega_\kappa I_\kappa(b_0 - \alpha_\kappa')$$
$$+ 2\omega_\kappa^2 R_\kappa^2(a_1 - \tau_\kappa') + 2\omega_\kappa^2 R_\kappa^2(1 - \sigma_\kappa') - 2\omega_\kappa^2 R_\kappa(b_1 - \beta_\kappa') = 0$$

$$\sum_{\kappa=0}^{m} -2\omega_\kappa^2 R_\kappa^2(1 - \sigma_\kappa') + 2\omega_\kappa^3 R_\kappa I_\kappa(a_1 - \tau_\kappa') + 2\omega_\kappa^2 R_\kappa(b_0 - \alpha_\kappa')$$
$$- 2\omega_\kappa^3 I_\kappa R_\kappa(a_1 - \tau_\kappa') - \omega_\kappa^2 I_\kappa^2(1 - \sigma_\kappa') + \omega_\kappa^3 I_\kappa(b_1 - \beta_\kappa') = 0$$

$$\sum_{\kappa=0}^{m} \omega_\kappa^3 I_\kappa(b_0 - \alpha_\kappa') - \omega_\kappa^4 R_\kappa(b_1 - \beta_\kappa') + \omega_\kappa^4(R_\kappa^2 + I_\kappa^2)(a_1 - \tau_\kappa') = 0$$

Simplification yields

$$\sum_{\kappa=0}^{m} b_0 - \alpha_\kappa + R_\kappa \sigma_\kappa' + \omega_\kappa I_\kappa a_1 - \omega_\kappa \tau_\kappa' = \sum_{\kappa=0}^{m} R_\kappa$$

$$\sum_{\kappa=0}^{m} \omega_\kappa^2(b_1 - \beta_\kappa') + \omega_\kappa I_\kappa \sigma_\kappa' - \omega_\kappa^2 R_\kappa(a_1 - \tau_\kappa') = \sum_{\kappa=0}^{m} \omega_\kappa I_\kappa$$

$$\sum_{\kappa=0}^{m} \omega_\kappa^2 R_\kappa \sigma_\kappa' + \omega_\kappa^3 I_\kappa(a_1 - \tau_\kappa') + \omega_\kappa^2(b_0 - \alpha_\kappa') = \sum_{\kappa=0}^{m} \omega_\kappa^2 R_\kappa$$

$$\sum_{\kappa=0}^{m} \omega_\kappa^4 R_\kappa(a_1 - \tau_\kappa') + \omega_\kappa^3 I_\kappa \sigma_\kappa + \omega_\kappa^4(b_1 - \beta_\kappa') = \sum_{\kappa=0}^{m} \omega_\kappa^3 I_\kappa$$

$$\begin{matrix} \cdot & & \cdot \\ \cdot & & \cdot \\ \cdot & & \cdot \end{matrix} \qquad\qquad (12a)$$

$$\sum_{\kappa=0}^{m} \omega_\kappa I_\kappa(b_0 - \alpha_\kappa') - \omega_\kappa^2 R_\kappa(b_1 - \beta_\kappa') + \omega_\kappa^2(R_\kappa^2 + I_\kappa^2)(a_1 - \tau_\kappa') = 0$$

$$\sum_{\kappa=0}^{m} \omega_\kappa^2 R_\kappa(b_0 - \alpha_\kappa') + \omega_\kappa^3 I_\kappa(b_1 - \beta_\kappa') + \omega_\kappa^2(R_\kappa^2 + I_\kappa^2)\sigma_\kappa' = \sum_{\kappa=0}^{m} (R_\kappa^2 + I_\kappa^2)$$

$$\sum_{\kappa=0}^{m} \omega_\kappa^3 I_\kappa(b_0 - \alpha_\kappa') - \omega_\kappa^4 R_\kappa(b_1 - \beta_\kappa') + \omega_\kappa^4(R_\kappa^2 + I_\kappa^2)(a_1 - \tau_\kappa') = 0$$

$$\begin{matrix} \cdot & & \cdot \\ \cdot & & \cdot \\ \cdot & & \cdot \end{matrix}$$

Equations (12a) may now be condensed by using the following formulas:

$$\lambda_h = \sum_{\kappa=0}^{m} \omega_\kappa^h$$

$$S_h = \sum_{\kappa=0}^{m} \omega_\kappa^h R_\kappa$$

$$T_h = \sum_{\kappa=0}^{m} \omega_\kappa^h I_\kappa \qquad\qquad (13)$$

$$U_h = \sum_{\kappa=0}^{m} \omega_\kappa^h(R_\kappa^2 + I_\kappa^2), \qquad h = 0, 1, 2, 3, \dots$$

Applying the definitions of α'_κ, β'_κ, σ'_κ, and τ'_κ as given in (11), we obtain the following set of linear algebraic equations:

$$\lambda_0 b_0 - \lambda_2 b_2 + \lambda_4 b_4 - \lambda_6 b_6 + \cdots$$
$$+ T_1 a_1 + S_2 a_2 - T_3 a_3 - S_4 a_4 + T_5 a_5 + \cdots = S_0$$

$$\lambda_2 b_1 - \lambda_4 b_3 + \lambda_6 b_5 - \lambda_8 b_8 + \cdots$$
$$- S_2 a_1 + T_3 a_2 + S_4 a_3 - T_5 a_4 - S_6 a_5 + \cdots = T_1$$

$$\lambda_2 b_0 - \lambda_4 b_2 + \lambda_6 b_4 - \lambda_8 b_6 + \cdots$$
$$+ T_3 a_1 + S_4 a_2 - T_5 a_3 - S_6 a_4 + T_7 a_5 + \cdots = S_2$$

$$\lambda_4 b_1 - \lambda_6 b_3 + \lambda_8 b_5 - \lambda_{10} b_7 + \cdots$$
$$- S_4 a_1 + T_5 a_2 + S_6 a_3 - T_7 a_4 - S_6 a_5 + \cdots = T_3$$

$$\vdots$$

$$\tag{14}$$

$$T_1 b_0 - S_2 b_1 - T_3 b_2 + S_4 b_4 - \cdots$$
$$+ U_2 a_1 - U_4 a_3 + U_6 a_5 - U_8 a_7 + \cdots = 0$$

$$S_2 b_0 + T_3 b_1 - S_4 b_2 - T_5 b_3 + \cdots$$
$$+ U_4 a_2 - U_6 a_4 + U_8 a_6 - U_{10} a_8 + \cdots = U_2$$

$$T_3 b_0 - S_4 b_1 - T_5 b_2 + S_6 b_3 + \cdots$$
$$+ U_4 a_1 - U_6 a_3 + U_8 a_5 - U_{10} a_7 + \cdots = 0$$

The numerical value of the unknowns $a_1, a_2, \ldots, b_0, b_1, \ldots$ may now be determined once the coefficients of the equations have been evaluated.

4. A Digital Computer Program

A digital computer program is written to solve for the unknowns of a system function consisting of a tenth degree numerator and a tenth degree denominator (see Program 7).

5. Illustrative Example

The frequency response shown in Fig. 4-44 is generated from

$$F(j\omega) = \frac{1}{3.0 + 2.4(j\omega) + 4.21(j\omega)^2 + (j\omega)^3 + (j\omega)^4}$$

$$= \frac{0.33333}{1 + 0.8(j\omega) + 1.4033(j\omega)^2 + 0.33333(j\omega)^3 + 0.33333(j\omega)^4}$$

the 45 data points being used. The coefficients are computed and are given to five significant figures, as shown on page 231.

Program 7

```
C       TRANSFER FUNCTION SYNTHESIS  BY LEVYS METHOD
C       TYPE NO. OF DATA POINTS ON FIRST CARD
C       ENTER FREQUENCY RESPONSE DATA IN 3E15.8
C       SW 1 ON TO PUNCH OUT MATRIX ELEMENTS
C       SW2 AND 3 ON TO RE-ENTER ELEMENTS
*2004
        DIMENSION W(90),XM(90),PH(90),R(90)
        DIMENSION X(90),A(15,16),CF(90)
        EQUIVALENCE  (XM,CF),  (PH,A)
        IF(SENSE SWITCH 2)19,200
  200 READ 1,N
    1 FORMAT(I3)
        READ 2,(W(I),XM(I),PH(I),I=1,N)
    2 FORMAT(3E15.8)
        DO 3 I=1,N
        ANG=PH(I)*3.14159265/180.
        R(I)=XM(I)*COS(ANG)
    3 X(I)=XM(I)*SIN(ANG)
        DO 51 I=1,N
   51 CF(I)=1.0
        IT=1
    4 DO 5 I=1,15
        DO 5 J=1,16
    5 A(I,J)=0.
   52 DO 6 I=1,N
        A(1,1)=A(1,1)+CF(I)
        A(1,3)=A(1,3)-CF(I)*W(I)**2
        A(1,5)=A(1,5)+CF(I)*W(I)**4
        A(1,7)=A(1,7)-CF(I)*W(I)**6
        A(1,9)=A(1,9)+CF(I)*X(I)*W(I)
        A(1,10)=A(1,10)+CF(I)*R(I)*W(I)**2
        A(1,11)=A(1,11)-CF(I)*X(I)*W(I)**3
        A(1,12)=A(1,12)-CF(I)*R(I)*W(I)**4
        A(1,13)=A(1,13)+CF(I)*X(I)*W(I)**5
        A(1,14)=A(1,14)+CF(I)*R(I)*W(I)**6
    6 A(1,16)=A(1,16)+CF(I)*R(I)
        KF=8
        DO 7 K=2,8,2
        KG=KF-1
        L=K-1
        DO 8 I=1,N
        A(K,8)=A(K,8)-CF(I)*W(I)**KF
        A(K,15)=A(K,15)+CF(I)*R(I)*W(I)**KF
    8 A(L,15)=A(L,15)-CF(I)*X(I)*W(I)**KG
    7 KF=KF+2
        A(2,16)=A(1,9)
        A(3,16)=A(1,10)
        A(4,16)=-A(1,11)
        A(5,16)=-A(1,12)
        A(6,16)=A(1,13)
        A(7,16)=A(1,14)
        A(8,16)=-A(1,15)
        DO 9 I=1,7
        K=I+1
```

Program 7 (CONT.)

```
     L=I+8
  9  A(9,I)=A(I,L)*(-1.0)**K
     DO 10  J=2,8
     K=J+7
 10  A(K,8)=A(J,15)
     DO 11  I=1,N
     F1=R(I)**2+X(I)**2
     A(9,9)=A(9,9)+CF(I)*F1*W(I)**2
     A(9,11)=A(9,11)-CF(I)*F1*W(I)**4
     A(9,13)=A(9,13)+CF(I)*F1*W(I)**6
     A(9,15)=A(9,15)-CF(I)*F1*W(I)**8
     A(11,15)=A(11,15)-CF(I)*F1*W(I)**10
     A(13,15)=A(13,15)-CF(I)*F1*W(I)**12
 11  A(15,15)=A(15,15)-CF(I)*F1*W(I)**14
     A(10,16)=A(9,9)
     A(12,16)=-A(9,11)
     A(14,16)=A(9,13)
     DO 12  J=2,8
     M=J-1
     DO 12  L=1,7
     K=L+1
 12  A(J,L)=A(M,K)*(-1.0)**M
     DO 13  J=2,8
     M=J-1
     DO 13  L=9,14
     K=L+1
 13  A(J,L)=A(M,K)*(-1.0)**M
     DO 14  J=10,15
     M=J-1
     DO 14  L=1,7
     K=L+1
 14  A(J,L)=A(M,K)*(-1.0)**M
     DO 15  J=10,15
     M=J-1
     DO 15  L=9,14
     K=L+1
 15  A(J,L)=A(M,K)*(-1.0)**M
     IF(SENSE SWITCH 1)17,20
 17  PUNCH 18,((A(I,J),J=1,16),I=1,15)
 18  FORMAT(2E19.12)
 19  READ 18,((A(I,J),J=1,16),I=1,15)
 20  TYPE 21
 21  FORMAT(14HTYPE TOLERANCE)
     ACCEPT 22,TOLER
 22  FORMAT(F14.12)
     DO 120  I=1,15
     IDG=I
     DIAG=A(I,I)
107  IF(ABS(DIAG)-TOLER)124,124,108
108  DO 109  J=1,16
109  A(I,J)=A(I,J)/DIAG
     K=I
110  IF(K-I)111,113,111
111  FELMT=A(K,I)
```

Program 7 (CONCL.)

```
      DO 112 J=1,16
112   A(K,J)=A(K,J)-FELMT*A(I,J)
113   K=K+1
      IF(K-15)110,110,120
120   CONTINUE
      GO TO 30
124   TOLER=.1*TOLER
      TYPE 125,IDG,TOLER
125   FORMAT(21HTOLERANCE ON DIAGONALI3
     113H ADJUSTED TO F14.12)
      GO TO 107
30    TYPE1211
1211  FORMAT(30HTRANSFER FUNCTION COEFFICIENTS)
      TYPE121,IT
121   FORMAT(13HFOR ITERATIONI3)
      DO 33 I=1,8
      K=I-1
33    TYPE 34,K,A(I,16)
34    FORMAT(1HAI3,2H =E15.8)
      BO=1.
      TYPE 35,BO
35    FORMAT(6HB  O =E15.8)
      DO 36 I=9,15
      K=I-8
36    TYPE 37,K,A(I,16)
37    FORMAT(1HBI3,2H =F15.8)
      IF(SENSE SWITCH 3)132,130
130   DO 131 I=1,N
      XIDIO=A(9,16)-A(11,16)*W(I)**2
      XIDI=XIDIO+A(13,16)+W(I)**4-A(15,16)*W(I)**6
      RDO=1.0-A(10,16)*W(I)**2
      RD=RDO+A(12,16)*W(I)**4-A(14,16)*W(I)**6
      XID=W(2)*XIDI
131   CF(I)=1.0/(RD**2+XID**2)
      IT=IT+1
      GO TO 4
132   PAUSE
      END
```

$$b_0 = 3.3333 \cdot 10^{-1}$$
$$b_1 = 2.0574 \cdot 10^{-7}$$
$$b_2 = -1.4917 \cdot 10^{-5}$$
$$b_3 = -3.3257 \cdot 10^{-11}$$
$$a_0 = 1.$$
$$a_1 = 8.0000 \cdot 10^{-1}$$
$$a_2 = 1.4033$$
$$a_3 = 3.3333 \cdot 10^{-1}$$
$$a_4 = 3.3327 \cdot 10^{-1}$$
$$a_5 = -1.4747 \cdot 10^{-5}$$

It is noted that coefficients of higher ordered terms are negligible and that to five significant figures, $M(j\omega)$ is almost identical to the original $F(j\omega)$.

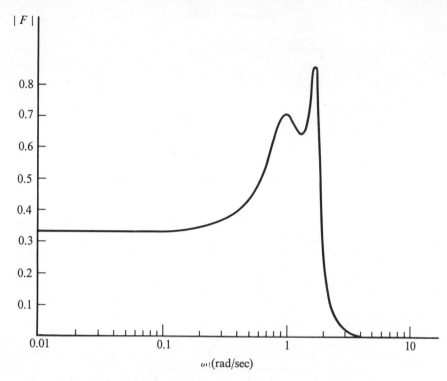

Fig. 4-44. Example for illustrating Levy's curve fitting technique.

6. Summary

This chapter contains two main topics: how to analyze a signal and how to recognize a system. In other words, how an analytical, mathematical expression can be constructed for a signal or a system which can become the starting point for carrying out further analyses.

The basic tool for analyzing a periodic function is the Fourier series; for an aperiodic function, it is the Fourier integral. These were treated in the first two sections. Next were presented several methods for identifying a system from its frequency response.

The most popular, although approximate, technique employs Bode's diagram. The most powerful method that is oriented to the use of a digital computer is Levy's curve fitting, which has been used in industry for several years.

For a nonoscillatory system, Wiener-Lee's method is very useful; for highly oscillatory systems, Bush's decomposition technique is recommended.

In general, this chapter deals with the identification problem in the frequency domain. Although the steps are presented in graphic form, they are actually the most convenient approach for application to a digital computer.

PROBLEMS

4-1. Taking 12 samples from each period, find the Fourier coefficients of the wave forms shown in Fig. P4-1.

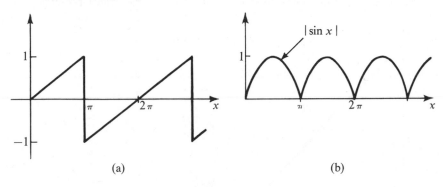

(a)   (b)

Fig. P4-1

4-2. The "real part" response and the magnitude response of a system are shown in Table 8.
(a) Use Bode's decomposition technique to approximate the transfer function.
(b) Use Wiener-Lee's decomposition technique to approximate the transfer function.

TABLE 8

ω	Real Part Response	Magnitude Response
1×10^{-6}	-1.52	1×10^{6}
0.1	-1.5011	9.9378
0.2	-1.4467	4.8785
0.3	-1.3630	3.1574
0.4	-1.2585	2.2760
0.5	-1.1425	1.7354
0.6	-1.0228	1.3688
0.7	-0.9057	1.1045
0.8	-0.7954	0.9061
0.9	-0.6944	0.7530
1.0	-0.6038	0.6323
1.1	-0.5236	0.5357
1.2	-0.4535	0.4573
1.3	-0.3925	0.3931
1.4	-0.3399	0.3400
1.5	-0.2946	0.2957
1.6	-0.2557	0.2585
1.7	-0.2223	0.2271
1.8	-0.1936	0.2004
1.9	-0.1690	0.1776
2.0	-0.1478	0.1580
2.5	-0.0783	0.0927
3.0	-0.0438	0.0584
5.0	-0.0066	0.0044
∞	0.0	0.0

4-3. A unit impulse is applied to the system

$$M(s) = \frac{2}{s^2 + 1.8s + 1.2}$$

Tabulate the transient response $m(t)$, and give the trapezoidal expression for $m(t)$.

(a) Evaluate the frequency response, $M(\omega)$ vs. ω and $\alpha(\omega)$ vs. ω.

(b) Use the result obtained from (a) to fit the transfer function by Levy's method.

(c) Use the result obtained from (a) to evaluate the transfer function by

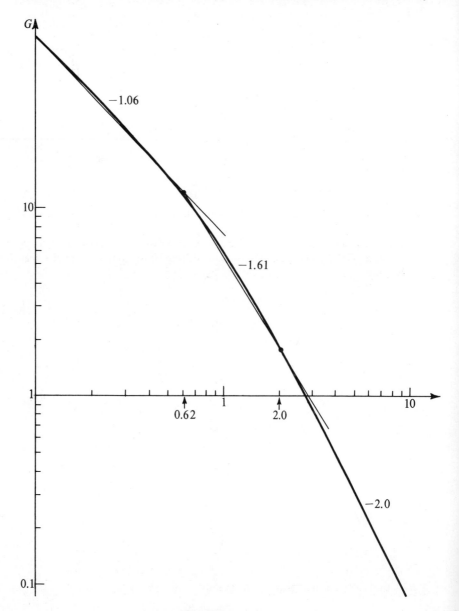

Fig. P4-4(a)

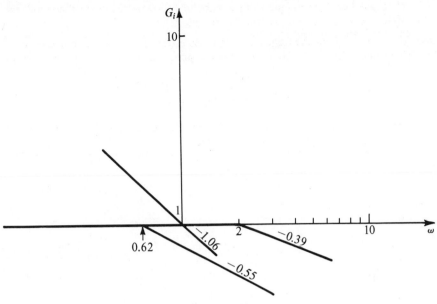

Fig. P4-4(b)

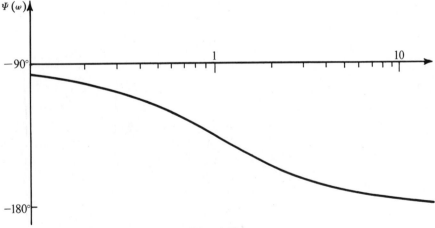

Fig. P4-4(c)

4-4. Bode's decomposition technique can be generalized by approximating a magnitude curve, log $G(\omega)$ vs. log ω, by a group of straight lines with arbitrary slopes as shown in Fig. P4-4(a). The component lines are shown in 4-4(b). Derive an approximate formula for finding the phase curve of $G(j\omega)$, and verify that the phase curve of the problem is as shown in 4-4(c).

4-5. Generate the frequency response of the following transfer function:

$$M(s) = \frac{s^2 + 4.5s + 7.8}{s^3 + 5.6s^2 + 8.2s + 5.3}$$

Then use Levy's technique to fit a transfer function. Compare the result you obtained with the original transfer function.

4-6. If the real part of the frequency response $g_1(\omega)$ vs. ω is given and the corresponding transient response $g(t)$ can be obtained by using Eq. (16) of section II:

$$g(t) = \frac{2}{\pi} \int_0^\infty g_1(\omega) \cos t\omega \, d\omega$$

Devise a procedure to produce an approximation formula for the conversion, and verify your approach by an example.

PART TWO

State Variable Analysis

CHAPTER 5

Fundamentals of Matrices

Any system of engineering importance needs to be described in a mathematical model containing more than a single variable. Many displacements, velocities, currents, voltages, and combinations of these may be required to give an adequate description. It therefore becomes necessary to have a systematic and orderly way to deal with a large set of variables.

Matrix theory is a powerful and elegant scheme for "mass production" mathematics. Its unifying concepts, simplicity of notation, and high degree of generalization make it the control engineer's most helpful analytical tool.

In this chapter we shall be introduced to some basic matrix theory as a necessary foundation for the state variable analyses of subsequent chapters.

I. Matrix Algebra

1. Linear Transformation

A set of simultaneous equations

$$
\begin{aligned}
y_1 &= a_{11}x_1 + a_{12}x_2 + \cdots + a_{1n}x_n \\
y_2 &= a_{21}x_1 + a_{22}x_2 + \cdots + a_{2n}x_n \\
&\ \cdot\ \cdot\ \cdot\ \cdot\ \cdot\ \cdot\ \cdot\ \cdot\ \cdot\ \cdot\ \cdot\ \cdot \\
y_n &= a_{n1}x_1 + a_{n2}x_2 + \cdots + a_{nn}x_n
\end{aligned}
\tag{1}
$$

can be interpreted as follows: The set of variable y_i is derived from the set of x_i by a linear transformation. Equation (1) can be represented by a simpler formula:

$$[y] = [A][x] \tag{2}$$

Equation (2) is not only a qualitative expression; it can be derived from (1) by using certain rules.

Let us rearrange (1) as follows:

$$
\begin{bmatrix} y_1 \\ y_2 \\ \cdot \\ \cdot \\ \cdot \\ y_n \end{bmatrix} = \begin{bmatrix} a_{11} & a_{12} & \cdots & a_{1n} \\ a_{21} & a_{22} & \cdots & a_{2n} \\ \cdot & \cdot & \cdots & \cdot \\ \cdot & \cdot & \cdots & \cdot \\ \cdot & \cdot & \cdots & \cdot \\ a_{n1} & a_{n2} & \cdots & a_{nn} \end{bmatrix} \begin{bmatrix} x_1 \\ x_2 \\ \cdot \\ \cdot \\ \cdot \\ x_n \end{bmatrix} \tag{3}
$$

Equation (3) is comparable to (2).

$$
[y] = \begin{bmatrix} y_1 \\ y_2 \\ \cdot \\ \cdot \\ \cdot \\ y_n \end{bmatrix}, \qquad [x] = \begin{bmatrix} x_1 \\ x_2 \\ \cdot \\ \cdot \\ \cdot \\ x_n \end{bmatrix}, \qquad [A] = \begin{bmatrix} a_{11} & a_{12} & \cdots & a_{1n} \\ a_{21} & a_{22} & \cdots & a_{2n} \\ \cdot & \cdot & \cdots & \\ \cdot & \cdot & \cdots & \\ a_{n1} & a_{n2} & \cdots & a_{nn} \end{bmatrix}
$$

$[y]$, $[x]$, or $[A]$ is called a matrix.

A collection of elements arranged in rows and columns as the three given above is called a matrix. Matrix $[A]$ is called a square matrix; it can be denoted by a_{ij}, where the suffixes i and j are understood to range from 1 to n.

2. Multiplication of Matrices

Multiplying the first row of $[A]$ by the column matrix $[x]$ of Eq. (3) gives the first equation of (1). Similarly, the second row of $[A]$ times $[x]$ of Eq. (3) gives the second equation of (1), etc.

We may now define the process of multiplication:

$$
y_i = \sum_{j=1}^{n} a_{ij} x_j
$$

Example 1. Find the product of $[A]$ and $[x]$, if

$$
[A] = \begin{bmatrix} 1 & 7 & 8 \\ 3 & 2 & 0 \\ 5 & 1 & 4 \end{bmatrix}, \qquad [x] = \begin{bmatrix} 1 \\ 6 \\ 4 \end{bmatrix}
$$

$$
[A][x] = \begin{bmatrix} 1 & 7 & 8 \\ 3 & 2 & 0 \\ 5 & 1 & 4 \end{bmatrix} \begin{bmatrix} 1 \\ 6 \\ 4 \end{bmatrix} = \begin{bmatrix} (1 \times 1) + (7 \times 6) + (8 \times 4) \\ (3 \times 1) + (2 \times 6) + (0 \times 4) \\ (5 \times 1) + (1 \times 6) + (4 \times 4) \end{bmatrix} = \begin{bmatrix} 75 \\ 15 \\ 27 \end{bmatrix}
$$

In multiplying one square matrix by another, the following formula is used:

$$
c_{ij} = \sum a_{ir} b_{rj}
$$

Example 2. Find $[A] \times [B]$, if

$$[A] = \begin{bmatrix} 1 & 3 \\ 5 & 2 \end{bmatrix}, \qquad [B] = \begin{bmatrix} 2 & 8 \\ 7 & 9 \end{bmatrix}$$

$$[C] = [A][B] = \begin{bmatrix} 1 & 3 \\ 5 & 2 \end{bmatrix} \begin{bmatrix} 2 & 8 \\ 7 & 9 \end{bmatrix} = \begin{bmatrix} (1 \times 2) + (3 \times 7) & (1 \times 8) + (3 \times 9) \\ (5 \times 2) + (2 \times 7) & (5 \times 8) + (2 \times 9) \end{bmatrix}$$

$$= \begin{bmatrix} 23 & 35 \\ 24 & 58 \end{bmatrix}$$

Example 3. Find $[B] \times [A]$.

$$[B] \times [A] = \begin{bmatrix} 2 & 8 \\ 7 & 9 \end{bmatrix} \times \begin{bmatrix} 1 & 3 \\ 5 & 2 \end{bmatrix}$$

$$= \begin{bmatrix} (2 \times 1) + (8 \times 5) & (2 \times 3) + (8 \times 2) \\ (7 \times 1) + (9 \times 5) & (7 \times 3) + (9 \times 2) \end{bmatrix}$$

$$= \begin{bmatrix} 42 & 22 \\ 52 & 39 \end{bmatrix}$$

We see that $[A] \times [B] \neq [B] \times [A]$. In order to avoid confusion, $[A] \times [B]$ is said, "$[B]$ is premultiplied by $[A]$," and $[B] \times [A]$ is said, "$[B]$ is post-multiplied by $[A]$."

3. Transposition of Matrices

The transpose of a matrix $[A]$, or $[A]^T$, is defined to be the matrix which has rows identical with the columns of $[A]$. In other words, if $[A] = a_{ij}$, then $[A]^T = a_{ji}$.

Example 1. If

$$[A] = \begin{bmatrix} 1 & 2 \\ 3 & 4 \end{bmatrix}$$

then

$$[A]^T = \begin{bmatrix} 1 & 3 \\ 2 & 4 \end{bmatrix}$$

Example 2. If

$$[b] = \begin{bmatrix} 5 \\ 7 \end{bmatrix}$$

then

$$[b]^T = [5 \quad 7]$$

We call $[b]^T$ a row matrix.

4. Special Matrices

(a) DIAGONAL MATRICES

If the elements of a square matrix other than those in the principal diagonal are zero, then the matrix is called a diagonal matrix.

Example.

$$\begin{bmatrix} 1 & 0 & 0 \\ 0 & 5 & 0 \\ 0 & 0 & 8 \end{bmatrix}$$

is a diagonal matrix.

(b) IDENTITY MATRIX

The identity matrix or unit matrix is defined to be the diagonal matrix which has units for all its principal diagonal elements. It is denoted by $[I]$.

Example.

$$\begin{bmatrix} 1 & 0 & 0 \\ 0 & 1 & 0 \\ 0 & 0 & 1 \end{bmatrix}$$

is a unit matrix of order 3.

(c) SYMMETRICAL MATRIX

When $a_{ij} = a_{ji}$ the matrix $[A]$ is said to be symmetrical.

Example.

$$\begin{bmatrix} 1 & -6 & 7 \\ -6 & 2 & -8 \\ 7 & -8 & 3 \end{bmatrix}$$

is a symmetrical matrix.

(d) NULL MATRIX

A matrix in which the elements are all zero is called a null matrix.

5. Addition and Subtraction of Matrices.

The sum of two matrices $[A]$ and $[B]$ is defined as follows.

$$c_{ij} = a_{ij} + b_{ij}$$

The difference of $[A]$ and $[B]$ is defined as

$$e_{ij} = a_{ij} - b_{ij}$$

Examples.

$$\begin{bmatrix} 4 & 2 \\ 3 & 7 \end{bmatrix} + \begin{bmatrix} 2 & 1 \\ 6 & 10 \end{bmatrix} = \begin{bmatrix} 6 & 3 \\ 9 & 17 \end{bmatrix}$$

$$\begin{bmatrix} 4 & 2 \\ 3 & 7 \end{bmatrix} - \begin{bmatrix} 2 & 1 \\ 6 & 10 \end{bmatrix} = \begin{bmatrix} 2 & 1 \\ -3 & -3 \end{bmatrix}$$

6. Determinant of a Square Matrix

The determinant of $[A]$, denoted by $|A|$, or det $[A]$, means using the elements of $[A]$ to form a determinant.

Examples.

$$\det \begin{bmatrix} 1 & 2 \\ 3 & 4 \end{bmatrix} = \begin{vmatrix} 1 & 2 \\ 3 & 4 \end{vmatrix} = (1 \times 4) - (2 \times 3) = -2$$

$$\det \begin{bmatrix} 1 & 2 & 4 \\ 5 & 6 & 3 \\ 8 & 4 & 9 \end{bmatrix} = 1 \begin{vmatrix} 6 & 3 \\ 4 & 9 \end{vmatrix} - 2 \begin{vmatrix} 5 & 3 \\ 8 & 9 \end{vmatrix} + 4 \begin{vmatrix} 5 & 6 \\ 8 & 4 \end{vmatrix}$$

$$= 42 - (2 \times 21) + \{4 \times (-28)\}$$

$$= -112$$

7. Reciprocal Matrices

Given a scalar equation $ab = c$, one can multiply both sides by a^{-1} and obtain $b = c/a$.

In matrix algebra, given

$$[A][B] = [I] \tag{4}$$

one can solve for the matrix $[B]$ by multiplying both sides of (4) by $[A]^{-1}$ and thus obtain

$$[B] = [A]^{-1}[I] \tag{5}$$

But, how do we find $[A]^{-1}$?

Let $[A]$ be a given square matrix of the nth order and $[I]$ be the identity matrix of nth order; then if a matrix $[B]$ can be determined so that

$$[A][B] = [I]$$

then $[B]$ is called the inverse of $[A]$.

Using a second order square matrix to explain the inverting procedure, we can easily generalize the procedure for the case of nth order square matrix.

A particular example of (4) is written as

$$\begin{bmatrix} a_{11} & a_{12} \\ a_{21} & a_{22} \end{bmatrix} \begin{bmatrix} b_{11} & b_{12} \\ b_{21} & b_{22} \end{bmatrix} = \begin{bmatrix} 1 & 0 \\ 0 & 1 \end{bmatrix} \tag{6}$$

where a_{ij} is known, b_{ij} is to be determined.

Decomposing (6) into four scalar equations

$$a_{11}b_{11} + a_{12}b_{21} = 1$$
$$a_{21}b_{11} + a_{22}b_{21} = 0$$
$$a_{11}b_{12} + a_{12}b_{22} = 0$$
$$a_{21}b_{12} + a_{22}b_{22} = 1$$

and using Cramer's rule, we obtain

$$b_{11} = \frac{F_{11}}{|A|}$$

$$b_{12} = \frac{F_{21}}{|A|}$$

$$b_{21} = \frac{F_{12}}{|A|} \tag{7}$$

$$b_{22} = \frac{F_{22}}{|A|}$$

where $|A|$ is the determinant of A. The quantities F_{ij} are the cofactors of the elements a_{ij} in the determinant $|A|$. Equation (7) can be written as

$$[B] = \frac{1}{|A|} \begin{bmatrix} F_{11} & F_{21} \\ F_{12} & F_{22} \end{bmatrix} = [A]^{-1} \tag{8}$$

or

$$[A]^{-1} = [B] = \begin{bmatrix} a_{22} & -a_{12} \\ -a_{21} & a_{11} \end{bmatrix} \frac{1}{|A|}$$

A practical four step technique to carry out the process is easier to remember. We will use an example to illustrate. Suppose the inverse of $[A]$ is required, where

$$[A] = \begin{bmatrix} a_{11} & a_{12} & a_{13} \\ a_{21} & a_{22} & a_{23} \\ a_{31} & a_{32} & a_{33} \end{bmatrix}$$

Step 1. Organize the corresponding "minor matrix" of A.

$$[M] = \begin{bmatrix} \begin{vmatrix} a_{22} & a_{23} \\ a_{32} & a_{33} \end{vmatrix}, & \begin{vmatrix} a_{21} & a_{23} \\ a_{31} & a_{33} \end{vmatrix}, & \begin{vmatrix} a_{21} & a_{22} \\ a_{31} & a_{32} \end{vmatrix} \\ \begin{vmatrix} a_{12} & a_{13} \\ a_{32} & a_{33} \end{vmatrix}, & \begin{vmatrix} a_{11} & a_{13} \\ a_{31} & a_{33} \end{vmatrix}, & \begin{vmatrix} a_{11} & a_{12} \\ a_{31} & a_{32} \end{vmatrix} \\ \begin{vmatrix} a_{12} & a_{13} \\ a_{22} & a_{23} \end{vmatrix}, & \begin{vmatrix} a_{11} & a_{13} \\ a_{21} & a_{23} \end{vmatrix}, & \begin{vmatrix} a_{11} & a_{12} \\ a_{21} & a_{22} \end{vmatrix} \end{bmatrix} \tag{9}$$

Step 2. Change to a "cofactor matrix."

$$[F] = \begin{bmatrix} m_{11} & -m_{12} & m_{13} \\ -m_{21} & m_{22} & -m_{23} \\ m_{31} & -m_{32} & m_{33} \end{bmatrix} \tag{10}$$

where

$$m_{11} = \begin{vmatrix} a_{22} & a_{23} \\ a_{32} & a_{33} \end{vmatrix}, \qquad m_{12} = \begin{vmatrix} a_{21} & a_{23} \\ a_{31} & a_{33} \end{vmatrix} \qquad \text{etc.}$$

Step 3. Transpose the cofactor matrix.

$$[F]^T = \begin{bmatrix} m_{11} & -m_{21} & m_{31} \\ -m_{12} & m_{22} & -m_{32} \\ m_{13} & -m_{23} & m_{33} \end{bmatrix}$$

Step 4. Divide by the determinant of $[A]$.

$$[A]^{-1} = \frac{\begin{bmatrix} m_{11} & -m_{21} & m_{31} \\ -m_{12} & m_{22} & -m_{32} \\ m_{13} & -m_{23} & m_{33} \end{bmatrix}}{|A|} = \frac{\begin{bmatrix} F_{11} & F_{21} & F_{31} \\ F_{12} & F_{22} & F_{32} \\ F_{13} & F_{23} & F_{33} \end{bmatrix}}{|A|} = \frac{[F]^T}{|A|} \tag{11}$$

where $[F]^T$ is called the *adjoint matrix* of $[A]$. In finding $[A]$, we use det $[A]$ as a divisor; therefore, det $[A]$ cannot be zero. If det $[A] = 0$, $[A]$ is called a *singular matrix*.

8. Solution of Simultaneous Equations

Consider the following simultaneous equations:

$$\begin{aligned} 3x_1 + 4x_2 &= 11 \\ 5x_1 - x_2 &= 3 \end{aligned} \tag{12}$$

We can write them into a matrix form:

$$\begin{bmatrix} 3 & 4 \\ 5 & -1 \end{bmatrix} \begin{bmatrix} x_1 \\ x_2 \end{bmatrix} = \begin{bmatrix} 11 \\ 3 \end{bmatrix} \tag{12a}$$

Multiplying both sides by

$$\begin{bmatrix} 3 & 4 \\ 5 & -1 \end{bmatrix}^{-1}$$

we have

$$\begin{bmatrix} 3 & 4 \\ 5 & -1 \end{bmatrix}^{-1} \begin{bmatrix} 3 & 4 \\ 5 & -1 \end{bmatrix} \begin{bmatrix} x_1 \\ x_2 \end{bmatrix} = \begin{bmatrix} 3 & 4 \\ 5 & -1 \end{bmatrix}^{-1} \begin{bmatrix} 11 \\ 3 \end{bmatrix}$$

by using the four-step technique we find

$$\begin{bmatrix} 3 & 4 \\ 5 & -1 \end{bmatrix}^{-1} = \begin{bmatrix} \frac{1}{23} & \frac{4}{23} \\ \frac{5}{23} & -\frac{3}{23} \end{bmatrix}$$

Therefore we obtain

$$\begin{bmatrix} 1 & 0 \\ 0 & 1 \end{bmatrix} \begin{bmatrix} x_1 \\ x_2 \end{bmatrix} = \begin{bmatrix} \frac{1}{23} & \frac{4}{23} \\ \frac{5}{23} & -\frac{3}{23} \end{bmatrix} \begin{bmatrix} 11 \\ 3 \end{bmatrix}$$

$$\begin{bmatrix} x_1 \\ x_2 \end{bmatrix} = \begin{bmatrix} \frac{11}{23} + \frac{12}{23} \\ \frac{55}{23} - \frac{9}{23} \end{bmatrix}$$

$$\begin{bmatrix} x_1 \\ x_2 \end{bmatrix} = \begin{bmatrix} 1 \\ 2 \end{bmatrix}$$

II. Gauss' Reduction and Premultiplication

1. Gauss' Reduction

A set of simultaneous equations is given.

$$a_{11}x_1 + a_{12}x_2 + \cdots + a_{1n}x_n = y_1$$
$$a_{21}x_1 + a_{22}x_2 + \cdots + a_{2n}x_n = y_2$$

. . . .   (1)

$$a_{n1}x_1 + a_{n2}x_2 + \cdots + a_{nn}x_n = y_n$$

Eliminate x_1 from all equations beginning with the second in the following manner: Multiply and divide the first equation by two numbers such that when the resulting equation is subtracted from the second equation, the new equation will have zero coefficient for the first term. This then becomes the new second equation. Again, multiply and divide all terms in the first equation by two numbers such that when this resulting modified equation is subtracted from the third given equation, a new equation is formed that will have lost its first term. This becomes the new third equation. Continuing this process, a new equivalent system of equations will be obtained.

$$a_{11}x_1 + a_{12}x_2 + a_{13}x_3 + \cdots + a_{1n}x_n = y_1$$
$$a_{22}^{(1)}x_2 + a_{23}^{(1)}x_3 + \cdots + a_{2n}^{(1)}x_n = y_2^{(1)}$$
$$a_{32}^{(1)}x_2 + a_{33}^{(1)}x_3 + \cdots + a_{3n}^{(1)}x_n = y_3^{(1)}$$ (2)

. . . .

$$a_{n2}^{(1)}x_2 + a_{n3}^{(1)}x_3 + \cdots + a_{nn}^{(1)}x_n = y_n^{(1)}$$

The superscripts indicate that the values of the numbers so marked are new numbers, having been changed by the preceding operations. What has been done can be stated in the following algorithmic form:

$$a_{ij}^{(1)} = a_{ij} - \frac{a_{i1}}{a_{11}} a_{1j}$$

$$y_i^{(1)} = y_i - \frac{a_{i1}}{a_{11}} y_1 \qquad (i, j = 2, \ldots, n) \tag{3}$$

Then x_2 is eliminated in the same way from the last $(n - 2)$ equations.

$$a_{11}x_1 + a_{12}x_2 + a_{13}x_3 + \cdots + a_{1n}x_n = y_1$$
$$a_{22}^{(1)}x_2 + a_{23}^{(1)}x_3 + \cdots + a_{2n}^{(1)}x_n = y_2^{(1)}$$
$$a_{33}^{(2)}x_3 + \cdots + a_{3n}^{(2)}x_n = y_3^{(2)} \tag{4}$$

$$\cdots \cdots \cdots \cdots \cdots$$

$$a_{n3}^{(2)}x_3 + \cdots + a_{nn}^{(2)}x_n = y_n^{(2)}$$

This time the algorithm is slightly different but methodically similar.

$$a_{ij}^{(2)} = a_{ij}^{(1)} - \frac{a_{i2}^{(1)}}{a_{22}^{(1)}} a_{2j}^{(1)}$$

$$y_i^{(2)} = y_i^{(1)} - \frac{a_{i2}^{(1)}}{a_{22}^{(1)}} y_2^{(1)} \qquad (i, j = 3, \ldots, n) \tag{5}$$

One continues the algorithm $(n - 1)$ steps from the original system to the final.

$$a_{11}x_1 + a_{12}x_2 + a_{13}x_3 + \cdots + a_{1n}x_n = y_1$$
$$a_{22}^{(1)}x_2 + a_{23}^{(1)}x_3 + \cdots + a_{2n}^{(1)}x_n = y_2^{(1)}$$
$$a_{33}^{(2)}x_3 + \cdots + a_{3n}^{(2)}x_n = y_3^{(2)} \tag{6}$$
$$a_{nn}^{(n-1)}x_n = y_n^{(n-1)}$$

Expressing (6) as the matrix form, one has

$$
\begin{bmatrix}
a_{11} & a_{12} & a_{13} & \cdots & a_{1n} \\
0 & a_{22}^{(1)} & a_{23}^{(1)} & \cdots & a_{2n}^{(1)} \\
0 & 0 & a_{33}^{(2)} & \cdots & a_{3n}^{(2)} \\
0 & 0 & 0 & \cdots & \cdot \\
0 & 0 & 0 & \cdots & \cdot \\
0 & 0 & 0 & \cdots & a_{nn}^{(n-1)}
\end{bmatrix}
\begin{bmatrix}
x_1 \\ x_2 \\ x_3 \\ \cdot \\ \cdot \\ x_n
\end{bmatrix}
=
\begin{bmatrix}
y_1 \\ y_2^{(1)} \\ y_3^{(2)} \\ \cdot \\ \cdot \\ y_n^{(n-1)}
\end{bmatrix}
\tag{6a}
$$

The left-hand square matrix of (6a) is called a triangular matrix. Changing a regular square matrix into a triangular matrix can be called triangularization and the technique mentioned above is called the Gauss reduction.

After the triangularization, one proceeds with the back substitution; that is, the last equation is solved for x_n. This value is substituted into the $(n - 1)$ equation to obtain x_{n-1}. These are substituted into the next equation and so on until all the values of x_i are obtained.

Example

$$3x_1 + 2x_2 + x_3 = 17 \tag{7}$$

$$2x_1 + 4x_2 + 5x_3 = 41 \tag{8}$$

$$4x_1 + x_2 + x_3 = 16 \tag{9}$$

This problem will be solved by means of Gaussian reduction. The "first round" yields the following:

$$3x_1 \qquad + 2x_2 \qquad + x_3 = 17$$

$$(2 - \tfrac{3}{3}\cdot 2)x_1 + (4 - \tfrac{2}{3}\cdot 2)x_2 + (5 - \tfrac{1}{3}\cdot 2)x_3 = (41 - \tfrac{17}{3}\cdot 2) \tag{10}$$

$$(4 - \tfrac{3}{3}\cdot 4)x_1 + (1 - \tfrac{2}{3}\cdot 4)x_2 + (1 - \tfrac{1}{3}\cdot 4)x_3 = (16 - \tfrac{17}{3}\cdot 4)$$

or

$$3x_1 + 2x_2 + x_3 = 17$$

$$\tfrac{8}{3}x_2 + \tfrac{13}{3}x_3 = \tfrac{89}{3} \tag{11}$$

$$-\tfrac{5}{3}x_2 - \tfrac{1}{3}x_3 = -\tfrac{20}{3}$$

It is noted that the first coefficient in the new second equation came from the following substitution:

$$a_{ij}^{(1)} = a_{ij} - \frac{a_{i1}}{a_{11}}a_{1j} \tag{12}$$

$$\tfrac{8}{3} = 4 - \tfrac{2}{3}\cdot 2$$

The right-hand side of the second equation was obtained from

$$y_i^{(1)} = y_i - \frac{a_{i1}}{a_{11}}y_1 \tag{13}$$

$$\tfrac{89}{3} = 41 - \tfrac{17}{3}\cdot 2$$

The "second round" gives the following:

$$3x_1 + 2x_2 + x_3 = 17$$

$$\tfrac{8}{3}x_2 + \tfrac{13}{3}x_3 = \tfrac{89}{3} \tag{14}$$

$$[-\tfrac{5}{3} + (\tfrac{8}{3}/\tfrac{8}{3})\cdot\tfrac{5}{3}]x_2 + [-\tfrac{1}{3} + (\tfrac{13}{3}/\tfrac{8}{3})\cdot\tfrac{5}{3}]x_3 = -\tfrac{20}{3} + (\tfrac{89}{3}/\tfrac{8}{3})\cdot\tfrac{5}{3}$$

or it can be rewritten as

$$3x_1 + 2x_2 + x_3 = 17$$

$$\tfrac{8}{3}x_2 + \tfrac{13}{3}x_3 = \tfrac{89}{3} \tag{15}$$

$$\tfrac{57}{24}x_3 = \tfrac{285}{24}$$

and from the third equation of (15) one obtains

$$x_3 = 5 \tag{16}$$

Then substituting the value of x_3 into the second equation of (15), we have

$$x_2 = 3 \tag{17}$$

and substituting (16) and (17) into the first equation of (15) gives

$$x_1 = 2 \tag{18}$$

2. Premultiplication

Gauss' reduction offers a very simple procedure for solving a set of simultaneous equations. However the technique is nothing but a systematic elimination of variables.

If we use matrix notation to do Gauss' reduction, the procedure looks even simpler. The operation corresponding to the "first round" reduction of coefficients to zero is as follows. By premultiplying (1) by the matrix (19), Eq. (2) is obtained directly.

$$\begin{bmatrix} 1 & 0 & 0 & \cdots & 0 \\ -\dfrac{a_{21}}{a_{11}} & 1 & 0 & \cdots & 0 \\ -\dfrac{a_{31}}{a_{11}} & 0 & 1 & \cdots & 0 \\ \cdot & \cdot & \cdot & & \cdot \\ \cdot & \cdot & \cdot & & \cdot \\ \cdot & \cdot & \cdot & & \cdot \\ \dfrac{a_{n1}}{a_{11}} & 0 & 0 & \cdots & 1 \end{bmatrix} \tag{19}$$

Similarly, if one premultiplies (2) by the following matrix

$$\begin{bmatrix} 1 & 0 & 0 & \cdots & 0 \\ 0 & 1 & 0 & \cdots & 0 \\ 0 & -\dfrac{a_{32}^{(1)}}{a_{22}^{(1)}} & 1 & \cdots & 0 \\ 0 & -\dfrac{a_{42}^{(1)}}{a_{22}^{(1)}} & 0 & 1 & \\ \cdot & \cdot & \cdot & & \cdot \\ \cdot & \cdot & \cdot & & \cdot \\ 0 & -\dfrac{a_{n2}^{(1)}}{a_{22}^{(1)}} & 0 & \cdots & 1 \end{bmatrix} \tag{20}$$

one obtains Eq. (4).

We use the foregoing procedure to illustrate how to solve the set of Eqs. (7) through (9). The matrix equation of the example is

$$\begin{bmatrix} 3 & 2 & 1 \\ 2 & 4 & 5 \\ 4 & 1 & 1 \end{bmatrix} \begin{bmatrix} x_1 \\ x_2 \\ x_3 \end{bmatrix} = \begin{bmatrix} 17 \\ 41 \\ 16 \end{bmatrix} \tag{21}$$

Performing the premultiplication,

$$\begin{bmatrix} 1 & 0 & 0 \\ -\frac{2}{3} & 1 & 0 \\ -\frac{4}{3} & 0 & 1 \end{bmatrix} \begin{bmatrix} 3 & 2 & 1 \\ 2 & 4 & 5 \\ 4 & 1 & 1 \end{bmatrix} \begin{bmatrix} x_1 \\ x_2 \\ x_3 \end{bmatrix} = \begin{bmatrix} 1 & 0 & 0 \\ -\frac{2}{3} & 1 & 0 \\ -\frac{4}{3} & 0 & 1 \end{bmatrix} \begin{bmatrix} 17 \\ 41 \\ 16 \end{bmatrix}$$

one obtains

$$\begin{bmatrix} 3 & 2 & 1 \\ 0 & 4-\frac{4}{3} & 5-\frac{2}{3} \\ 0 & 1-\frac{8}{3} & 1-\frac{4}{3} \end{bmatrix} \begin{bmatrix} x_1 \\ x_2 \\ x_3 \end{bmatrix} = \begin{bmatrix} 1 & 0 & 0 \\ -\frac{2}{3} & 1 & 0 \\ -\frac{4}{3} & 0 & 1 \end{bmatrix} \begin{bmatrix} 17 \\ 41 \\ 16 \end{bmatrix} \tag{22}$$

or

$$\begin{bmatrix} 3 & 2 & 1 \\ 0 & \frac{8}{3} & \frac{13}{3} \\ 0 & -\frac{5}{3} & -\frac{1}{3} \end{bmatrix} \begin{bmatrix} x_1 \\ x_2 \\ x_3 \end{bmatrix} = \begin{bmatrix} 17 \\ \frac{89}{3} \\ -\frac{20}{3} \end{bmatrix} \tag{23}$$

Premultiplying (23) by the matrix

$$\begin{bmatrix} 1 & 0 & 0 \\ 0 & 1 & 0 \\ 0 & \frac{5}{8} & 1 \end{bmatrix}$$

finishes the triangularization.

$$\begin{bmatrix} 1 & 0 & 0 \\ 0 & 1 & 0 \\ 0 & \frac{5}{8} & 1 \end{bmatrix} \begin{bmatrix} 3 & 2 & 1 \\ 0 & \frac{8}{3} & \frac{13}{3} \\ 0 & -\frac{5}{3} & -\frac{1}{3} \end{bmatrix} \begin{bmatrix} x_1 \\ x_2 \\ x_3 \end{bmatrix} = \begin{bmatrix} 1 & 0 & 0 \\ 0 & 1 & 0 \\ 0 & \frac{5}{8} & 1 \end{bmatrix} \begin{bmatrix} 17 \\ \frac{89}{3} \\ -\frac{20}{3} \end{bmatrix} \tag{24}$$

$$\begin{bmatrix} 3 & 2 & 1 \\ 0 & \frac{8}{3} & \frac{13}{3} \\ 0 & 0 & \frac{57}{24} \end{bmatrix} \begin{bmatrix} x_1 \\ x_2 \\ x_3 \end{bmatrix} = \begin{bmatrix} 17 \\ \frac{89}{3} \\ \frac{285}{24} \end{bmatrix} \tag{25}$$

Referring to Eq. (15), we note that the results are the same; however, the use of the matrix operations is easier. A digital computer program based on the Gauss reduction is shown in Program 8.

The following observations concerning premultiplications are obvious from the above example.

A. If a set of equations is expressed in a matrix form and if both sides of the matrix equation are premultiplied by a unit matrix, nothing will change.

$$\begin{bmatrix} 1 & 0 & 0 \\ 0 & 1 & 0 \\ 0 & 0 & 1 \end{bmatrix} \begin{bmatrix} a_{11} & a_{12} & a_{13} \\ a_{21} & a_{22} & a_{23} \\ a_{31} & a_{32} & a_{33} \end{bmatrix} \begin{bmatrix} x_1 \\ x_2 \\ x_3 \end{bmatrix} = \begin{bmatrix} 1 & 0 & 0 \\ 0 & 1 & 0 \\ 0 & 0 & 1 \end{bmatrix} \begin{bmatrix} y_1 \\ y_2 \\ y_3 \end{bmatrix} \tag{26}$$

$$\begin{bmatrix} a_{11} & a_{12} & a_{13} \\ a_{21} & a_{22} & a_{23} \\ a_{31} & a_{32} & a_{33} \end{bmatrix} \begin{bmatrix} x_1 \\ x_2 \\ x_3 \end{bmatrix} = \begin{bmatrix} y_1 \\ y_2 \\ y_3 \end{bmatrix}$$

Program 8. SOLVING SIMULTANEOUS EQUATIONS BY GAUSS' ELIMINATION

```
C      THE MATRIX IS FIRST PUT IN BEST ORDER
C      EACH DATA CARD IS TO CONTAIN 5 COEFFICIENTS OF X
C      AND ONE Y(X) IN E10.3
       DIMENSION A(5,5),X(5),Y(5),ORDER(5)
    7  DO 5 I=1,5
    5  READ6,A(I,1),A(I,2),A(I,3),A(I,4),A(I,5),Y(I)
    6  FORMAT(6E10.3)
       VMAX=ABS(A(1,1))
       ROW=1
       COL=1
       DO 107 I=1,5
       ORDER(I)=1
  107  CONTINUE
       DO 91 M=1,5
       DO 105 N=1,5
       IF(VMAX-ABS(A(M,N)))70,105,105
   70  VMAX=ABS(A(M,N))
       ROW = M
       COL = N
  105  CONTINUE
   91  CONTINUE
       M=ROW
       N=COL
       STORE = ORDER(1)
       ORDER(1)=ORDER(N)
       ORDER(N)=STORE
       DO 92 L=1,5
       STORE=A(L,1)
       A(L,1) = A(L,N)
       A(L,N)=STORE
   92  CONTINUE
       DO 106 L=1,5
       STORE = A(1,L)
       A(1,L)=A(M,L)
       A(M,L)=STORE
  106  CONTINUE
       STORE=Y(1)
       Y(1)=Y(M)
       Y(M)=STORE
       PRINT57
   57  FORMAT(16HTHE COEFFICIENTS)
       DO 18 I=1,5
       PRINT12, A(I,1),A(I,2),A(I,3),A(I,4),A(I,5),Y(I)
   18  CONTINUE
       DO 4 K=1,4
       M=K+1
       DO 2 I=M,5
       DO 3 J=M,5
       A(I,J) = A(I,J) -(A(K,J)*A(I,K))/ A(K,K)
    3  CONTINUE
       Y(I) = Y(I) - A(I,K)*Y(K)/A(K,K)
    2  CONTINUE
   80  IF(M-5)85,81,85
   81  IF(A(M,M))4,82,4
   82  PRINT83
   83  FORMAT(30HINDEPENDENT EQUATION DISPARITY
       GO TO 7
   85  VMAX=ABS(A(M,M))
       SUBS = M
       N=M+1
       DO 88 L = N,5
       IF(VMAX-ABS(A(M,L)))87,88,88
```

```
 87  VMAX=ABS(A(M,L))
     SUBS = L
 88  CONTINUE
     N=SUBS
     STORE = ORDER(M)
     ORDER(M)=ORDER(N)
     ORDER(N)=STORE
     IF(VMAX)89,82,89
 89  DO 90 L=1,5
     STORE=A(L,M)
     A(L,M)=A(L,N)
     A(L,N) = STORE
 90  CONTINUE
  4  CONTINUE
     DO 16 I = 2,5
     M=I-1
     DO 19 J=1,M
     A(I,J)=0.0
 19  CONTINUE
 16  CONTINUE
     PRINT21
 21  FORMAT(9HREDUCTION)
     DO 15 I=1,5
     PRINT12,A(I,1),A(I,2),A(I,3),A(I,4),A(I,5),Y(I)
 12  FORMAT(5E10.3,3H = E10.3
 15  CONTINUE
     PRINT 22
 22  FORMAT(8HSOLUTION)
     M=5+1
 58  DO61 K=1,5
     M1=M-K
     X(M1)=Y(M1)/A(M1,M1)
     IF((M-K)-5)59,60,10
 10  PRINT50
 50  FORMAT(7HTRIVIAL)
     GO TO 7
 59  DO 25 L=1,5
     M2=M-L
     X(M1) = X(M1) - A(M1,M2)*X(M2)/A(M1,M1)
     IF((M-L-1)-(M-K))53,60,53
 53  CONTINUE
 25  CONTINUE
 60  CONTINUE
 61  CONTINUE
109  J=0
     DO 101 I=1,5
     N=ORDER(I)
     IF(N-1)108,101,108
108  STORE = X(N)
     X(N)=X(I)
     X(I)=STORE
     STORE = ORDER(I)
     ORDER(I)=ORDER(N)
     ORDER(N)=STORE
     J=J+1
101  CONTINUE
     IF(J)109,110,109
110  CONTINUE
     PRINT52,X(1),X(2),X(3),X(4),X(5)
 52  FORMAT(3HX1=E10.3,4H X2=E10.3,4H X3=E10.3
    14H X4=E10.3,4H X5=E10.3)
     GO TO 7
     END
```

Flow Chart for Program 8

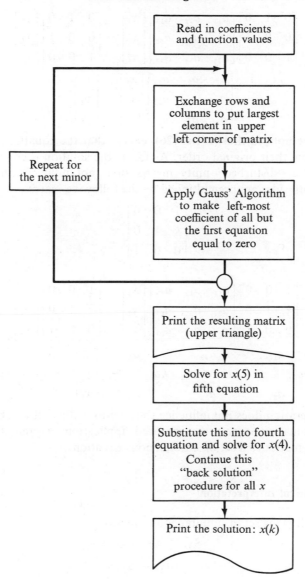

B. If both sides are premultiplied by

$$\begin{bmatrix} 0 & 1 & 0 \\ 0 & 0 & 1 \\ 1 & 0 & 0 \end{bmatrix}$$

the following is obtained:

$$\begin{bmatrix} 0 & 1 & 0 \\ 0 & 0 & 1 \\ 1 & 0 & 0 \end{bmatrix} \begin{bmatrix} a_{11} & a_{12} & a_{13} \\ a_{21} & a_{22} & a_{23} \\ a_{31} & a_{32} & a_{33} \end{bmatrix} \begin{bmatrix} x_1 \\ x_2 \\ x_3 \end{bmatrix} = \begin{bmatrix} 0 & 1 & 0 \\ 0 & 0 & 1 \\ 1 & 0 & 0 \end{bmatrix} \begin{bmatrix} y_1 \\ y_2 \\ y_3 \end{bmatrix} \tag{27}$$

$$\begin{bmatrix} a_{21} & a_{22} & a_{23} \\ a_{31} & a_{32} & a_{33} \\ a_{11} & a_{12} & a_{13} \end{bmatrix} \begin{bmatrix} x_1 \\ x_2 \\ x_3 \end{bmatrix} = \begin{bmatrix} y_2 \\ y_3 \\ y_1 \end{bmatrix}$$

It will be noticed that (27) is (26), except that the equations have been changed from their original order. As far as the solution is concerned, premultiplying a "disturbed" unity matrix does not influence the solution.

C. If both sides are premultiplied by the following matrix

$$\begin{bmatrix} 1 & 0 & 0 \\ k & 1 & 0 \\ 0 & 0 & 1 \end{bmatrix}$$

one obtains

$$\begin{bmatrix} 1 & 0 & 0 \\ k & 1 & 0 \\ 0 & 0 & 1 \end{bmatrix} \begin{bmatrix} a_{11} & a_{12} & a_{13} \\ a_{21} & a_{22} & a_{23} \\ a_{31} & a_{32} & a_{33} \end{bmatrix} \begin{bmatrix} x_1 \\ x_2 \\ x_3 \end{bmatrix} = \begin{bmatrix} 1 & 0 & 0 \\ k & 1 & 0 \\ 0 & 0 & 1 \end{bmatrix} \begin{bmatrix} y_1 \\ y_2 \\ y_3 \end{bmatrix} \tag{28}$$

$$\begin{bmatrix} a_{11} & a_{12} & a_{13} \\ ka_{11} + a_{21} & ka_{12} + a_{22} & ka_{13} + a_{23} \\ a_{31} & a_{32} & a_{33} \end{bmatrix} \begin{bmatrix} x_1 \\ x_2 \\ x_3 \end{bmatrix} = \begin{bmatrix} y_1 \\ ky_1 + y_2 \\ y_3 \end{bmatrix}$$

This operation does not influence the solution either. It can be seen that Gauss' elimination is only the repeated application of premultiplying by this type of matrix to a set of simultaneous equations.

3. An Electrical Interpretation

If the following set of simultaneous equations

$$5x_1 - 4x_2 + 0 = y_1$$
$$-4x_1 + 10x_2 - 5x_3 = y_2 \tag{29}$$
$$0 - 5x_2 + 6x_3 = y_3$$

is simulated by a multiloop circuit, shown in Fig. 5–1, in which y is the emf, x is the current, and resistors are the constant coefficients, the model in Fig. 5-1 can be obtained. If it is to be shown that the solution of the problem by circuit techniques is the same as Gauss' reduction, then the final equation format should be that of Eq. (30), which is the Gauss reduction of the original equation.

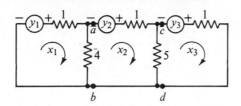

Fig. 5-1. A circuit interpretation of Eq. (30).

$$5x_1 - 4x_2 = y_1$$
$$\tfrac{34}{5}x_2 - 5x_3 = y_2 + \tfrac{4}{5}y_1 \tag{30}$$
$$\tfrac{79}{34}x_3 = y_3 + \tfrac{25}{34}y_2 + \tfrac{20}{34}y_1$$

To solve the problem by circuit techniques, cut the circuit at a and b and replace the right-hand part by a source $4x_2$, in series with the 4 ohm resistor. The equation of the new circuit is

$$5x_1 - 4x_2 = y_1 \tag{31}$$

This is exactly the same as the first equation of (30).

Then draw another circuit as shown in Fig. 5-2(b). This is constructed by drawing the Thevenin equivalent by looking into the left-hand side of Fig. 5-1 from a, b. The equivalent source is $\tfrac{4}{5}y_1$ and the equivalent resistor is $\tfrac{4}{5}$. On the right-hand branch of Fig 5-2(b), we put a source $5x_3$ in series with the 5 ohm resistor, to represent the remaining right-hand side. The cir-

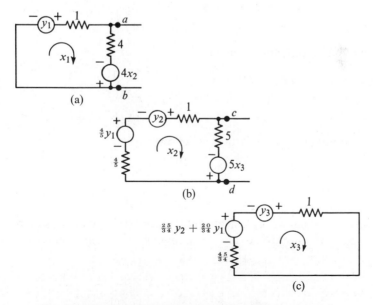

Fig. 5-2. The equivalent circuits corresponding to the equations obtained by Gauss' elimination.

cuit equation is then

$$\left(\frac{1 \times 4}{1 + 4} + 1 + 5\right)x_2 - 5x_3 = y_2 + \tfrac{4}{5}y_1$$

or

$$\tfrac{34}{5}x_2 - 5x_3 = y_2 + \tfrac{4}{5}y_1 \tag{32}$$

Following similar reasoning, we find the Thevenin equivalent of the first two loops; it appears as Fig. 5-2(c). The circuit equation is

$$x_3 + \frac{\left(\frac{1 \times 4}{1 + 4} + 1\right)5}{\left(\frac{1 \times 4}{1 + 4} + 1 + 5\right)}x_3 = y_3 + \frac{y_2 + \tfrac{4}{5}y_1}{6 + \tfrac{4}{5}}5$$

$$\tfrac{79}{34}x_3 = y_3 + \tfrac{25}{34}y_2 + \tfrac{20}{34}y_1 \tag{33}$$

It is noted that Equations (31) through (33) are the three equations of (30), respectively. Therefore, we conclude that using Thevenin's theorem to simplify the electric circuit is directly analogous to using Gaussian reduction to solve the set of simultaneous equations.

III.   Elementary Transformations

1. Postmultiplication
A system is given:

$$3x_1 + 2x_2 + 5x_3 = 39$$
$$4x_1 + 7x_2 + x_3 = 38 \tag{1}$$
$$x_1 + 6x_2 + 8x_3 = 49$$

Applying Gauss' reduction, one obtains

$$3x_1 + 2x_2 + 5x_3 = 39$$
$$\tfrac{13}{3}x_2 - \tfrac{17}{3}x_3 = -14 \tag{1a}$$
$$\tfrac{519}{39}x_3 = \tfrac{692}{13}$$

The corresponding matrix form is as follows:

$$\begin{bmatrix} 3 & 2 & 5 \\ 0 & \tfrac{13}{3} & -\tfrac{17}{3} \\ 0 & 0 & \tfrac{519}{39} \end{bmatrix}\begin{bmatrix} x_1 \\ x_2 \\ x_3 \end{bmatrix} = \begin{bmatrix} 39 \\ -14 \\ \tfrac{692}{13} \end{bmatrix} \tag{1b}$$

It is desirable to make the elements of the upper triangle of the square matrix equal to zero. In order to do this, the following transformation is used.

$$\begin{bmatrix} x_1 \\ x_2 \\ x_3 \end{bmatrix} = \begin{bmatrix} 1 & -\tfrac{2}{3} & -\tfrac{5}{3} \\ 0 & 1 & 0 \\ 0 & 0 & 1 \end{bmatrix}\begin{bmatrix} y_1 \\ y_2 \\ y_3 \end{bmatrix} \tag{2}$$

The reason for using $-\tfrac{2}{3}$, $-\tfrac{5}{3}$ is self-explanatory.

Substituting (2) into (1b), we have

$$\begin{bmatrix} 3 & 2 & 5 \\ 0 & \frac{13}{3} & \frac{17}{3} \\ 0 & 0 & \frac{519}{39} \end{bmatrix} \begin{bmatrix} 1 & -\frac{2}{3} & -\frac{5}{3} \\ 0 & 1 & 0 \\ 0 & 0 & 1 \end{bmatrix} \begin{bmatrix} y_1 \\ y_2 \\ y_3 \end{bmatrix} = \begin{bmatrix} 39 \\ -14 \\ \frac{692}{13} \end{bmatrix} \tag{3}$$

As stated previously, the square matrix of (1b) is to be postmultiplied by the square matrix of (2).

Performing the multiplication, we obtain

$$\begin{bmatrix} 3 & 0 & 0 \\ 0 & \frac{13}{3} & -\frac{17}{3} \\ 0 & 0 & \frac{519}{39} \end{bmatrix} \begin{bmatrix} y_1 \\ y_2 \\ y_3 \end{bmatrix} = \begin{bmatrix} 39 \\ -14 \\ \frac{692}{13} \end{bmatrix} \tag{3a}$$

In order to create another zero on the third column, second row, another transformation is used.

$$\begin{bmatrix} y_1 \\ y_2 \\ y_3 \end{bmatrix} = \begin{bmatrix} 1 & 0 & 0 \\ 0 & 1 & \frac{17/3}{13/3} \\ 0 & 0 & 1 \end{bmatrix} \begin{bmatrix} z_1 \\ z_2 \\ z_3 \end{bmatrix} \tag{4}$$

Substitution of (4) into (3a) gives

$$\begin{bmatrix} 3 & 0 & 0 \\ 0 & \frac{13}{3} & -\frac{17}{3} \\ 0 & 0 & \frac{519}{39} \end{bmatrix} \begin{bmatrix} 1 & 0 & 0 \\ 0 & 1 & \frac{17}{13} \\ 0 & 0 & 1 \end{bmatrix} \begin{bmatrix} z_1 \\ z_2 \\ z_3 \end{bmatrix} = \begin{bmatrix} 39 \\ -14 \\ \frac{692}{13} \end{bmatrix}$$

or

$$\begin{bmatrix} 3 & 0 & 0 \\ 0 & \frac{13}{3} & 0 \\ 0 & 0 & \frac{519}{39} \end{bmatrix} \begin{bmatrix} z_1 \\ z_2 \\ z_3 \end{bmatrix} = \begin{bmatrix} 39 \\ -14 \\ \frac{692}{13} \end{bmatrix} \tag{5}$$

or

$$\begin{bmatrix} z_1 \\ z_2 \\ z_3 \end{bmatrix} = \begin{bmatrix} \frac{1}{3} & 0 & 0 \\ 0 & \frac{3}{13} & 0 \\ 0 & 0 & \frac{39}{519} \end{bmatrix} \begin{bmatrix} 39 \\ -14 \\ \frac{692}{13} \end{bmatrix} \tag{6}$$

Equation (6) is the solution. In order to find the values for x_1, x_2, and x_3, the inverse transformation is performed. In other words, the substitution (6) into (2) and (4) is made.

$$\begin{bmatrix} x_1 \\ x_2 \\ x_3 \end{bmatrix} = \begin{bmatrix} 1 & -\frac{2}{3} & -\frac{5}{3} \\ 0 & 1 & 0 \\ 0 & 0 & 1 \end{bmatrix} \begin{bmatrix} 1 & 0 & 0 \\ 0 & 1 & \frac{17}{13} \\ 0 & 0 & 1 \end{bmatrix} \begin{bmatrix} \frac{1}{3} & 0 & 0 \\ 0 & \frac{3}{13} & 0 \\ 0 & 0 & \frac{39}{519} \end{bmatrix} \begin{bmatrix} 39 \\ -14 \\ \frac{692}{13} \end{bmatrix} = \begin{bmatrix} 5 \\ 2 \\ 4 \end{bmatrix}$$

From this example it can be seen that postmultiplicating a matrix to the square matrix of a matrix equation is easily understood by using the idea of change of variables.

2. Geometrical Interpretation of Postmultiplication

In general, (2), or (4) should be expressed as

$$
\begin{bmatrix} y_1 \\ y_2 \\ y_3 \\ \cdot \\ \cdot \\ \cdot \\ y_n \end{bmatrix} = \begin{bmatrix} a_{11} & a_{12} & a_{13} & \cdots & a_{1n} \\ a_{21} & a_{22} & a_{23} & \cdots & a_{2n} \\ a_{31} & a_{32} & a_{33} & \cdots & a_{3n} \\ \cdot & \cdot & \cdot & & \cdot \\ \cdot & \cdot & \cdot & \cdots & \cdot \\ \cdot & \cdot & \cdot & & \cdot \\ a_{n1} & a_{n2} & a_{n3} & \cdots & a_{nn} \end{bmatrix} \begin{bmatrix} x_1 \\ x_2 \\ x_3 \\ \cdot \\ \cdot \\ \cdot \\ x_n \end{bmatrix}
\tag{7}
$$

This transformation can be interpreted in two ways. It can be considered as transforming one n-dimensional vector $(x_1, x_2, \ldots, x_n)$ into another vector $(y_1, y_2, y_3, \ldots, y_n)$ in the same coordinate system. On the other hand, if $x_1, x_2, \ldots, x_n$ and $y_1, y_2, \ldots, y_n$ are regarded as the components of the same vector with respect to two different choices of the axes, then Eq. (7) gives the transformation law of the vector components, in going from one coordinate system to the other.

The second interpretation will be used in working a simple problem as an illustration.

Consider the following equations:

$$
0.8x_1 + 0.6x_2 = y_1
$$
$$
-0.6x_1 + 0.8x_2 = y_2
$$

or

$$
\begin{bmatrix} 0.8 & 0.6 \\ -0.6 & 0.8 \end{bmatrix} \begin{bmatrix} x_1 \\ x_2 \end{bmatrix} = \begin{bmatrix} y_1 \\ y_2 \end{bmatrix}
$$

x_1, x_2, and y_1, y_2 are the coordinate axes in the x and y planes, respectively.

There is a vector A (shown in Fig. 5-3). It can be referred to the x plane or the y plane.

If A is referred to the x plane, it can be read

$$
\begin{bmatrix} x_1 \\ x_2 \end{bmatrix} = \begin{bmatrix} 0.4 \\ 0.2 \end{bmatrix}
$$

If it is referred to the y plane, it will be expressed as

$$
\begin{bmatrix} y_1 \\ y_2 \end{bmatrix} = \begin{bmatrix} 0.44 \\ -0.08 \end{bmatrix}
$$

It can be seen that the matrix

$$
\begin{bmatrix} 0.8 & 0.6 \\ -0.6 & 0.8 \end{bmatrix}
$$

gives the transformation law of the components.

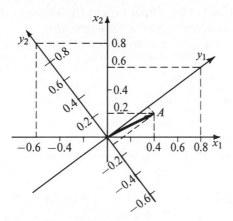

Fig. 5-3. Geometrical interpretation of change of variables.

3. Elementary Transformations

The special types of constant matrices that have been used with illustration of premultiplication and postmultiplication can be classified into the following categories.

(a) OPERATION OF TYPE I

Let $[J]$ denote the unit matrix $[I]$ with the ith and the jth rows interchanged. Then $[J][A]$ is the matrix obtained when the ith and jth rows of $[A]$ are interchanged. And $[A][J]$ is the matrix obtained from $[A]$ by interchange of the ith and jth columns.

Example 1.

$$\begin{bmatrix} 1 & 0 & 0 \\ 0 & 0 & 1 \\ 0 & 1 & 0 \end{bmatrix} \begin{bmatrix} a_{11} & a_{12} & a_{13} \\ a_{21} & a_{22} & a_{23} \\ a_{31} & a_{32} & a_{33} \end{bmatrix} = \begin{bmatrix} a_{11} & a_{12} & a_{13} \\ a_{31} & a_{32} & a_{33} \\ a_{21} & a_{22} & a_{23} \end{bmatrix}$$

Example 2.

$$\begin{bmatrix} a_{11} & a_{12} & a_{13} \\ a_{21} & a_{22} & a_{23} \\ a_{31} & a_{32} & a_{33} \end{bmatrix} \begin{bmatrix} 1 & 0 & 0 \\ 0 & 0 & 1 \\ 0 & 1 & 0 \end{bmatrix} = \begin{bmatrix} a_{11} & a_{13} & a_{12} \\ a_{21} & a_{23} & a_{22} \\ a_{31} & a_{33} & a_{32} \end{bmatrix}$$

(b) OPERATION OF TYPE II

Let $[L]$ be the unit matrix modified by the introduction of the element l in the ith row and jth column, where i and j are unequal. Then $[L][A]$ is the matrix obtained from $[A]$ by taking l times the elements of the jth row and adding the corresponding terms to the ith row. For postmultiplication $[A]$ by $[L]$, change the word "row" to "column" and interchange i and j.

Example 1. $(i = 3, j = 2)$

$$\begin{bmatrix} 1 & 0 & 0 \\ 0 & 1 & 0 \\ 0 & l & 1 \end{bmatrix} \begin{bmatrix} a_{11} & a_{12} & a_{13} \\ a_{21} & a_{22} & a_{23} \\ a_{31} & a_{32} & a_{33} \end{bmatrix} \begin{bmatrix} a_{11} & a_{12} & a_{13} \\ a_{21} & a_{22} & a_{23} \\ a_{31} + la_{21} & a_{32} + la_{22} & a_{33} + la_{23} \end{bmatrix}$$

Example 2. $(i = 3, j = 2)$

$$\begin{bmatrix} a_{11} & a_{12} & a_{13} \\ a_{21} & a_{22} & a_{23} \\ a_{31} & a_{32} & a_{32} \end{bmatrix} \begin{bmatrix} 1 & 0 & 0 \\ 0 & 1 & 0 \\ 0 & l & 1 \end{bmatrix} = \begin{bmatrix} a_{11} & a_{12} + la_{13} & a_{13} \\ a_{21} & a_{22} + la_{23} & a_{23} \\ a_{31} & a_{32} + la_{33} & a_{33} \end{bmatrix}$$

(c) OPERATION OF TYPE III

Let $[H]$ denote the unit matrix with h ($\neq 0$) substituted for unity in the ith element in the diagonal. Then $[H][A]$ and $[A][H]$ are shown as examples.

Example 1.

$$\begin{bmatrix} 1 & 0 & 0 \\ 0 & 1 & 0 \\ 0 & 0 & h \end{bmatrix} \begin{bmatrix} a_{11} & a_{12} & a_{13} \\ a_{21} & a_{22} & a_{23} \\ a_{31} & a_{32} & a_{33} \end{bmatrix} = \begin{bmatrix} a_{11} & a_{12} & a_{13} \\ a_{21} & a_{22} & a_{23} \\ ha_{31} & ha_{32} & ha_{33} \end{bmatrix}$$

Example 2.

$$\begin{bmatrix} a_{11} & a_{12} & a_{13} \\ a_{21} & a_{22} & a_{23} \\ a_{31} & a_{32} & a_{33} \end{bmatrix} \begin{bmatrix} 1 & 0 & 0 \\ 0 & 1 & 0 \\ 0 & 0 & h \end{bmatrix} = \begin{bmatrix} a_{11} & a_{12} & ha_{13} \\ a_{21} & a_{22} & ha_{23} \\ a_{31} & a_{32} & ha_{33} \end{bmatrix}$$

In general, the relationship between $[B]$ and $[A]$ can be expressed by the following equation:

$$[B] = [P][A][Q] \tag{8}$$

where $[P]$ and $[Q]$ are nonsingular matrices.

4. Inverse from Elementary Matrices

We can reduce a given matrix $[A]$ to a unit matrix $[I]$ by elementary transformations.

$$[P_m]\dots[P_2][P_1][A][Q_1][Q_2]\dots[Q_r] = [I] \tag{9}$$

or simply

$$[P][A][Q] = [I] \tag{9a}$$

Rewriting gives

$$[A] = [P]^{-1}[Q]^{-1} \tag{10}$$

Inverting yields

$$[A]^{-1} = [Q][P]$$

$$= [Q_1][Q_2]\ldots[Q_r][P_m]\ldots[P_2][P_1] \tag{11}$$

$[P_1], [P_2], \ldots, [P_m]$ are constructed by inspection according to the well-known rules of Gauss' elimination, and $[Q_1], [Q_2], \ldots, [Q_r]$ are constructed in a similar way. Equation (11) then becomes a basic formula for obtaining the inverse of a matrix, as shown in the following example.

$$[A] = \begin{bmatrix} 1 & -2 & 0 \\ 3 & 4 & -1 \\ 5 & 0 & -3 \end{bmatrix}$$

Its inverse is required.

Recalling the rules of Gauss' elimination, we construct $[P_1]$ as follows:

$$[P_1] = \begin{bmatrix} 1 & 0 & 0 \\ -3 & 1 & 0 \\ -5 & 0 & 1 \end{bmatrix} \tag{12}$$

Then,

$$[P_1][A] = \begin{bmatrix} 1 & -2 & 0 \\ 0 & 10 & -1 \\ 0 & 10 & -3 \end{bmatrix} \tag{13}$$

From (13), we construct $[P_2]$.

$$[P_2] = \begin{bmatrix} 1 & 0 & 0 \\ 0 & 1 & 0 \\ 0 & -1 & 1 \end{bmatrix} \tag{14}$$

And

$$[P_2][P_1][A] = \begin{bmatrix} 1 & -2 & 0 \\ 0 & 10 & -1 \\ 0 & 0 & -2 \end{bmatrix} \tag{15}$$

which is a triangularized matrix.

We then try to construct $[Q_1]$:

$$[Q_1] = \begin{bmatrix} 1 & 2 & 0 \\ 0 & 1 & 0 \\ 0 & 0 & 1 \end{bmatrix} \tag{16}$$

and then

$$[Q_2] = \begin{bmatrix} 1 & 0 & 0 \\ 0 & 1 & \frac{1}{10} \\ 0 & 0 & 1 \end{bmatrix} \tag{17}$$

Substituting $[P_1]\ldots[Q_1]\ldots$ into (11), we have

$$[A]^{-1} = [Q_1][Q_2][Q_3][P_2][P_1]$$

$$= \begin{bmatrix} 1 & 2 & 0 \\ 0 & 1 & 0 \\ 0 & 0 & 1 \end{bmatrix} \begin{bmatrix} 1 & 0 & 0 \\ 0 & 1 & \frac{1}{10} \\ 0 & 0 & 1 \end{bmatrix} \begin{bmatrix} 1 & 0 & 0 \\ 0 & \frac{1}{10} & 0 \\ 0 & 0 & -\frac{1}{2} \end{bmatrix} \begin{bmatrix} 1 & 0 & 0 \\ 0 & 1 & 0 \\ 0 & -1 & 1 \end{bmatrix} \begin{bmatrix} 1 & 0 & 0 \\ -3 & 1 & 0 \\ -5 & 0 & 1 \end{bmatrix}$$

$$= \begin{bmatrix} \frac{3}{5} & \frac{3}{10} & -\frac{1}{10} \\ -\frac{1}{5} & \frac{3}{20} & -\frac{1}{20} \\ 1 & \frac{1}{2} & -\frac{1}{2} \end{bmatrix} \tag{18}$$

For inverting a matrix, thus far, we have two methods: namely, the adjoint matrix method mentioned in the first section of this chapter, and the linear transformation method shown in this section. The former is suitable for small problems, whereas the latter is good for larger matrix inversion. If a digital computer is available, the linear transformation method is usually adopted for most cases.

IV.   Quadratic Forms

1. Three Expressions

A homogeneous polynomial of the type

$$q = a_{11}x_1^2 + 2a_{12}x_1x_2 + 2a_{13}x_1x_3 + \cdots + 2a_{1n}x_1x_n$$
$$+ a_{22}x_2^2 \qquad + 2a_{23}x_2x_3 + \cdots + 2a_{2n}x_2x_n$$
$$+ a_{33}x_3^2 \qquad + \cdots + 2a_{3n}x_3x_n$$
$$\cdot$$
$$\cdot$$
$$\cdot$$
$$+ a_{nn}x_n^2 \tag{1}$$

is called a quadratic form in the variables x_1, x_2, ..., x_n. We can write (1) in the following compact way.

$$q = \sum_{i=1}^{n} \sum_{j=1}^{n} a_{ij}x_ix_j \tag{1a}$$

In matrix notation a quadratic form is written as

$$q = [x_1, x_2, \ldots, x_n] \begin{bmatrix} a_{11} & a_{12} & \cdots & a_{1n} \\ a_{12} & a_{22} & \cdots & a_{2n} \\ \cdot & \cdot & \cdots & \cdot \\ \cdot & \cdot & \cdots & \cdot \\ \cdot & \cdot & \cdots & \cdot \\ a_{1n} & a_{2n} & \cdots & a_{nn} \end{bmatrix} \begin{bmatrix} x_1 \\ x_2 \\ \cdot \\ \cdot \\ \cdot \\ x_n \end{bmatrix} \tag{1b}$$

The symmetric matrix $[a_{ij}]$ is called the matrix of the quadratic form. For example

$$q = x_1^2 + 3x_2^2 + 8x_3^2 - 4x_1x_2 + 10x_1x_3$$

is a quadratic form in the variables x_1, x_2, and x_3. Its matrix expression is

$$q = [x_1, x_2, x_3] \begin{bmatrix} 1 & -2 & 5 \\ -2 & 3 & 0 \\ 5 & 0 & 8 \end{bmatrix} \begin{bmatrix} x_1 \\ x_2 \\ x_3 \end{bmatrix} \tag{2}$$

2. Positive Semidefinite

A real, symmetric, quadratic form, $q = [x]^T[a_{ij}][x]$ in n variables, is called positive semidefinite if and only if

$$\sum_{i,j=1}^{n} a_{ij} x_i x_j \geq 0$$

for all finite values of the real variables x_1, x_2, $\ldots$, x_n.

Example.

$$q = (x_1 + x_2)^2 + x_3^2$$

$$q = [x_1, x_2, x_3] \begin{bmatrix} 1 & 1 & 0 \\ 1 & 1 & 0 \\ 0 & 0 & 1 \end{bmatrix} \begin{bmatrix} x_1 \\ x_2 \\ x_3 \end{bmatrix}$$

is positive semidefinite.

3. Positive Definite

A real symmetric quadratic form $q = [x]^T[b_{ij}][x]$ in n variables is said to be positive definite if and only if $[x]^T[b_{ij}][x]$ is positive semidefinite, and in addition

$$\sum_{i,j=1}^{n} b_{ij} x_i x_j = 0 \tag{3}$$

if and only if $x_i = 0$ for $i = 1, 2, 3, \ldots, n$.

Example.

$$q = x_1^2 + x_2^2 + x_3^2$$

is positive definite. The matrix $[A] = [a_{ij}]$ of a real quadratic form

$$q = [x]^T[A][x]$$

is called semidefinite or definite depending upon whether the quadratic form is semidefinite or definite.

4. Lagrange's Reduction

How do we examine the sign-definiteness of a quadratic form? A logical starting point is to change the quadratic form into one in which only the squared terms of the variables appear. This procedure is called a reduction technique.

Lagrange's reduction consists essentially of repeated completing of the square. Given a quadratic form as follows:

$$q = x_1^2 + 3x_2^2 + 5x_3^2 + 4x_1x_2 + 8x_1x_3 + x_2x_3 \tag{4}$$

we combine the terms in the following manner.

$$\begin{aligned}
q &= x_1^2 + 4x_1(x_2 + 2x_3) + 2^0(x_2 + 2x_3)^2 - 2^2(x_2 + 2x_3)^2 + 5x_3^2 + x_2x_3 + 3x_2^2 \\
&= [x_1 + 2(x_2 + 2x_3)]^2 - 4x_2^2 - 16x_2x_3 - 16x_3^2 + 5x_3^2 + x_2x_3 + 3x_2^2 \\
&= [x_1 + 2(x_2 + 2x_3)]^2 - x_2^2 - 11x_3^2 - 15x_2x_3 \\
&= [x_1 + 2(x_2 + 2x_3)]^2 - [x_2 + 7.5x_3]^2 + 45.25x_3^2
\end{aligned}$$

which is a sign-indefinite quadratic form because the middle term is with a minus sign.

5. Transformation of Quadratic Forms

Can we change the variables from one set to another without changing the nature of the quadratic form?

Consider the following:

$$q = [x_1, x_2, x_3, \ldots, x_n] \begin{bmatrix} a_{11} & a_{12} & \cdots & a_{1n} \\ a_{21} & a_{22} & \cdots & a_{2n} \\ \cdot & \cdot & \cdots & \cdot \\ \cdot & \cdot & \cdots & \cdot \\ \cdot & \cdot & \cdots & \cdot \\ a_{n1} & a_{n2} & \cdots & a_{nn} \end{bmatrix} \begin{bmatrix} x_1 \\ x_2 \\ \cdot \\ \cdot \\ \cdot \\ x_n \end{bmatrix} \tag{5}$$

or

$$q = [x]^T[A][x] \tag{5a}$$

Suppose we have to change the variables $x_1, x_2, \ldots, x_n$ into $y_1, y_2, \ldots, y_n$. The vector y is defined by

$$[P][y] = [x] \tag{6}$$

in which $[P]$ is regarded as a transformation matrix.

Substitution of (6) into (5a) gives

$$q = [y]^T [P]^T [A][P][y] \tag{7}$$

If we write the result as

$$q = [y]^T [B][y] \tag{8}$$

Then

$$[B] = [P]^T [A][P] \tag{9}$$

Equation (9) is usually called a congruent transformation of the matrix $[A]$.

It is easily seen that the nature of a given quadratic form is invariant to a congruent transformation of its matrix.

6. Kronecker's Reduction

This method is based on repeated congruent transformations. It enables us systematically to reduce a quadratic form into one in which only the squared terms of the variables appear. An example will illustrate the procedure.

Given a quadratic form

$$q = x_1^2 + 3x_2^2 + 5x_3^2 + 4x_1x_2 + 8x_1x_3 + x_2x_3$$

First of all, we change it into a matrix form

$$q = [x_1, x_2, x_3] \begin{bmatrix} 1 & 2 & 4 \\ 2 & 3 & \frac{1}{2} \\ 4 & \frac{1}{2} & 5 \end{bmatrix} \begin{bmatrix} x_1 \\ x_2 \\ x_3 \end{bmatrix} \tag{10}$$

Recall the linear transformation described previously in connection with Gauss' elimination. Perform a transformation by using the following matrix equation:

$$\begin{bmatrix} x_1 \\ x_2 \\ x_3 \end{bmatrix} = \begin{bmatrix} 1 & -2 & -4 \\ 0 & 1 & 0 \\ 0 & 0 & 1 \end{bmatrix} \begin{bmatrix} y_1 \\ y_2 \\ y_3 \end{bmatrix} \tag{11}$$

The reason for choosing (11) is the same as that in treating a simultaneous equation by Gauss' elimination.

Substituting (10) and (11) into (7) gives

$$q = [y_1, y_2, y_3] \begin{bmatrix} 1 & 0 & 0 \\ -2 & 1 & 0 \\ -4 & 0 & 1 \end{bmatrix} \begin{bmatrix} 1 & 2 & 4 \\ 2 & 3 & \frac{1}{2} \\ 4 & \frac{1}{2} & 5 \end{bmatrix} \begin{bmatrix} 1 & -2 & -4 \\ 0 & 1 & 0 \\ 0 & 0 & 1 \end{bmatrix} \begin{bmatrix} y_1 \\ y_2 \\ y_3 \end{bmatrix} \qquad (12)$$

$$q = [y_1 \, y_2, y_3] \begin{bmatrix} 1 & 0 & 0 \\ 0 & -1 & -7\frac{1}{2} \\ 0 & -7\frac{1}{2} & -11 \end{bmatrix} \begin{bmatrix} y_1 \\ y_2 \\ y_3 \end{bmatrix} \qquad (13)$$

Note that we had produced four zeros in (13), but the nature of the quadratic form (10) did not change.

We need to produce more zeros as elements by using another congruent transformation.

Let

$$\begin{bmatrix} y_1 \\ y_2 \\ y_3 \end{bmatrix} = \begin{bmatrix} 1 & 0 & 0 \\ 0 & 1 & -7\frac{1}{2} \\ 0 & 0 & 1 \end{bmatrix} \begin{bmatrix} z_1 \\ z_2 \\ z_3 \end{bmatrix} \qquad (14)$$

Substitution of (14) into (7) gives

$$q = [z_1, z_2, z_3] \begin{bmatrix} 1 & 0 & 0 \\ 0 & 1 & 0 \\ 0 & -7\frac{1}{2} & 1 \end{bmatrix} \begin{bmatrix} 1 & 0 & 0 \\ 0 & -1 & -7\frac{1}{2} \\ 0 & -7\frac{1}{2} & -11 \end{bmatrix} \begin{bmatrix} 1 & 0 & 0 \\ 0 & 1 & -7\frac{1}{2} \\ 0 & 0 & 1 \end{bmatrix} \begin{bmatrix} z_1 \\ z_2 \\ z_3 \end{bmatrix}$$

$$= [z_1, z_2, z_3] \begin{bmatrix} 1 & 0 & 0 \\ 0 & -1 & 0 \\ 0 & 0 & 45.25 \end{bmatrix} \begin{bmatrix} z_1 \\ z_2 \\ z_3 \end{bmatrix} \qquad (15)$$

The result in (15) is the same as that obtained by Lagrange's reduction.

7. Sylvester's Theorem

Since we often wish to examine whether a quadratic form $[x]^T[A][x]$ is positive definite without using repeated reduction processes, a set of conditions will be developed that will enable one to make such examination. The conditions so developed are based on Sylvester's theorem.

There is a quadratic form:

$$q = [x]^T[A][x] \qquad (16)$$

The necessary and sufficient conditions for the quadratic form to be positive definite are

$$a_{11} > 0, \quad \begin{vmatrix} a_{11} & a_{12} \\ a_{12} & a_{22} \end{vmatrix} > 0, \quad \begin{vmatrix} a_{11} & a_{12} & a_{13} \\ a_{12} & a_{22} & v_{23} \\ a_{13} & a_{23} & a_{33} \end{vmatrix} > 0, \ldots, |A| > 0 \qquad (17)$$

The segments in (17) are usually called the leading principal minors or naturally ordered principal minors.

This extremely important theorem can be proven by the following inductive procedure:

Starting with a two variable quadratic form:

$$v = a_{11}x^2 + 2a_{12}x_1x_2 + a_{22}x_2^2 \tag{18}$$

rearrangement gives

$$v = a_{11}\left(x_1 + \frac{a_{12}}{a_{11}}x_2\right)^2 - a_{11}\left(\frac{a_{12}}{a_{11}}x_2\right)^2 + a_{22}x_2^2$$

$$= a_{11}\left(x_1 + \frac{a_{12}}{a_{11}}x_2\right)^2 + \left(a_{22} - \frac{a_{12}^2}{a_{11}}\right)x_2^2 \tag{18a}$$

The necessary and sufficient conditions for v to be positive definite are

$$a_{11} > 0$$

$$a_{22} - \frac{a_{12}^2}{a_{11}} > 0 \tag{19}$$

These conditions can be written in determinant forms:

$$a_{11} > 0$$

$$\begin{vmatrix} a_{11} & a_{12} \\ a_{12} & a_{22} \end{vmatrix} > 0 \tag{19a}$$

Continuing with a three-variable form, we have

$$w = a_{11}x_1^2 + 2a_{12}x_1x_2 + a_{22}x_2^2 + 2a_{23}x_2x_3 + 2a_{13}x_1x_3 + a_{33}x_3^2 \tag{20}$$

Combination of related terms by means of Lagrange's reduction gives

$$w = a_{11}\left(x_1 + \frac{a_{12}}{a_{11}}x_2 + \frac{a_{13}}{a_{11}}x_3\right)^2 - \left[\left(\frac{a_{12}}{a_{11}}x_2\right)^2 + \left(\frac{a_{13}}{a_{11}}x_3\right)^2\right]a_{11}$$

$$- 2\frac{a_{12}a_{13}}{a_{11}}x_2x_3 + a_{22}x_2^2 + a_{33}x_3^2 + 2a_{23}x_2x_3 \tag{21}$$

$$= a_{11}\left(x_1 + \frac{a_{12}}{a_{11}}x_2 + \frac{a_{13}}{a_{11}}x_3\right)^2 + \left(a_{22} - \frac{a_{12}^2}{a_{11}}\right)x_2^2 + \left(a_{33} - \frac{a_{13}^2}{a_{11}}\right)x_3^2$$

$$+ 2\left(a_{23} - \frac{a_{12}a_{13}}{a_{11}}\right)x_2x_3 \tag{22}$$

$$= a_{11}\left(x_1 + \frac{a_{12}}{a_{11}}x_2 + \frac{a_{13}}{a_{11}}x_3\right)^2 + \left(a_{22} - \frac{a_{12}^2}{a_{11}}\right)$$

$$\times \left[x_2^2 + \frac{2\left(a_{23} - \dfrac{a_{12}a_{13}}{a_{11}}\right)}{\left(a_{22} - \dfrac{a_{12}^2}{a_{11}}\right)}x_2x_3 + \frac{\left(a_{23} - \dfrac{a_{12}a_{13}}{a_{11}}\right)^2}{\left(a_{22} - \dfrac{a_{12}^2}{a_{11}}\right)^2}x_3^2\right]$$

$$- \frac{\left(a_{23} - \dfrac{a_{12}a_{13}}{a_{11}}\right)^2}{\left(a_{22} - \dfrac{a_{12}^2}{a_{11}}\right)}x_3^2 + \left(a_{33} - \frac{a_{13}^2}{a_{11}}\right)x_3^2 \tag{23}$$

$$= a_{11}\left(x_1 + \frac{a_{12}}{a_{11}}x_2 + \frac{a_{13}}{a_{11}}x_3\right)^2 + \left(a_{22} - \frac{a_{12}^2}{a_{11}}\right)$$

$$\times \left[x_2 + \frac{\left(a_{23} - \dfrac{a_{12}a_{13}}{a_{11}}\right)}{\left(a_{22} - \dfrac{a_{12}^2}{a_{11}}\right)}x_3\right]^2 + \left[\left(a_{33} - \frac{a_{13}^2}{a_{11}}\right) - \frac{\left(a_{23} - \dfrac{a_{12}a_{13}}{\cdot a_{11}}\right)^2}{\left(a_{22} - \dfrac{a_{12}^2}{a_{11}}\right)}\right]x_3^2$$

$$(24)$$

Similar to the two-variable case, we have the following necessary and sufficient conditions for w to be positive definite:

$$a_{11} > 0 \tag{25}$$

$$a_{22} - \frac{a_{12}^2}{a_{11}} > 0 \tag{26}$$

and

$$\left[\left(a_{33} - \frac{a_{13}^2}{a_{11}}\right) - \frac{\left(a_{23} - \dfrac{a_{12}a_{13}}{a_{11}}\right)^2}{\left(a_{22} - \dfrac{a_{12}^2}{a_{11}}\right)}\right] > 0 \tag{27}$$

The first two conditions are the same as for the two-variable case. The third one can be rearranged into the following form.

$$\begin{vmatrix} a_{22} - \dfrac{a_{12}^2}{a_{11}} & a_{23} - \dfrac{a_{12}a_{13}}{a_{11}} \\[2ex] a_{23} - \dfrac{a_{12}a_{13}}{a_{11}} & a_{33} - \dfrac{a_{13}^2}{a_{11}} \end{vmatrix} > 0 \tag{27a}$$

It can also be written

$$\begin{vmatrix} a_{11} & a_{12} & a_{13} \\[2ex] 0 & a_{22} - \dfrac{a_{12}^2}{a_{11}} & a_{23} - \dfrac{a_{12}a_{13}}{a_{11}} \\[2ex] 0 & a_{23} - \dfrac{a_{13}a_{12}}{a_{11}} & a_{33} - \dfrac{a_{13}^2}{a_{11}} \end{vmatrix} > 0 \tag{27b}$$

The above form can be considered as having resulted from applying Gauss' algorithm twice to the form below

$$\begin{vmatrix} a_{11} & a_{12} & a_{13} \\ a_{12} & a_{22} & a_{23} \\ a_{13} & a_{23} & a_{33} \end{vmatrix} > 0 \tag{27c}$$

Consequently, it is an inductive proof of Sylvester's theorem.

By using Sylvester's theorem, we can also examine whether a quadratic function is positive semidefinite; thus, we can restate the theorems as a conclusion.

(1) *A matrix [A] is positive definite if and only if all leading principal minors of A are positive.*

(2) *A matrix [A] is positive semidefinite if and only if all the leading principal minors of [A] are nonnegative.*

For example,

$$\begin{bmatrix} 3 & 1 \\ 1 & 4 \end{bmatrix} \tag{28}$$

is a positive definite matrix because

$$3 > 0, \quad \begin{vmatrix} 3 & 1 \\ 1 & 4 \end{vmatrix} = 11 > 0$$

and

$$\begin{bmatrix} 1 & -1 & 0 \\ -1 & 1 & 0 \\ 0 & 0 & 1 \end{bmatrix} \tag{29}$$

is a positive semidefinite matrix because one of the leading principal minors is zero, which is nonnegative.

8. Negative Definite and Negative Semidefinite

From the reduction techniques developed in the above section, the definitions of positive definite and positive semidefinite can be given as follows: reduce the matrix to a form such that $a_{ij} = 0$, $(i \neq j)$; then examine:

1. $q = [x]^T[A][x]$ is positive definite if and only if all the diagonal elements of $[A]$ are positive.

2. $q = [x]^T[A][x]$ is positive semidefinite if and only if all the diagonal elements of $[A]$ are nonnegative and at least one of the values vanishes.

3. $q = [x]^T[A][x]$ is sign-indefinite if and only if the diagonal elements of $[A]$ have both positive and negative values.

Following similar reasoning, we have the following definitions.

4. $q = [x]^T[A][x]$ is negative definite if and only if all the diagonal elements of $[A]$ are negative.

Example.

$$q = -2x_1^2 + (-3x_2^2) + (-5x_3^2), \quad n = 3$$

is negative definite.

5. $q = [x]^T[A][x]$ is negative semidefinite if and only if all the diagonal elements of $[A]$ are nonpositive, and at least one of them vanishes.

Example.

$$q = -x_1^2 - 4x_1 x_2 - 2x_2^2, \quad n = 2$$
$$= -(x_1 + 2x_2)^2$$

is negative semidefinite.

V. Vectors

1. Two-dimensional Vectors

The concept of vectors in mathematics is different in many aspects from that in engineering analysis. However, the foundation is the same.

In elementary physics a vector is defined as a quantity which has both magnitude and direction. Figure 5–4 shows a vector **a**. The length of line OA indicates the magnitude of the quantity. The angle θ indicates its direction from some reference direction.

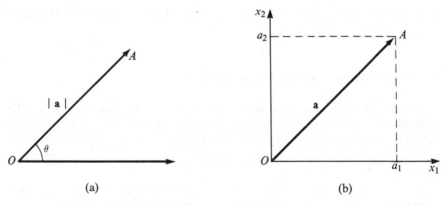

(a)   (b)

Fig. 5-4. Two expressions of a two-dimensional vector: (a) polar form; (b) rectangular form.

If rectangular coordinates (x_1, x_2) are introduced, the magnitude of vector **a** can be expressed in terms of two projections, Oa_1 and Oa_2, on the axes. Note that the point (a_1, a_2) on the x_1, x_2 plane gives complete information about **a**.

When the coordinates are set up with the origin at one end, a vector and the coordinates of a point (the other end) have a one-to-one correspondence. This is the fundamental idea from which many of the mathematical aspects of vectors are developed.

2. Three-dimensional Vectors

A three-dimensional vector is shown in Fig. 5-5. Because of the correspondence between a vector and a point, we can write the vector as follows:

$$\begin{bmatrix} x_1 \\ x_2 \\ x_3 \end{bmatrix} = \begin{bmatrix} 2.0 \\ 2.5 \\ 3.7 \end{bmatrix} \qquad (1)$$

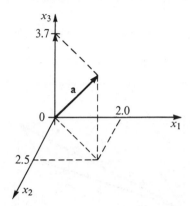

Fig. 5-5. A three-dimensional vector.

It means that there is a vector from the origin to the point $(2, 2.5, 3.7)$ with coordinates (x_1, x_2, x_3).

3. Higher-dimensional Vectors

As mentioned before, (a_1, a_2) can uniquely define a vector and they can be called the components of that vector. The numbers a_i are the coordinates of the point where the shadows of the vector terminate. Similarly, (a_1, a_2, a_3) define a vector in three-dimensional space. We can extend this idea logically (even though we cannot easily draw a picture of the result) to an n-dimensional space in which the termination of the vector is at the point $(a_1, a_2, a_3, \ldots, a_n)$ with the coordinates $(x_1, x_2, x_3, \ldots, x_n)$.

An n-dimensional vector, **a**, is a set of n ordered numbers, in which a_i $(i = 1, 2, \ldots, n)$ are called the components of the vector.

$$\mathbf{a} = \begin{bmatrix} a_1 \\ a_2 \\ \cdot \\ \cdot \\ \cdot \\ a_n \end{bmatrix} \qquad (2)$$

When (2) is written as above, it can be seen that it is the same as a column matrix.

4. Addition

The rules for operating on two-dimensional vectors in mechanics can be generalized. The parallelogram law for addition is shown in Fig. 5-6.

If

$$\mathbf{a} = (a_1, a_2) \quad \text{and} \quad \mathbf{b} = (b_1, b_2) \tag{3}$$

then

$$\mathbf{a} + \mathbf{b} = (a_1 + b_1, a_2 + b_2) \tag{4}$$

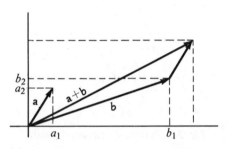

Fig. 5-6. The parallelogram law for addition.

This can be extended to the domain of n-dimensional vectors:

$$\mathbf{a} = (a_1, a_2, a_3, \ldots, a_n)$$
$$\mathbf{b} = (b_1, b_2, b_3, \ldots, b_n)$$

Then

$$\mathbf{a} + \mathbf{b} = (a_1 + b_1, a_2 + b_2, a_3 + b_3, \ldots, a_n + b_n) \tag{5}$$

5. Multiplication by a Scalar

If a two-dimensional vector $\mathbf{a}$ is multiplied by an ordinary (scalar) quantity, we write

$$k\mathbf{a} = (ka_1, ka_2)$$

The operation is easily generalized:

$$k\mathbf{a} = (ka_1, ka_2, ka_3, \ldots, ka_n) \tag{6}$$

6. Scalar Product of Two Vectors

For the vectors $\mathbf{f}$ and $\mathbf{v}$ in two-dimensional space, shown in Fig. 5–7, the product is a scalar quantity:

$$p = |\mathbf{f}| \cdot |\mathbf{v}| \cos \theta \tag{7}$$

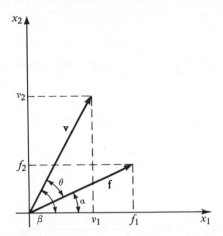

Fig. 5-7. Two vectors and their components.

On the other hand, consider the following quantity from Fig. 5–7:

$$f_1 v_1 + f_2 v_2 = |\mathbf{f}| \cos \alpha \, |\mathbf{v}| \cos \beta + |\mathbf{f}| \sin \alpha \, |\mathbf{v}| \sin \beta$$
$$= |\mathbf{f}| \cdot |\mathbf{v}| \cos (\beta - \alpha)$$
$$= |\mathbf{f}| \cdot |\mathbf{v}| \cos \theta \tag{8}$$

Comparing (7) and (8), we have

$$f_1 v_1 + f_2 v_2 = p = |\mathbf{f}| \cdot |\mathbf{v}| \cos \theta \tag{9}$$

In elementary physics we learn that (7) is the equation for calculating power from the product of the external force, the resultant velocity, and the angle between their two directions. It is a scalar quantity. The fact that it is equatable to (8), as shown in (9), gives us a basis for defining the *n*-dimensional product of vectors.

$$q = a_1 b_1 + a_2 b_2 + \cdots + a_n b_n = \sum_{i=1}^{n} a_i b_i \tag{10}$$

where q is a scalar quantity.

For convenience, we may write **a** as a row, and **b** as a column. Then (10) becomes

$$q = [a_1, a_2, \ldots, a_n] \begin{bmatrix} b_1 \\ b_2 \\ \cdot \\ \cdot \\ \cdot \\ b_n \end{bmatrix} \tag{11}$$

This can be interpreted as the product of a row matrix times a column matrix; or the transposition of a column matrix times another column matrix.

7. Distance

In two-dimensional space, the distance between two points (a_1, a_2) and (b_1, b_2) is defined as

$$|\mathbf{a} - \mathbf{b}| \triangleq \sqrt{(a_1 - b_1)^2 + (a_2 - b_2)^2} \tag{12}$$

In higher-dimensional space the distance between two vectors $\mathbf{a}$ and $\mathbf{b}$ is generalized from (12):

$$|\mathbf{a} - \mathbf{b}| = \sqrt{(a_1 - b_1)^2 + (a_2 - b_2)^2 + (a_3 - b_3)^2 + \cdots + (a_n - b_n)^2}$$
$$= \sqrt{\sum_{i=1}^{n} (a_i - b_i)^2} \tag{13}$$

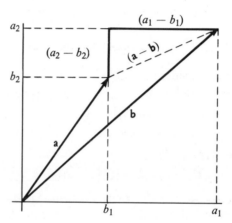

Fig. 5-8. The distance between two points.

8. Euclidean Space

An n-dimensional Euclidean space is defined as a collection of all points (vectors) for which the operations (5) and (6) are valid and for which the distance between any two vectors is defined by (13). It is interesting to note that the distance between points is the fundamental quantity upon which a geometry is erected.

A two-dimensional Euclidean space is usually denoted by E^2, a three-dimensional space is denoted by E^3, and an n-dimensional space by E^n.

9. Sets

A group of vectors in E^n is usually to be interpreted as a group of points in E^n. The collection of points is called a set. For example, all the points lying inside and on the circumference of a circle with a radius of unity constitute a set which conforms to the relationship

$$x_1^2 + x_2^2 \leq 1$$

This is described very concisely by the following notation:

$$X = \{[x_1, x_2] : x_1^2 + x_2^2 \leq 1\} \tag{14}$$

A set is "open" if it contains only the interior points, whereas it is "closed" if it contains all the boundary points also. Figure 5-9, then, represents a closed set.

A sphere in E^3 with center at $\mathbf{a}$ and radius r is described by

$$|\mathbf{x} - \mathbf{a}| = r$$

All the points inside the sphere can be described by

$$X = \{\mathbf{x} : |\mathbf{x} - \mathbf{a}| < r\} \tag{15}$$

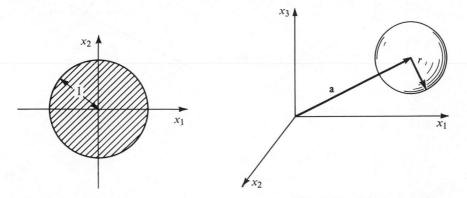

Fig. 5-9. A closed set is defined by (14).   Fig. 5-10. An open set is defined by (15).

10. Hypersphere

Extending this idea to E^n space, we define a hypersphere: A hypersphere in E^n with center at $\mathbf{a}$ and radius r is defined as the set of points

$$X = (\mathbf{x} : |\mathbf{x} - \mathbf{a}| = r)$$

The equation of a hypersphere in E^n is

$$|\mathbf{x} - \mathbf{a}| = r$$

or

$$\sum_{i=1}^{n} (x_i - a_i)^2 = r^2 \tag{16}$$

11. Hyperplane

In E^2, if we write

$$a_1 x_1 + a_2 x_2 = c \tag{17}$$

where c is a constant, the equation represents a straight line (Fig. 5-11).
Similarly,

$$a_1 x_1 + a_2 x_2 + a_3 x_3 = c \tag{18}$$

represents a plane in E^3 as shown in Fig. 5-12.

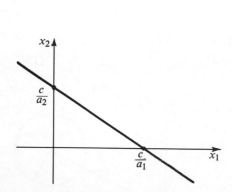

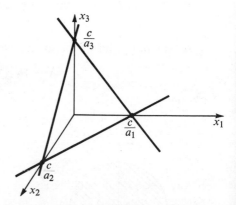

Fig. 5-11. A geometrical interpretation of $a_1 x + a_2 x = c$.

Fig. 5-12. A geometrical interpretation of $a_1 x_1 + a_2 x_2 + a_3 x_3 = c$.

A hyperplane in E^n is defined as

$$a_1 x_1 + a_2 x_2 + a_3 x_3 + \cdots + a_n x_n = c \tag{19}$$

or, in matrix notation,

$$[a_1, a_2, a_3, \ldots, a_n] \begin{bmatrix} x_1 \\ x_2 \\ x_3 \\ \cdot \\ \cdot \\ \cdot \\ x_n \end{bmatrix} = c \tag{19a}$$

or, in vector notation,

$$\mathbf{a}\mathbf{x} = c$$

It is evident that when $c = 0$, the plane passes through the origin:

$$\mathbf{a}\mathbf{x} = 0 \tag{20}$$

If (20) is interpreted as the scalar product, $\mathbf{a}$ is said to be *normal* to the vector $\mathbf{x}$.

VI.   State Variable Concept

1. State Variable Approach to Dynamic Systems

Any group of components, whether they be mechanical devices, electric circuits, or a chemical apparatus which operate dependently to perform a specific function can be considered under the general term "system." Interactions among the elements of a system can usually be detected and described by the measurement of certain quantities such as mechanical displacement, electric current, chemical concentration.

A system is said to be dynamic if the quantities which describe it vary in some way as time goes by.

A very convenient way to describe a dynamic system is to set up equations showing the time rate of change of these variables.

For a simple system with only one variable involved we could write:

$$\frac{dx}{dt} = f(x, t) \tag{1}$$

The function $f(x, t)$ is determined by the system in question.

Any system of engineering importance requires more than a single quantity to describe it. In fact, there may be a great number of velocities, accelerations, currents, voltages, as well as combinations of them that are necessary for a complete description .In such cases the whole group of differential equations is best set up in matrix form. The equation for a complicated system then becomes

$$\frac{d}{dt} [x] = [f(x, t)] \tag{2}$$

where $[x]$ is a column matrix of variables that are to be determined in terms of the variable t and, where $f_1 \ldots f_n$ are given functions of $x_1, x_2, \ldots, x_n, t$.

These variables, $x_1, x_2, \ldots, x_n$, are called the *state variables*.

Essential properties of a set of state variables are that the left-hand members be first derivatives of the variables and the right-hand members be functions of the variables $x_1, x_2, \ldots, x_n$ and the variable t, but are not involved with any of the derivatives. Geometrically speaking, the matrix $[x]$, or

$$[x] = \begin{bmatrix} x_1 \\ x_2 \\ x_3 \\ \cdot \\ \cdot \\ \cdot \\ x_n \end{bmatrix} \tag{3}$$

can be interpreted as a moving point in Euclidean space. The coordinates of the point are $x_1, x_2, \ldots, x_n$. From the origin of the coordinates to the point, one could construct a vector. This vector is expanding, shrinking, swinging, or rotating in the Euclidean space as time goes on. The position of the point or the length and attitude of the vector at any given instant represents the condition of the system at that time and is known as the state of the system.

Formally defined, the state of a dynamic system is a set of variables which contain as much information regarding the past history of the system as is required for the determination of its future behavior.

2. Classification

Although the description of system behavior by (2) is most convenient, the form is ordinarily too general to permit a detailed study. In order to classify the differential equations, we use the following form instead.

$$\frac{d}{dt}[x] = [A(t)][x] + [g(x_1, x_2, \ldots, x_n\ t)] + [\mu(t)] \tag{4}$$

The corresponding component equations are written as

$$\frac{dx_1}{dt} = a_{11}(t)x_1 + \cdots + a_{1n}(t)x_n + g_1(x_1, x_2, \ldots, x_n, t) + \mu_1(t)$$

$$\frac{dx_2}{dt} = a_{21}(t)x_1 + \cdots + a_{nn}(t)x_n + g_2(x_1, x_2, \ldots, x_n, t) + \mu_2(t)$$

$$\tag{4a}$$

$$\cdot \quad \cdot \quad \cdot \quad \cdot \quad \cdot \quad \cdot \quad \cdot \quad \cdot \quad \cdot \quad \cdot$$

$$\frac{dx_n}{dt} = a_{n1}(t)x_1 + \cdots + a_{nn}(t)x_n + g_n(x_1, x_2, \ldots, x_n, t) + \mu_n(t)$$

On the basis of Eq. (4), we can classify the differential equations which we usually encounter.

Special Case 1. If $[A]$ in Eq. (4) is a constant matrix, $[g]$ is zero, and $[\mu(t)]$ is zero, the system is called a homogeneous linear system with constant coefficients.

Special Case 2. If $[A]$ in Eq. (4) is a constant matrix, $[g]$ is zero, and $[\mu(t)]$ is not zero, the system is called a nonhomogeneous linear system with constant coefficients.

Special Case 3. If the elements of $[A]$ are constants and g does not explicitly depend on time, the system is called a stationary system.

Special Case 4. If any one of the conditions in special case 3 is violated, the system is called a *nonstationary* system.

Special Case 5. If a system is stationary and $\mu(t)$ is zero, the system is referred to as an autonomous system.

3. Example 1. Planetary Orbit

The first scientist to use state variable formulation was probably Lagrange. He discussed a planetary orbit problem for which he had the following set of equations.

$$\frac{dx_1}{dt} = +a_{11}y_1 + \cdots + a_{1n}y_n$$

$$\cdots \cdots \cdots \cdots \cdots$$

$$\frac{dx_n}{dt} = +a_{n1}y_1 + \cdots + a_{nn}y_n$$

$$\frac{dy_1}{dt} = -a_{11}x_1 - \cdots - a_{1n}x_n \tag{5}$$

$$\cdots \cdots \cdots \cdots \cdots$$

$$\frac{dy_n}{dt} = -a_{n1}x_1 - \cdots - a_{nn}x_n$$

where the a_{ij} are real constants and $a_{ij} = a_{ji}$.

Laplace proved that the x_i and the y_j do not increase with t beyond very narrow limits. Upon multipying the first n equations by $x_1, \ldots, x_n$, respectively, and the last n equations by $y_1, \ldots, y_n$, respectively and adding these products, the complete derivative is obtained. The integral of this is

$$x_1^2 + \cdots + x_n^2 + y_1^2 + \cdots + y_n^2 = \text{constant} = c$$

Therefore no x_i^2 or y_i^2 can exceed the positive constant c that is determined by the initial conditions.

It is seen that (5) is the linear homogeneous system with constant coefficients.

4. Example 2. Simple Pendulum

A simple pendulum swinging freely in a vacuum is usually described by

$$\frac{d^2\theta}{dt^2} + k \sin \theta = 0 \tag{6}$$

Let us make the following transformation.

$$\theta = x_1$$

$$\frac{d\theta}{dt} = x_2 \tag{7}$$

Then by appropriate substitution

$$\frac{dx_1}{dt} = x_2$$

$$\frac{dx_2}{dt} = -k \sin x_1 \tag{8}$$

which is a state variable formulation of the simple pendulum.

Equation (8) is a set of nonlinear autonomous differential equations.

5. Example 3. *RLC* Circuit

The differential equation of an *RLC* series circuit is well-known.

$$L\frac{d^2 x}{dt^2} + R\frac{dx}{dt} + \frac{1}{C}x = M \cos \omega t \tag{9}$$

By using the following transformation

$$\frac{dx_1}{dt} = x_2 \tag{10}$$

we finally have

$$\frac{dx_1}{dt} = x_2$$

$$\frac{dx_2}{dt} = -\frac{1}{CL}x_1 - \frac{R}{L}x_2 + \frac{M}{L}\cos \omega t \tag{11}$$

Equation (11) is a linear, nonhomogeneous system.

6. Example 4. Bessel Equations

A typical Bessel equation reads

$$\frac{d^2 x}{dt^2} + \frac{1}{t}\frac{dx}{dt} + \left(1 - \frac{1}{t}\right)x = 0 \tag{12}$$

which appears many times in various branches of electrical and mechanical engineering. The state variable equations are obtained as follows:

$$\frac{dx_1}{dt} = x_2$$

$$\frac{dx_2}{dt} = -\left(1 - \frac{1}{t}\right)x_1 - \frac{1}{t}x_2 \tag{13}$$

It is a linear nonstationary system.

7. Higher Order Equations vs. State Variable Equations

From the above examples, we see that a higher order differential equation can be reduced to a set of state variable equations through a transformation. In our cases, we used

$$\frac{dx_1}{dt} = x_2$$

$$\frac{dx_2}{dt} = x_3$$

$$\vdots \tag{14}$$

$$\frac{d^{n-1}x}{dt^{n-1}} = x_n$$

as a transformation.

This is by no means the only transformation. In the next chapter, we will develop a systematic study of the formulation problem from many viewpoints.

A dualistic question arises: Can we reduce a set of state variable equations into a higher order differential equation?

In order to illustrate the procedure, suppose we have three state equations:

$$\frac{dx_1}{dt} = f_1(x_1, x_2, x_3, t)$$

$$\frac{dx_2}{dt} = f_2(x_1, x_2, x_3, t) \tag{15}$$

$$\frac{dx_3}{dt} = f_3(x_1, x_2, x_3, t)$$

The procedure is to eliminate two of the variables, say x_1 and x_2, along with their first derivatives.

Since four unknowns cannot generally be eliminated from three equations, it is necessary to deduce three additional equations:

$$\frac{d^2 x_1}{dt^2} = \frac{\partial f_1}{\partial x_1}\frac{dx_1}{dt} + \frac{\partial f_1}{\partial x_2}\frac{dx_2}{dt} + \frac{\partial f_1}{\partial x_3}\frac{dx_3}{dt} + \frac{\partial f_1}{\partial t}$$

$$\frac{d^2 x_2}{dt} = \frac{\partial f_2}{\partial x_1}\frac{dx_1}{dt} + \frac{\partial f_2}{\partial x_2}\frac{dx_2}{dt} + \frac{\partial f_2}{\partial x_3}\frac{dx_3}{dt} + \frac{\partial f_2}{\partial t} \tag{16}$$

$$\frac{d^2 x_3}{dt} = \frac{\partial f_3}{\partial x_1}\frac{dx_1}{dt} + \frac{\partial f_3}{\partial x_2}\frac{dx_2}{dt} + \frac{\partial f_3}{\partial x_3}\frac{dx_3}{dt} + \frac{\partial f_3}{\partial t}$$

Usually the six equations of (15) and (16) can be solved for the six quantities: x_1, x_2, and their first and second derivatives, in terms of x_3 and its first and second derivatives. Finally, we obtain

$$\frac{d^3 x_3}{dt^3} = f\left(x_3, \frac{dx_3}{dt}, \frac{d^2 x_3}{dt^2}, t\right) \tag{17}$$

which is the higher order differential equation corresponding to (15).

A very important expression which we will use frequently should be defined and illustrated as follows: Rewrite Eq. (16) into the matrix form

$$\frac{d}{dt}\begin{bmatrix} \dot{x}_1 \\ \dot{x}_2 \\ \dot{x}_3 \end{bmatrix} = \begin{bmatrix} \dfrac{\partial f_1}{\partial x_1} & \dfrac{\partial f_1}{\partial x_2} & \dfrac{\partial f_1}{\partial x_3} \\ \dfrac{\partial f_2}{\partial x_1} & \dfrac{\partial f_2}{\partial x_2} & \dfrac{\partial f_2}{\partial x_3} \\ \dfrac{\partial f_3}{\partial x_1} & \dfrac{\partial f_3}{\partial x_2} & \dfrac{\partial f_3}{\partial x_3} \end{bmatrix} \begin{bmatrix} \dot{x}_1 \\ \dot{x}_2 \\ \dot{x}_3 \end{bmatrix} + \begin{bmatrix} \dfrac{\partial f_1}{\partial t} \\ \dfrac{\partial f_2}{\partial t} \\ \dfrac{\partial f_3}{\partial t} \end{bmatrix} \qquad (16a)$$

$$= [J][\dot{x}] + \cdots$$

where

$$[J] = \begin{bmatrix} \dfrac{\partial f_1}{\partial x_1} & \dfrac{\partial f_1}{\partial x_2} & \dfrac{\partial f_1}{\partial x_3} \\ \dfrac{\partial f_2}{\partial x_1} & \dfrac{\partial f_2}{\partial x_2} & \dfrac{\partial f_2}{\partial x_3} \\ \dfrac{\partial f_3}{\partial x_1} & \dfrac{\partial f_3}{\partial x_2} & \dfrac{\partial f_3}{\partial x_3} \end{bmatrix} \qquad (18)$$

is called the Jacobian matrix of the function f_i with respect to the variables x_j. Abbreviated symbols are frequently used:

$$[J] = \left[\frac{\partial f_i}{\partial x_j}\right] \quad \text{or} \quad [J] = \left[\frac{\partial f}{\partial x}\right] \qquad (19)$$

Form (19) is particularly concise and can be used to advantage in the chain rule application. In other words,

$$\left[\frac{\partial f}{\partial x}\right]\left[\frac{\partial x}{\partial y}\right] = \left[\frac{\partial f}{\partial y}\right] \qquad (20)$$

8. Summary

The complete analysis of a dynamic system using the state variable techniques consists in solving actually three problems:
1. The representation problem.
2. The solution problem.
3. The stability problem.
We will treat these problems, one by one, in the subsequent chapters.

PROBLEMS

5-1. Given

$$[A] = \begin{bmatrix} 1 & 2 & 4 \\ 6 & 7 & 0 \\ 8 & 3 & 10 \end{bmatrix}, \quad [B] = \begin{bmatrix} 5 & 5 & 4 \\ 0 & 9 & 7 \\ 11 & 8 & 3 \end{bmatrix}$$

(a) Compute

$$[A] + [B]$$
$$[A] - [B]$$
$$[A] \times [B]$$
$$[B] \times [A]$$

(b) Find $[A]^{-1}$ and $[B]^{-1}$.

(c) Verify that

$$[A]^{-1}[B]^{-1} = \{[B][A]\}^{-1}$$
$$[B]^{-1}[A]^{-1} = \{[A][B]\}^{-1}$$

5-2. Given

$$x_1 + x_2 + x_3 = 18$$
$$2x_1 + 3x_2 + 7x_3 = 77$$
$$x_1 + 5x_2 + 2x_3 = 49$$

(a) Solve by using the Gauss elimination method.

(b) Solve by using the linear transformation method.

(c) Solve by the method which involves inverting a matrix with four steps.

5-3. Write the following quadratic quantities in matrix expressions:

(a) $q = x_1^2 + 6x_1x_2 + 5x_2^2$

(b) $q = 3x_1^2 + 4x_2^2 + x_3^2 - 6x_1x_2 + 7x_2x_3 + 9x_1x_3$

(c) $q = 3x_1^2 + 4x_2^2 + 10x_3^2 + 6x_1x_2$

5-4. Reduce the following quadratic forms by Lagrange's reduction:

(a) $x_1^2 + 6x_1x_2 + 5x_2^2$

(b) $3x_1^2 + 6x_1x_2 + 4x_3^2$

(c) $3x_1^2 - 6x_1x_2 + 4x_2^2$

(d) $x_1^2 + 5x_1x_2 + x_2^2 + 3x_1x_2 + x_3^2$

5-5. Reduce the quadratic forms shown in Problem 5-4 by Kronecker's method.

5-6. Indicate whether each of the following is positive definite, positive semidefinite, negative definite, negative semidefinite or sign-indefinite:

(a) $x_1^2 + x_2^2 + x_3^2,$ $n = 3$

(b) $x_1^2 + x_2^2 + x_3^2,$ $n = 4$

(c) $(x_1 + x_2)^2 + x_3^2,$ $n = 3$

(d) $-x_1^2 - x_2^2 - x_3^2 - x_4^2,$ $n = 4$

(e) $x_1^2 + x_2^2 + x_3^2 - x_4^2,$ $n = 4$

5-7. Use Sylvester's theorem to determine the sign-definiteness of each of the following:

(a)

$$q = [x_1, x_2, x_3] \begin{bmatrix} 1 & 0 & 0 \\ 0 & 3 & 0 \\ 0 & 0 & 5 \end{bmatrix} \begin{bmatrix} x_1 \\ x_2 \\ x_3 \end{bmatrix}$$

(b)

$$q = [x_1, x_2, x_3] \begin{bmatrix} 1 & 3 & 0 \\ 3 & 2 & 1 \\ 0 & 1 & 4 \end{bmatrix} \begin{bmatrix} x_1 \\ x_2 \\ x_3 \end{bmatrix}$$

(c)

$$q = [x_1, x_2, x_3] \begin{bmatrix} 0 & 0 & 0 \\ 0 & 0 & 0 \\ 0 & 0 & -7 \end{bmatrix} \begin{bmatrix} x_1 \\ x_2 \\ x_3 \end{bmatrix}$$

(d)

$$q = [x_1, x_2, x_3] \begin{bmatrix} 2 & 4 & 1 \\ 4 & 8 & 0 \\ 1 & 0 & 5 \end{bmatrix} \begin{bmatrix} x_1 \\ x_2 \\ x_3 \end{bmatrix}$$

5-8. Show graphically the regions represented by the following sets:

(a) $X = \{\mathbf{x}: x_1^2 + x_2^2 \leq 2\}$

(b) $X = \{\mathbf{x}: x_1^2 + x_2^2 < 2\}$

(c) $X = \{\mathbf{x}: x_1^2 + x_2^2 \geq 2\}$

CHAPTER 6

The Representation Problem

As was explained in Chapter 1, the first problem that confronts an engineer in analyzing or designing a control system is that of representing the system with a suitable mathematical model. In this chapter it will be shown why the state variable representation is such a powerful one. Also, the various conventional vehicles which may be converted to the state variable representation will be described, and specific examples for formulating the state equations from well-known, classical representations will be demonstrated.

In state variable analysis, a system is represented by a set of *first order differential equations*:

$$\dot{x}_1 = f_1(x_1, x_2, x_3, x_4, \ldots, x_n)$$
$$\dot{x}_2 = f_2(x_1, x_2, x_3, x_4, \ldots, x_n)$$
$$\dot{x}_3 = f_3(x_1, x_2, x_3, x_4, \ldots, x_n) \qquad (1)$$
$$\cdots\cdots\cdots\cdots\cdots\cdots\cdots\cdots$$
$$\dot{x}_n = f_n(x_1, x_2, x_3, x_4, \ldots, x_n)$$

The set of differential equations is easily expressed in matrix notation:

$$\frac{d[x]}{dt} = [f(x)] \qquad (1a)$$

where $[x]$ is defined as a state vector. For a linear system with constant coefficients, Equation (1) is reduced to the following form:

$$
\begin{bmatrix} \dot{x}_1 \\ \dot{x}_2 \\ \dot{x}_3 \\ \cdot \\ \cdot \\ \cdot \\ \dot{x}_n \end{bmatrix}
=
\begin{bmatrix}
a_{11} & a_{12} & a_{13} & \cdots & a_{1n} \\
a_{21} & a_{22} & a_{23} & \cdots & a_{2n} \\
a_{31} & a_{32} & a_{33} & \cdots & a_{3n} \\
\cdot & \cdot & \cdot & \cdots & \cdots \\
\cdot & \cdot & \cdot & \cdots & \cdots \\
\cdot & \cdot & \cdot & \cdots & \cdots \\
a_{n1} & a_{n2} & a_{n3} & \cdots & a_{nn}
\end{bmatrix}
\begin{bmatrix} x_1 \\ x_2 \\ x_3 \\ \cdot \\ \cdot \\ \cdot \\ x_n \end{bmatrix}
\qquad (2)
$$

or

$$\frac{d[x]}{dt} = [A][x] \qquad\qquad (2a)$$

where $[A]$ is a constant square matrix.

With the system represented by these state equations, much light is shed on the problem they represent because analyses have already been made on the same equations representing phenomena in other fields, such as:

1. Theoretical physics: classical and quantum mechanics.
2. Pure mathematics: linear algebra and group theory.
3. Numerical analysis: digital computer techniques.

The third item on the above list is mainly responsible for the great importance of the state variable representation in our day. Because of digital computer techniques which are directly applicable, it has a most promising future. Moreover, because it supplies a common notation, it makes communication between researchers in a diversity of disciplines very easy, thus portending accelerated progress. Control engineers, especially, are attracted by the simplicity of the state equations.

Conventional ways of representing an engineering problem may be classified in three categories.

1. The circuit, or topological, approach, when we represent a circuit by loop or node equations following Kirchhoff's laws.
2. The input-output, or transfer function, approach, using poles-zeros of the input-output relationship to represent the system.
3. The energy approach, or Lagrange's equation, formulating a set of differential equations based on the energy content considerations.

The first and third approaches give us a set of second order differential equations in general. The second approach gives us a high order polynomial of s variables in the complex domain. Since the state equations are a set of *first order* differential equations, the question naturally arises as to how we can change the representations developed in the three schools listed above to serve in state variable representation. This chapter will treat each in turn.

I. Energy Method

1. Lagrange's Equation

Lagrange derived a general equation to describe a conservative or lossless system. Rayleigh extended his equation to the nonconservative system. We have derived their equations in Chapter 1.

Now we rewrite their symbols:

T = kinetic energy of the entire system
V = potential energy of the entire system
D = dissipation function of the entire system

and Lagrange's equation as

$$\frac{d}{dt}\left(\frac{\partial T}{\partial \dot{q}_n}\right) - \frac{\partial T}{\partial q_n} + \frac{\partial D}{\partial \dot{q}_n} + \frac{\partial V}{\partial q_n} = Q_n \tag{1}$$

where $n = 1, 2, 3, \ldots$ are integers; n means the number of degrees of freedom of the system.

Q_n = the generalized force applied at the coordinate n
q_n = generalized coordinates
$\dot{q}_n$ = generalized velocities

2. Application of Lagrange's Equation

Let us take an example and substitute into Lagrange's equation for a review.

Two loop currents i_1 and i_2 are shown in Fig. 6-1. The resulting charges q_1 and q_2 are the variables making two degrees of freedom.

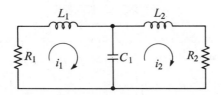

Fig. 6-1. A circuit with two loops.

The magnetic energy associated with the coils constitutes the kinetic energy of the system:

$$T = \tfrac{1}{2}L_1 i_1^2 + \tfrac{1}{2}L_2 i_2^2 \tag{2}$$

The electric energy stored in the capacitors constitutes the potential energy of the system.

$$V = \frac{1}{2} \cdot \frac{1}{C_1}(q_1 - q_2)^2 \tag{3}$$

The dissipation function of the system is

$$D = \tfrac{1}{2}R_1 i_1^2 + \tfrac{1}{2}R_2 i_2^2 \tag{4}$$

which is half the value of power.

There are two degrees of freedom, so we will substitute into (1) twice. For the first variable $\dot{q}_1$ we make the following substitution:

$$\frac{d}{dt}\left(\frac{\partial T}{\partial \dot{q}_1}\right) - \frac{\partial T}{\partial q_1} + \frac{\partial D}{\partial \dot{q}_1} + \frac{\partial V}{\partial q_1} = Q_1 \tag{5}$$

or

$$\frac{d}{dt}(L_1 i_1) - 0 + R_1 i_1 + \frac{1}{C_1}(q_1 - q_2) = 0 \tag{5a}$$

For the second variable, $\dot{q}_2$,

$$\frac{d}{dt}\left(\frac{\partial T}{\partial \dot{q}_2}\right) - \frac{\partial T}{\partial q_2} + \frac{\partial D}{\partial \dot{q}_2} + \frac{\partial V}{\partial q_2} = Q_2 \tag{6}$$

or

$$\frac{d}{dt}(L_2 i_2) - 0 + R_2 i_2 + \frac{1}{C_1}(q_1 - q_2)(-1) = 0 \tag{6a}$$

3. State Equations

Rewrite (5a) and (6a) as follows:

$$L_1 \frac{d^2 q_1}{dt^2} + R_1 \frac{dq_1}{dt} + \frac{1}{C_1}(q_1 - q_2) = 0 \tag{5b}$$

$$L_2 \frac{d^2 q_2}{dt^2} + R_2 \frac{dq_2}{dt} + \frac{1}{C_1}(q_2 - q_1) = 0 \tag{6b}$$

These must be changed to a set of first order differential equations. This can be done by first changing them to matrix form:

$$\begin{bmatrix} L_1 & 0 \\ 0 & L_2 \end{bmatrix}\begin{bmatrix} \ddot{q}_1 \\ \ddot{q}_2 \end{bmatrix} + \begin{bmatrix} R_1 & 0 \\ 0 & R_2 \end{bmatrix}\begin{bmatrix} \dot{q}_1 \\ \dot{q}_2 \end{bmatrix} + \begin{bmatrix} \frac{1}{C_1} & -\frac{1}{C_1} \\ -\frac{1}{C_1} & \frac{1}{C_1} \end{bmatrix}\begin{bmatrix} q_1 \\ q_2 \end{bmatrix} = \begin{bmatrix} 0 \\ 0 \end{bmatrix} \tag{7}$$

Then let $q_1 = x_1$, $q_2 = x_2$, etc. $\qquad(8)$

$$\begin{bmatrix} \dot{x}_1 \\ \dot{x}_2 \\ \dot{x}_3 \\ \dot{x}_4 \end{bmatrix} = \begin{bmatrix} 0 & 0 & 1 & 0 \\ 0 & 0 & 0 & 1 \\ -\frac{1}{L_1 C_1} & \frac{1}{L_1 C_1} & -\frac{R_1}{L_1} & 0 \\ \frac{1}{L_2 C_2} & -\frac{1}{L_2 C_1} & 0 & -\frac{R_2}{L_2} \end{bmatrix}\begin{bmatrix} x_1 \\ x_2 \\ x_3 \\ x_4 \end{bmatrix} \tag{9}$$

Obtaining (9) from (7) is based on the following reasoning: The Lagrange equations for a dynamical system in m generalized coordinates, q_1, q_2, q_m are of the type:

$$A\ddot{q} + B\dot{q} + Cq = 0 \tag{10}$$

If A, B, and C are scalars, we can make the following substitutions:

$$q = y_1 \tag{11}$$

$$\dot{y}_1 = y_2 \tag{12}$$

Then it follows that

$$\dot{y}_2 = \frac{d}{dt}y_1 = \frac{d^2 y_1}{dt} = \frac{d^2 q}{dt^2} = -\frac{C}{A}y_1 - \frac{B}{A}y_2$$

Substituting (11) and (12) into (10) gives

$$y_1 = y_2$$

$$y_2 = -\frac{C}{A}y_1 - \frac{B}{A}y_2$$

This can be rewritten as

$$\begin{bmatrix} \dot{y}_1 \\ \dot{y}_2 \end{bmatrix} = \begin{bmatrix} 0 & 1 \\ -\dfrac{C}{A} & -\dfrac{B}{A} \end{bmatrix} \begin{bmatrix} y_1 \\ y_2 \end{bmatrix} \tag{13}$$

Comparing (13) and (9) shows that (9) is merely the matrix general form of (13). Therefore, if we are given a matrix equation

$$[A]\ddot{q} + [B]\dot{q} + [C]q = 0 \tag{14}$$

where $|A| \neq 0$, they can be replaced by a system of $2m$ first order equations:

$$[\dot{y}_m] = [u][y_m] \tag{15}$$

where

$$[y_m] = \begin{bmatrix} y_1 \\ y_2 \\ \vdots \\ y_m \end{bmatrix} = \begin{bmatrix} q \\ \dot{q} \\ \vdots \\ q^{(m-1)} \end{bmatrix} \tag{16}$$

and

$$[u] = \begin{bmatrix} 0 & I_m \\ -[A]^{-1}[C] & -[A]^{-1}[B] \end{bmatrix} \tag{17}$$

In the latter, I_m is the identity matrix with m elements of unity in its diagonal.

Note that (9) is a special application of (17), and (14) is the general form of (7).

4. Example—Electromechanical System

Figure 6-2 is a loudspeaker. The mass M is supported by springs, and its motion is damped by friction. Rigidly fastened to the mass M is a coil of n turns lying in a magnetic field of flux density β. The coil has a radius r and an inductance L and a resistance R. The kinetic energy of the entire system is

$$T = \tfrac{1}{2}Li^2 + \tfrac{1}{2}Mx^2 + Uxi \tag{18}$$

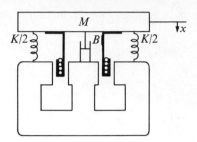

Fig. 6-2. A loudspeaker.

The first two terms are self-explanatory. The term Uxi describes the coupling between the mechanical and electrical parts of the system. It has the dimensions of energy and is equal to the total flux linkages cutting the coil multiplied by the current, or to the magnetic flux times the ampere turns ni. Since

$$\phi = 2\pi r\beta$$

the coupling term is

$$(2\pi r\beta n)xi = Uxi \tag{19}$$

where U is a constant and is equal to $2\pi r\beta n$.

The dissipation function of the entire system and the potential energy of the entire system are the following:

$$D = \tfrac{1}{2}Bx^2 + \tfrac{1}{2}Ri^2 \tag{20}$$

$$V = \tfrac{1}{2}Kx^2 \tag{21}$$

The two degrees of freedom are the displacement x and the current i. Therefore,

$$M\ddot{x} + B\dot{x} + Kx - Ui = 0 \tag{22}$$

$$L\ddot{q} + R\dot{q} + U\dot{x} = e(t) \tag{23}$$

The corresponding matrix form is

$$\begin{bmatrix} M & 0 \\ 0 & L \end{bmatrix}\begin{bmatrix} \ddot{x} \\ \ddot{q} \end{bmatrix} + \begin{bmatrix} B & -U \\ U & R \end{bmatrix}\begin{bmatrix} \dot{x} \\ \dot{q} \end{bmatrix} + \begin{bmatrix} K & 0 \\ 0 & 0 \end{bmatrix}\begin{bmatrix} x \\ q \end{bmatrix} = \begin{bmatrix} 0 \\ e(t) \end{bmatrix} \tag{24}$$

Substituting (24) into (15) gives

$$\begin{bmatrix} \dot{y}_1 \\ \dot{y}_2 \\ \dot{y}_3 \\ \dot{y}_4 \end{bmatrix} = \begin{bmatrix} 0 & 0 & 1 & 0 \\ 0 & 0 & 0 & 1 \\ \dfrac{-K}{M} & 0 & \dfrac{-B}{M} & \dfrac{U}{M} \\ 0 & 0 & \dfrac{-U}{L} & \dfrac{-R}{L} \end{bmatrix}\begin{bmatrix} y_1 \\ y_2 \\ y_3 \\ y_4 \end{bmatrix} + \begin{bmatrix} 0 \\ 0 \\ 0 \\ \dfrac{1}{L}e(t) \end{bmatrix} \tag{25}$$

In detail, the substitutions made were

$$-[A]^{-1}[C] = \begin{bmatrix} -\dfrac{1}{M} & 0 \\ 0 & -\dfrac{1}{L} \end{bmatrix} \begin{bmatrix} K & 0 \\ 0 & 0 \end{bmatrix} = \begin{bmatrix} -\dfrac{K}{M} & 0 \\ 0 & 0 \end{bmatrix}$$

$$-[A]^{-1}[B] = \begin{bmatrix} -\dfrac{1}{M} & 0 \\ 0 & -\dfrac{1}{L} \end{bmatrix} \begin{bmatrix} B & -U \\ U & R \end{bmatrix} = \begin{bmatrix} -\dfrac{B}{M} & \dfrac{U}{M} \\ -\dfrac{U}{L} & -\dfrac{R}{L} \end{bmatrix}$$

Therefore, equations in (25) are the state equations of the loudspeaker.

5. Hamilton's Equations

For a conservative, or lossless, system, Lagrange's equation can be written

$$\frac{d}{dt}\left[\frac{\partial(T-V)}{\partial \dot{q}}\right] - \frac{\partial(T-V)}{\partial q} = 0 \tag{26}$$

and $T - V = L$, where L is called the Lagrangian mentioned in Chapter 1. Equation (26) contains the variables $\dot{q}$ and q. (The Lagrangian is a function of $\dot{q}$ and q).

In substituting new variables to make a first order equation, Hamilton chose the following variables:

$$p_i = \frac{\partial L}{\partial \dot{q}_i} \tag{27}$$

and

$$q_i \tag{28}$$

On the basis of these variables, Hamilton defined a new function:

$$H(p, q) = \sum \dot{q}_i p_i - L(\dot{q}, q) \tag{29}$$

H is called the Hamiltonian. Equation (29) shows the relationship between $H(p, q)$ and $L(\dot{q}, q)$, and is called the Legendre transformation.

If a conservative system does not depend upon the time t, its Hamiltonian is the total energy of the system, i.e.,

$$H = T + V$$

For *any* H function, the following relationship always holds:

$$dH = \frac{\partial H}{\partial p}dp + \frac{\partial H}{\partial q}dq$$

Substituting (29) in the equation shown above, we have

$$dH = \sum (\dot{q}_i\, dp_i + p_i\, d\dot{q}_i) - \frac{\partial L}{\partial \dot{q}}d\dot{q} - \frac{\partial L}{\partial q}dq$$

A combination of this equation and (27) gives

$$dH = \sum (\dot{q}_i \, dp_i - \dot{p}_i \, dq_i)$$

and (27) can also be written

$$\frac{d}{dt} p_i = \frac{\partial L}{\partial q_i} \tag{28a}$$

We then have the following state equations:

$$\dot{q}_i = \frac{\partial H}{\partial p_i}, \qquad \dot{p}_i = -\frac{\partial H}{\partial q_i} \tag{30}$$

These equations are called Hamilton's equations for conservative systems.

6. Example

For the circuit shown in Fig. 6-3, the Lagrangian is

$$L = \frac{L_1}{2}\dot{q}_1^2 + \frac{L_2}{2}\dot{q}_2^2 - \frac{(q_1 - q_2)^2}{2C}$$

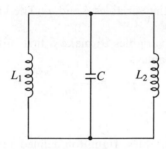

Fig. 6-3. An LC circuit.

Substituting the variables gives

$$p_1 = \frac{\partial L}{\partial \dot{q}_1} = L_1 \dot{q}_1 \tag{31}$$

$$p_2 = \frac{\partial L}{\partial \dot{q}_2} = L_2 \dot{q}_2 \tag{32}$$

The Hamiltonian is

$$\begin{aligned}
H &= \frac{L_1}{2}\dot{q}_1^2 + \frac{L_2}{2}\dot{q}_2^2 + \frac{(q_1 - q_2)^2}{2C} \\
&= \frac{1}{2}\frac{p_1^2}{L_1} + \frac{1}{2}\frac{p_2^2}{L_2} + \frac{(q_1 - q_2)^2}{2C}
\end{aligned} \tag{33}$$

Applying Eq. (30) to (33), we have

$$\dot{q}_1 = \frac{p_1}{L_1}$$

$$\dot{q}_2 = \frac{p_2}{L_2}$$

(34)

$$\dot{p}_1 = -\frac{q_1 - q_2}{C}$$

$$\dot{p}_2 = -\frac{q_1 - q_2}{C}(-1)$$

Using matrix notation, we have

$$\begin{bmatrix} \dot{q}_1 \\ \dot{q}_2 \\ \dot{p}_1 \\ \dot{p}_2 \end{bmatrix} = \begin{bmatrix} 0 & 0 & \frac{1}{L_1} & 0 \\ 0 & 0 & 0 & \frac{1}{L_2} \\ -\frac{1}{C} & \frac{1}{C} & 0 & 0 \\ \frac{1}{C} & -\frac{1}{C} & 0 & 0 \end{bmatrix} \begin{bmatrix} q_1 \\ q_2 \\ p_1 \\ p_2 \end{bmatrix}$$

(34a)

This is Hamilton's canonical equation for the circuit shown. It is interesting to note that (34a) is quite similar to (25).

II. Topological Approach

Drawing a circuit diagram, defining constraints by the diagram, and applying hypothetical forcing potentials to find resulting currents (and vice versa) with deductions based on Ohm's law and Kirchhoff's laws have given us a great heritage of representation methods. These can be transformed quite simply to state variable representations.

Usually equations derived from circuit diagrams are second order differential equations. We will show a typical transformation in the following example.

1. Node Method

Consider the circuit shown in Fig. 6-4. We name variables v_1, v_2, v_3 with respect to the ground. From node 1,

$$2(v_1 - v_2) + 1\frac{dv_1}{dt} + 0.3v_1 = 0$$

(1)

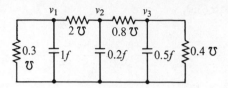

Fig. 6-4. Example illustrating the node method.

Similarly, for node 2,

$$2(v_2 - v_1) + 0.2\frac{dv_2}{dt} + 0.8(v_2 - v_3) = 0 \tag{2}$$

For node 3,

$$0.8(v_3 - v_2) + 0.5\frac{dv_3}{dt} + 0.4v_3 = 0 \tag{3}$$

Rearranging gives

$$\dot{v}_1 = 2.3v_1 + 2v_2$$
$$\dot{v}_2 = 10v_1 - 14v_2 + 4v_3$$
$$\dot{v}_3 = 1.6v_2 - 2.4v_3$$

The corresponding matrix form is

$$\begin{bmatrix} \dot{v}_1 \\ \dot{v}_2 \\ \dot{v}_3 \end{bmatrix} = \begin{bmatrix} -2.3 & 2 & 0 \\ 10 & -14 & 4 \\ 0 & 1.6 & -2.4 \end{bmatrix} \begin{bmatrix} v_1 \\ v_2 \\ v_3 \end{bmatrix} \tag{4}$$

2. Loop Method

Using Kirchhoff's second law to formulate the loop equation, we can illustrate the method of choosing state variables by the example shown in Fig. 6-5.

$$R_3 i_2 + \frac{1}{C} \int i_2 \, dt + R_1(i_1 + i_2) = 0 \tag{5}$$

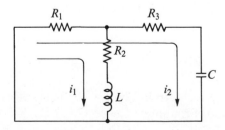

Fig. 6-5. Example illustrating the loop method.

$$R_2 i_1 + L\frac{di_1}{dt} = R_3 i_2 + \frac{1}{C}\int i_2 \, dt \tag{6}$$

Let

$$q_2 = \int i_2 \, dt = x_1 \tag{7}$$

$$i_1 = x_2 \tag{8}$$

Substitution gives

$$R_3 \dot{x}_1 + \frac{1}{C}x_1 + R_1(x_2 + \dot{x}_1) = 0$$

$$R_2 x_2 + L\dot{x}_2 = R_3 \dot{x}_1 + \frac{1}{C}x_1$$

The corresponding matrix equation is

$$
\begin{bmatrix} \dot{x}_1 \\ \dot{x}_2 \end{bmatrix}
=
\begin{bmatrix}
\dfrac{-1}{C(R_1 + R_3)} & \dfrac{-R_1}{R_1 + R_3} \\[2ex]
\dfrac{R_1}{LC(R_1 + R_3)} & -\dfrac{R_1 R_2 + R_2 R_3 + R_1 R_3}{L(R_1 + R_3)}
\end{bmatrix}
\begin{bmatrix} x_1 \\ x_2 \end{bmatrix}
\tag{9}
$$

3. Combined Loop and Node Method

Sometimes by using the combined loop and node method, we can get the simplest description.

The simple example shown in Fig. 6-6 demonstrates the method. Two node equations are

$$0.5\frac{dv_1}{dt} + 2v_1 + i = 0 \tag{10}$$

$$1\frac{dv_2}{dt} + 1v_2 = i \tag{11}$$

One loop equation is

$$1\frac{di}{dt} + v_2 = v_1 \tag{12}$$

Let

$$v_1 = x_1, \qquad v_2 = x_2, \quad \text{and} \quad i = x_3$$

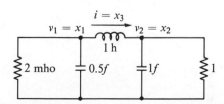

Fig. 6-6. Example illustrating the combined method.

Equations (10), (11), and (12) become

$$\dot{x}_1 = -4x_1 - 2x_3 \tag{10a}$$

$$\dot{x}_2 = -x_2 + x_3 \tag{11a}$$

$$\dot{x}_3 = x_1 - x_2 \tag{12a}$$

The matrix equation is

$$\begin{bmatrix} \dot{x}_1 \\ \dot{x}_2 \\ \dot{x}_3 \end{bmatrix} = \begin{bmatrix} -4 & 0 & -2 \\ 0 & -1 & +1 \\ 1 & -1 & 0 \end{bmatrix} \begin{bmatrix} x_1 \\ x_2 \\ x_3 \end{bmatrix} \tag{13}$$

4. A General Approach—2n Method

So far the methods discussed are borrowed from the classical second order equation description. A direct method which is based on $2n$ equations leads naturally to the formulation of a set of first order differential equations.

If a network is considered to consist of branches, each containing a single element—inductance, capacitance, or resistance—we can assume a voltage and a current for each branch.

For instance, given a network with five branches as shown in Fig. 6-7, five voltage unknowns, $v_1, v_2, v_3, v_4,$ and v_5, and five current unknowns, $i_1, i_2, i_3, i_4,$ and i_5, are labeled.

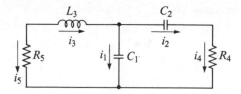

Fig. 6-7. Example illustrating the $2n$ method.

These are ten unknowns; we need ten equations. Immediately, we see five equations.

$$C_1 \frac{dv_1}{dt} = i_1 \tag{14}$$

$$C_2 \frac{dv_2}{dt} = i_2 \tag{15}$$

$$L_3 \frac{di_3}{dt} = v_3 \tag{16}$$

$$R_4 i_4 = v_4 \tag{17}$$

$$R_5 i_5 = v_5 \tag{18}$$

The equations formulated above are called *performance equations*, which are mathematical expressions of Ohm's law, Faraday's law, or the laws governing condensers.

We need five more equations in order to solve the "ten unknowns" problem. We must find the five equations according to the structure of the network and two laws of Kirchhoff. We usually call this kind of equation an *equation of constraint*.

The five constraint equations for this problem are

$$i_1 = i_3 - i_2 \tag{19}$$

$$i_2 = i_4 \tag{20}$$

$$i_3 = -i_5 \tag{21}$$

$$v_1 = v_2 + v_4 \tag{22}$$

$$v_5 = v_3 + v_1 \tag{23}$$

Writing Eqs. (14), (15), and (16) in matrix form gives

$$
\begin{bmatrix} C_1 \dfrac{dv_1}{dt} \\[2mm] C_2 \dfrac{dv_2}{dt} \\[2mm] L_3 \dfrac{di_3}{dt} \end{bmatrix}
=
\begin{bmatrix} 1 & & \\ & 1 & \\ & & 1 \end{bmatrix}
\begin{bmatrix} i_1 \\ i_2 \\ v_3 \end{bmatrix}
\tag{24}
$$

Then, using Eqs. (17) through (23), we write i_1, i_2, and v_3 in terms of v_1, v_2, and i_3 as follows:

$$i_1 = i_3 - \frac{1}{R_4}(v_1 - v_2) \tag{25}$$

$$i_2 = \frac{1}{R_4}(v_1 - v_2) \tag{26}$$

$$v_3 = -R_5 i_3 - v_1 \tag{27}$$

Equations (25) through (27) are simply a new combination of Eqs. (17) through (23).

The matrix form of Eqs. (25) through (27) is

$$
\begin{bmatrix} i_1 \\ i_2 \\ v_3 \end{bmatrix}
=
\begin{bmatrix} -\dfrac{1}{R_4} & \dfrac{1}{R_4} & 1 \\[2mm] \dfrac{1}{R_4} & -\dfrac{1}{R_4} & 0 \\[2mm] -1 & 0 & -R_5 \end{bmatrix}
\begin{bmatrix} v_1 \\ v_2 \\ i_3 \end{bmatrix}
\tag{28}
$$

Substituting Eq. (28) into (24), we have

$$
\begin{bmatrix} C_1 \dfrac{dv_1}{dt} \\[2mm] C_2 \dfrac{dv_2}{dt} \\[2mm] L_3 \dfrac{di_3}{dt} \end{bmatrix}
=
\begin{bmatrix} 1 & & \\ & 1 & \\ & & 1 \end{bmatrix}
\begin{bmatrix} -\dfrac{1}{R_4} & \dfrac{1}{R_4} & 1 \\[2mm] \dfrac{1}{R_4} & -\dfrac{1}{R_4} & 0 \\[2mm] -1 & 0 & -R_5 \end{bmatrix}
\begin{bmatrix} v_1 \\ v_2 \\ i_3 \end{bmatrix}
$$

or

$$\begin{bmatrix} C_1 \dfrac{dv_1}{dt} \\[2ex] C_2 \dfrac{dv_2}{dt} \\[2ex] L_3 \dfrac{di_3}{dt} \end{bmatrix} = \begin{bmatrix} -\dfrac{1}{R_4} & \dfrac{1}{R_4} & 1 \\[2ex] \dfrac{1}{R_4} & -\dfrac{1}{R_4} & 0 \\[2ex] -1 & 0 & -R_5 \end{bmatrix} \begin{bmatrix} v_1 \\[2ex] v_2 \\[2ex] i_3 \end{bmatrix} \qquad (29)$$

which is the state variable matrix equation of the network.

III. Transfer Function Decomposition— State Diagrams

If the transfer function of a system is given, what is the corresponding set of state variable equations? There is no unique answer to this question. The situation is much like that of programming a transfer function on an analog computer; it can be done in several ways.

As an operational definition at this point, we might define a state variable as the variable appearing at the output of the integrators of an analog computer program.

1. Direct Programming—Bush's Form

V. Bush developed a differential analyzer which handled general differential equations. His programming technique is sometimes still used on electronic analog computers. We shall employ it to change a transfer function into a set of state variables. Consider

$$G(s) = \frac{s^3 + 9s^2 + 23s + 15}{s^4 + 13s^3 + 50s^2 + 56s} \qquad (1)$$

Rewrite (1) as follows:

$$G(s) = \frac{15}{s^4 + 13s^3 + 50s^2 + 56s} + \frac{23s}{s^4 + 13s^3 + 50s^2 + 56s}$$
$$+ \frac{9s^2}{s^4 + 13s^3 + 50s^2 + 56s} + \frac{s^3}{s^4 + 13s^3 + 50s^2 + 56s} \qquad (2)$$

Program the beginning term of Eq. (2) first. In other words, attack the following transfer function:

$$\frac{15}{s^4 + 13s^3 + 50s^2 + 56s} \qquad (3)$$

Equation (3) can be rewritten as

$$\frac{15}{\{[(s + 13)s + 50]s + 56\}s} \tag{3a}$$

The corresponding relations between the transfer functions and block diagrams shown in Fig. 6-8 explain the thinking process.

The transfer function being considered	The corresponding program
$\dfrac{1}{s + 13}$	
$\dfrac{1}{(s + 13)\,s + 50}$	
$\dfrac{15}{(((s+13)s+50)s+56)\,s}$	

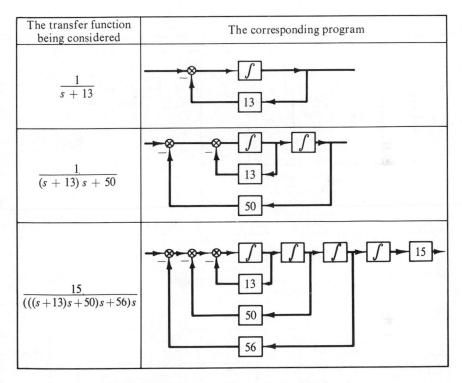

Fig. 6-8. Block diagrams interpreting transfer functions.

The second term of Eq. (2) is easily programmed by slightly modifying the first program. In other words, the program of the second term:

$$\frac{23s}{s^4 + 13s^3 + 50s^2 + 56s}$$

is shown in Fig. 6-9.

Similarly, we can get the program for the third and fourth terms. After combining, we obtain the whole program of Eq. (1) as shown in Fig. 6-10.

Examining the figure shown in Fig. 6-10 reveals that the values of elements used are directly related to the coefficients of the original polynomials.

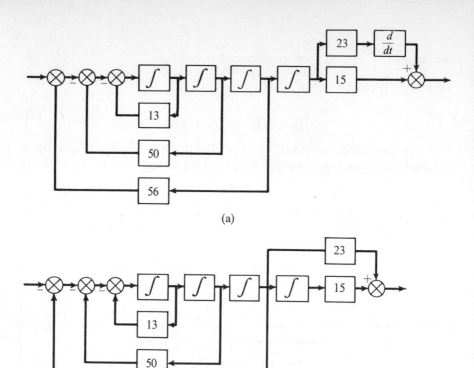

(a)

(b)

Fig. 6-9. (a) The block diagram for

$$\frac{23s}{s^4 + 13s^3 + 50s^2 + 56s};$$

(b) an alternate way to express it.

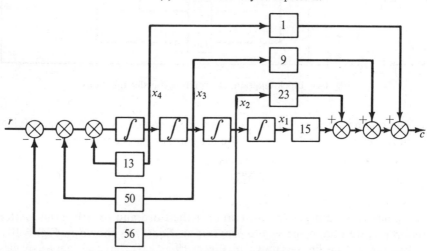

Fig. 6-10. The direct programming or Bush's form for

$$\frac{s^3 + 9s^2 + 23s + 15}{s^4 + 13s^3 + 50s^2 + 56s}.$$

Assigning a state variable name after each integrator gives x_1, x_2, x_3, and x_4 as the figure shown. Then the state equations can be written as

$$\dot{x}_1 = x_2$$
$$\dot{x}_2 = x_3$$
$$\dot{x}_3 = x_4 \tag{4}$$
$$\dot{x}_4 = -56x_2 - 50x_3 - 13x_4 + r$$
$$c = 15x_1 + 23x_2 + 9x_3 + x_4 \tag{5}$$

Equation (4) is the set of state equations of the system; (5) is an auxiliary equation which defines the output desired. The corresponding matrix forms of (4) and (5) can be written

$$\begin{bmatrix} \dot{x}_1 \\ \dot{x}_2 \\ \dot{x}_3 \\ \dot{x}_4 \end{bmatrix} = \begin{bmatrix} 0 & 1 & 0 & 0 \\ 0 & 0 & 1 & 0 \\ 0 & 0 & 0 & 1 \\ 0 & -56 & -50 & -13 \end{bmatrix} \begin{bmatrix} x_1 \\ x_2 \\ x_3 \\ x_4 \end{bmatrix} + \begin{bmatrix} 0 \\ 0 \\ 0 \\ r \end{bmatrix} \tag{4a}$$

$$c = [15, 23, 9, 1] \begin{bmatrix} x_1 \\ x_2 \\ x_3 \\ x_4 \end{bmatrix} \tag{5a}$$

Equation (4a) is called Bush's form, the phase variable form, or the companion matrix form.

2. Iterative Programming—Guillemin's Form

If $G(s)$ in Eq. (1) is factored out,

$$G(s) = \frac{s+1}{s+2} \cdot \frac{s+3}{s+4} \cdot \frac{s+5}{s+7} \cdot \frac{1}{s} \tag{6}$$

Based on (6), a program can be directly derived as shown in Fig. 6-11. The corresponding state variable equations are

$$\dot{x}_4 = r - 2x_4$$
$$\dot{x}_3 = \dot{x}_4 - 4x_3 + x_4$$
$$\dot{x}_2 = \dot{x}_3 + 3x_3 - 7x_2 \tag{7}$$
$$\dot{x}_1 = 5x_2 + \dot{x}_2$$

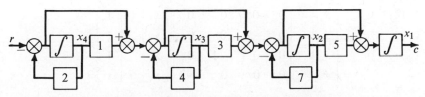

Fig. 6-11. The iterative programming, or Guillemin's form.

Rearrangement gives

$$
\begin{aligned}
\dot{x}_4 &= r - 2x_4 \\
\dot{x}_3 &= r - 2x_4 + x_4 - 4x_3 \\
\dot{x}_2 &= r - 2x_4 + x_4 - 4x_3 + 3x_3 - 7x_2 \\
\dot{x}_1 &= r - 2x_4 + x_4 - 4x_3 + 3x_3 - 7x_2 + 5x_2
\end{aligned}
\tag{7a}
$$

or

$$
\begin{aligned}
\dot{x}_1 &= -2x_2 - x_3 - x_4 + r \\
\dot{x}_2 &= -7x_2 - x_3 - x_4 + r \\
\dot{x}_3 &= -4x_3 - x_4 + r \\
\dot{x}_4 &= -2x_4 + r
\end{aligned}
\tag{7b}
$$

The corresponding matrix form is

$$
\begin{bmatrix} \dot{x}_1 \\ \dot{x}_2 \\ \dot{x}_3 \\ \dot{x}_4 \end{bmatrix}
=
\begin{bmatrix}
0 & -2 & -1 & -1 \\
0 & -7 & -1 & -1 \\
0 & 0 & -4 & -1 \\
0 & 0 & 0 & -2
\end{bmatrix}
\begin{bmatrix} x_1 \\ x_2 \\ x_3 \\ x_4 \end{bmatrix}
+
\begin{bmatrix} 1 \\ 1 \\ 1 \\ 1 \end{bmatrix} r
\tag{7c}
$$

Equation (7c) is called Guillemin's form or the pole-zero form. It can be noted that Guillemin's form gives us a triangular matrix which is one type of standard form in matrix analysis. This time, the auxiliary equation is not written because the desired output is the x_1 variable. In other words, the auxiliary equation here would be

$$
c = x_1
$$

If the roots of the denominator of Eq. (1) are complex and we do not want the coefficients of the state equations expressed in complex numbers, then it is necessary to use direct programming for the conjugate part.

3. Parallel Programming—Foster's Form

Equation (1) also can be written as a partial fraction form through the Heaviside expansion:

$$
G(s) = \frac{\frac{15}{56}}{s} + \frac{\frac{3}{20}}{s+2} + \frac{\frac{3}{24}}{s+4} + \frac{\frac{48}{105}}{s+7}
\tag{8}
$$

The programming is as shown in Fig. 6-12, and the state equation is

$$
\begin{bmatrix} \dot{x}_1 \\ \dot{x}_2 \\ \dot{x}_3 \\ \dot{x}_4 \end{bmatrix}
=
\begin{bmatrix}
0 & 0 & 0 & 0 \\
0 & -2 & 0 & 0 \\
0 & 0 & -4 & 0 \\
0 & 0 & 0 & -7
\end{bmatrix}
\begin{bmatrix} x_1 \\ x_2 \\ x_3 \\ x_4 \end{bmatrix}
+
\begin{bmatrix} 1 \\ 1 \\ 1 \\ 1 \end{bmatrix} r
\tag{9}
$$

$$c = \left[\tfrac{15}{56}, \tfrac{3}{20}, \tfrac{3}{24}, \tfrac{48}{105}\right] \begin{bmatrix} x_1 \\ x_2 \\ x_3 \\ x_4 \end{bmatrix} \tag{10}$$

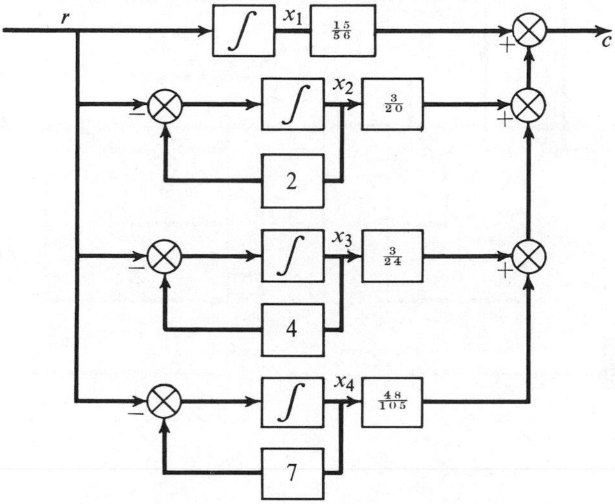

Fig. 6-12. The parallel programming, or Foster's form.

This form is similar to Foster's form in network analysis and is called Foster's form, Heaviside's form, the canonical form, or the partial fraction expansion form.

4. Continued Fraction Programming—Cauer's Form

If Eq. (1) is expanded into a continued fraction,

$$G(s) = \cfrac{1}{s + \cfrac{1}{\tfrac{1}{4} + \cfrac{1}{\tfrac{16}{9}s + \cfrac{1}{\tfrac{27}{52} + \cfrac{1}{\tfrac{169}{207}s + \cfrac{1}{\tfrac{4761}{1872} + \cfrac{1}{\tfrac{432}{3105}s}}}}}} \tag{11}$$

This is a typical form for a generalized feedback transformation. One can realize (8) by the block diagram shown in Fig. 6-13(a).

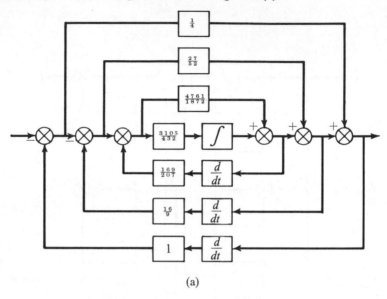

(a)

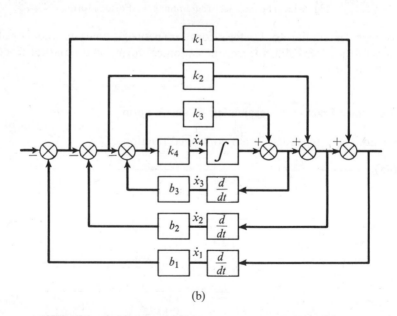

(b)

Fig. 6-13. The continuous fraction programming, or Cauer's form.

For convenience, each block is rewritten as a symbol instead of a number, giving the diagram shown in Fig. 6-13(b).

The corresponding state equation will be

$$
\begin{bmatrix} \dot{x}_1 \\ \dot{x}_2 \\ \dot{x}_3 \\ \dot{x}_4 \end{bmatrix} =
\begin{bmatrix}
-\dfrac{1}{b_1 k_1} & \dfrac{1}{b_1 k_1} & 0 & 0 \\
\dfrac{1}{b_2 k_1} & -\dfrac{1}{b_2 k_1}-\dfrac{1}{b_2 k_2} & \dfrac{1}{b_2 k_2} & 0 \\
0 & \dfrac{1}{b_3 k_2} & -\dfrac{1}{b_3 k_2}-\dfrac{1}{b_3 k_3} & \dfrac{1}{b_3 k_3} \\
0 & 0 & \dfrac{k_4}{k_3} & -\dfrac{k_4}{k_3}
\end{bmatrix}
\begin{bmatrix} x_1 \\ x_2 \\ x_3 \\ x_4 \end{bmatrix} +
\begin{bmatrix} \dfrac{r}{b_1} \\ 0 \\ 0 \\ 0 \end{bmatrix}
\tag{12}
$$

Equation (12) is called Cauer's form or the continuous fraction expansion form.

5. Repeated Roots—Jordan's Form

Some poles of a transfer function may be equal to each other; for example,

$$
G_1(s) = \frac{1}{(s+2)^3(s+4)} \tag{13}
$$

In this case, the partial fraction expansion is used:

$$
G_1(s) = \frac{\frac{1}{2}}{(s+2)^3} + \frac{-\frac{1}{4}}{(s+2)^2} + \frac{\frac{1}{8}}{s+2} + \frac{-\frac{1}{8}}{s+4} \tag{14}
$$

The block diagram shown in Fig. 6-14 is a little different from those

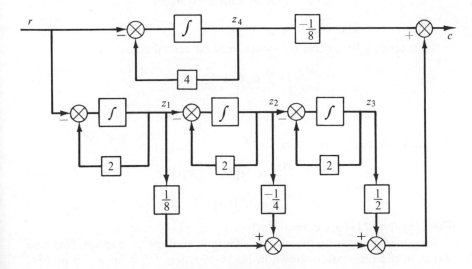

Fig. 6-14. Jordan's form.

shown in previous sections. Then we name a variable after each block. From inspection,

$$\dot{z}_4 = -4z_4 + r$$
$$\dot{z}_1 = -2z_1 + r$$
$$\dot{z}_2 = -z_1 - 2z_2$$
$$\dot{z}_3 = -z_2 - 2z_3$$

and

$$c = \tfrac{1}{8}z_1 + (-\tfrac{1}{4}z_2) + \tfrac{1}{2}z_3 - \tfrac{1}{8}z_4$$

The corresponding matrix equation is

$$
\begin{bmatrix} \dot{z}_1 \\ \dot{z}_2 \\ \dot{z}_3 \\ \dot{z}_4 \end{bmatrix}
=
\begin{bmatrix} -2 & 0 & 0 & 0 \\ -1 & -2 & 0 & 0 \\ 0 & -1 & -2 & 0 \\ 0 & 0 & 0 & -4 \end{bmatrix}
\begin{bmatrix} z_1 \\ z_2 \\ z_3 \\ z_4 \end{bmatrix}
+
\begin{bmatrix} 1 \\ 0 \\ 0 \\ 1 \end{bmatrix} r
\tag{15}
$$

$$
c = [\tfrac{1}{8}, -\tfrac{1}{4}, \tfrac{1}{2}, -\tfrac{1}{8}]
\begin{bmatrix} z_1 \\ z_2 \\ z_3 \\ z_4 \end{bmatrix}
\tag{16}
$$

Equation (15) is called Jordan's form.

IV.   General Linear Systems

In general, a linear control system may be described as

$$\frac{dx_i}{dt} = \sum_{j=1}^{n} a_{ij}x_j + \sum_{j=1}^{m} b_{ij}u_j \tag{1}$$

$$y_i = \sum_{j=1}^{n} c_{ij}x_j \tag{2}$$

In matrix notation, it becomes

$$\frac{d[x]}{dt} = [A][x] + [B][u] \tag{1a}$$

$$[y] = [C]^T[x] \tag{2a}$$

where $[x][u]$ and $[y]$ are column vectors.

Graphically, we can use block diagram analogs to express (1a) and (2a) as in the illustration shown in Fig. 6-15 where $[A]$ is an $n \times n$ matrix; $[B]$ is an $n \times m$ matrix; $[C]$ is an $n \times r$ matrix; $[C]^T$ is the transpose of $[C]$.

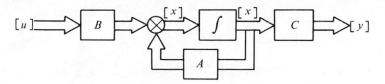

Fig. 6-15. A general matrix block diagram for linear systems.

1. Example—A Meter

Consider a meter system, shown in Fig. 6-16. According to the circuit approach, the system equations may be written as

$$e = iR + e_b + L\frac{di}{dt} \tag{3}$$

$$e_b = K_v\frac{d\theta}{dt} \tag{4}$$

$$K_t i = J\frac{d^2\theta}{dt^2} + K_s\theta \tag{5}$$

where e is the applied voltage which is to be measured by the meter.

     R is armature resistance.

     L is armature inductance.

     K_v is a constant relating the angular velocity and the back emf.

     i is the armature current.

     J is the polar moment of inertia of the rotating part.

     K_s is the spring constant and K_t is the torque constant.

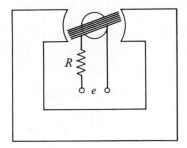

Fig. 6-16. A meter.

As was stated before, there is no unique way of assigning state variables. We will use the following procedure.

First, identify the input and output: The input is e and the output is θ . Since θ is the output, choose the first state variable:

$$x_1(t) = \theta(t) \tag{6}$$

Then let

$$\frac{d\theta}{dt} = x_2(t) \tag{7}$$

In Eq. (3) there is the term $L\, di/dt$. This suggests defining a third state variable as

$$i(t) = x_3(t) \tag{8}$$

Substitution of (6), (7), and (8) into (3) through (5) gives

$$e = Rx_3 + e_b + L\frac{dx_3}{dt}$$

$$e_b = K_v x_2$$

$$K_t x_3 = J\frac{d}{dt}x_2 + K_s x_1$$

Rearrangement gives

$$\frac{dx_1}{dt} = x_2 \tag{9}$$

$$\frac{dx_2}{dt} = \frac{K_t}{J}x_3 - \frac{K_s}{J}x_1 \tag{10}$$

$$\frac{dx_3}{dt} = \frac{e}{L} - \frac{R}{L}x_3 - K_v x_2 \tag{11}$$

The corresponding matrix form is

$$\begin{bmatrix} \dot{x}_1 \\ \dot{x}_2 \\ \dot{x}_3 \end{bmatrix} = \begin{bmatrix} 0 & 1 & 0 \\ \dfrac{-K_s}{J} & 0 & \dfrac{K_t}{J} \\ 0 & -K_v & \dfrac{-R}{L} \end{bmatrix} \begin{bmatrix} x_1 \\ x_2 \\ x_3 \end{bmatrix} + \begin{bmatrix} 0 \\ 0 \\ \dfrac{e}{L} \end{bmatrix} \tag{12}$$

In this case,

$$[\dot{x}] = [A][x] + [b]u$$

$$[y] = [C]^T[x]$$

where

$$[A] = \begin{bmatrix} 0 & 1 & 0 \\ \dfrac{-K_s}{J} & 0 & \dfrac{K_t}{J} \\ 0 & -K_v & \dfrac{-R}{L} \end{bmatrix} \qquad [b] = \begin{bmatrix} 0 \\ 0 \\ \dfrac{1}{L} \end{bmatrix}$$

$$[C]^T = [1, 0, 0]$$

2. Multi-input Considerations

If a mechanical torque is applied to the meter, $T(t)$ can be considered as a second source. The new matrix equations will be

$$\begin{bmatrix} \dot{x}_1 \\ \dot{x}_2 \\ \dot{x}_3 \end{bmatrix} = \begin{bmatrix} 0 & 1 & 0 \\ \dfrac{-K_s}{J} & 0 & \dfrac{K_t}{J} \\ 0 & -K_v & -\dfrac{R}{L} \end{bmatrix} \begin{bmatrix} x_1 \\ x_2 \\ x_3 \end{bmatrix} + \begin{bmatrix} 0 & 0 \\ -\dfrac{1}{J} & 0 \\ 0 & \dfrac{1}{L} \end{bmatrix} \begin{bmatrix} T(t) \\ e(t) \end{bmatrix} \qquad (13)$$

and the equation can be written as

$$[\dot{x}] = [A][x] + [B][u]$$
$$[y] = [C]^T[x]$$

where $[A]$ is the square matrix.

$$[B] = \begin{bmatrix} 0 & 0 \\ \dfrac{-1}{J} & 0 \\ 0 & L \end{bmatrix}, \qquad [u] = \begin{bmatrix} T(t) \\ e(t) \end{bmatrix}, \qquad [C]^T = [1, 0, 0]$$

The last term of (13) can also be decomposed into two terms:

$$\begin{bmatrix} \dot{x}_1 \\ \dot{x}_2 \\ \dot{x}_3 \end{bmatrix} = \begin{bmatrix} 0 & 1 & 0 \\ -\dfrac{K_s}{J} & 0 & \dfrac{K_t}{J} \\ 0 & -K_v & -\dfrac{R}{L} \end{bmatrix} \begin{bmatrix} x_1 \\ x_2 \\ x_3 \end{bmatrix} + \begin{bmatrix} 0 \\ -\dfrac{1}{J} \\ 0 \end{bmatrix} T(t) + \begin{bmatrix} 0 \\ 0 \\ \dfrac{1}{L} \end{bmatrix} e(t) \qquad (14)$$

$$[\dot{x}] = [A][x] + [n]T(t) + [m]e(t)$$
$$[y] = [C]^T[x] \qquad\qquad (15)$$

where

$$[m] = \begin{bmatrix} 0 \\ 0 \\ \dfrac{1}{L} \end{bmatrix}, \qquad [n] = \begin{bmatrix} 0 \\ -\dfrac{1}{J} \\ 0 \end{bmatrix}$$

The separation of the last term of (13) into the last two terms in (14) is for the purpose of further interpretation; that is, e can be considered as the normal output signal and $T(t)$ as undesirable noise. Therefore, (15) is a set of state equations which describes a linear system in general.

V. Nonlinear Feedback Systems

1. Introduction

The state variable representation of nonlinear systems is an immense subject, because there are an unlimited number of possible ways to express a system. However, if we restrict ourselves to the most common type, which is a feedback system involving a single nonlinearity in the forward loop, as shown in Fig. 6-17, the problem is straightforward.

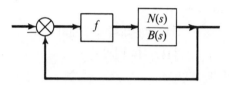

Fig. 6-17. A nonlinear feedback system.

As mentioned in the last chapter, the state variable formulation for a general nonlinear system is written as follows:

$$[\dot{x}] = [A][x] + [g(x)] \tag{1}$$

It is necessary to develop some procedures for converting Fig. 6-17 into Eq. (1). There are many ways of associating a set of state equations with a classical block diagram; two will be considered here.

2. Lure's Canonical Form

Lure's canonical form is based on the expansion of the linear transfer function into partial fractions.

$$\frac{N(s)}{B(s)} = \frac{k_1}{s - \lambda_1} + \frac{k_2}{s - \lambda_2} + \frac{k_3}{s - \lambda_3} + \cdots \tag{2}$$

The original block diagram then becomes one as shown in Fig. 6-18. Separate each fraction block into two cascaded parts and label them as shown in Fig. 6-19. After each integration block, assign names for state variables, such as $x_1, x_2, \ldots$, etc.

By definition,

$$\dot{x}_1 - \lambda_1 x_1 = f(e)$$
$$\dot{x}_2 - \lambda_2 x_2 = f(e) \tag{3}$$
$$\dot{x}_3 - \lambda_3 x_3 = f(e)$$
$$\cdots \cdots \cdots \cdots \cdots$$

and also

$$k_1 x_1 + k_2 x_2 + \cdots = r - e \tag{4}$$

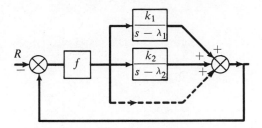

Fig. 6-18. Decomposing the linear part into a Foster form.

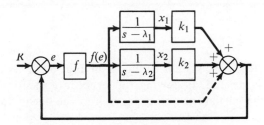

Fig. 6-19. Defining the state variables.

In matrix notation, (3) and (4) can be written as

$$
\begin{bmatrix} \dot{x}_1 \\ \dot{x}_2 \\ \dot{x}_3 \\ \cdot \\ \cdot \\ \cdot \\ \dot{x}_n \end{bmatrix} = \begin{bmatrix} \lambda_1 & & & & & \\ & \lambda_2 & & & & \\ & & \lambda_3 & & & \\ & & & \cdot & & \\ & & & & \cdot & \\ & & & & & \cdot \\ & & & & & \lambda_n \end{bmatrix} \begin{bmatrix} x_1 \\ x_2 \\ \cdot \\ \cdot \\ \cdot \\ \cdot \\ x_n \end{bmatrix}
$$
$$
+ \begin{bmatrix} f(r - k_1 x_1 - k_2 x_2 - k_3 x_3 - \cdots - k_n x_n) \\ f(r - k_1 x_1 - k_2 x_2 - k_3 x_3 - \cdots - k_n x_n) \\ \cdot \\ \cdot \\ \cdot \\ \cdot \\ f(r - k_1 x_1 - k_2 x_2 - k_3 x_3 - \cdots - k_n x_n) \end{bmatrix} \tag{5}
$$

A slight simplification gives

$$
\begin{bmatrix} \dot{x}_1 \\ \dot{x}_2 \\ \dot{x}_3 \\ \cdot \\ \cdot \\ \cdot \\ \dot{x}_n \end{bmatrix} = \begin{bmatrix} \lambda_1 & & & & & \\ & \lambda_2 & & & & \\ & & \lambda_3 & & & \\ & & & \cdot & & \\ & & & & \cdot & \\ & & & & & \cdot \\ & & & & & \lambda_n \end{bmatrix} \begin{bmatrix} x_1 \\ x_2 \\ \cdot \\ \cdot \\ \cdot \\ \cdot \\ x_n \end{bmatrix} + \begin{bmatrix} 1 \\ 1 \\ 1 \\ \cdot \\ \cdot \\ \cdot \\ 1 \end{bmatrix} f(e) \tag{5a}
$$

where

$$e = r - [k_1, k_2, k_3, \ldots, k_n] \begin{bmatrix} x_1 \\ x_2 \\ x_3 \\ \cdot \\ \cdot \\ \cdot \\ x_n \end{bmatrix}$$

Equation (5) is a set of state equations of the system shown in Fig. 6-17.

3. Example

A nonlinear feedback system is shown in Fig. 6-20. Its state equation in Lure's form is desired. Following the procedure mentioned in the last section, obtain the partial fraction expansion of the linear transfer function first.

$$\frac{1}{s^3 + 6s^2 + 11s + 6} = \frac{1}{(s+1)(s+2)(s+3)}$$

$$= \frac{\frac{1}{2}}{s+1} + \frac{-1}{s+1} + \frac{\frac{1}{2}}{s+3} \tag{6}$$

Constructing an equivalent block diagram for Fig. 6-20, we have that shown in Fig. 6-21. If we assume that there is no input ($r = 0$), the state equations can be written as follows:

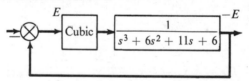

Fig. 6-20. A nonlinear system example.

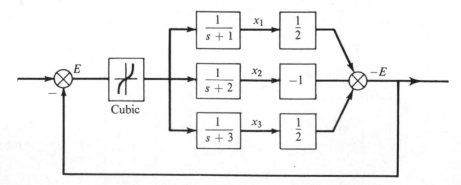

Fig. 6-21. The linear part is decomposed into Foster's form.

$$\begin{bmatrix} \dot{x}_1 \\ \dot{x}_2 \\ \dot{x}_3 \end{bmatrix} = \begin{bmatrix} -1 & & \\ & -2 & \\ & & -3 \end{bmatrix} \begin{bmatrix} x_1 \\ x_2 \\ x_3 \end{bmatrix} - \begin{bmatrix} (\frac{1}{2}x_1 - x_2 + \frac{1}{2}x_3)^3 \\ (\frac{1}{2}x_1 - x_2 + \frac{1}{2}x_3)^3 \\ (\frac{1}{2}x_1 - x_2 + \frac{1}{2}x_3)^3 \end{bmatrix} \qquad (7)$$

4. Bush's Canonical Form

The linear transfer function of Fig. 6-17 can also be decomposed into Bush's form. Let

$$\frac{N(s)}{B(s)} = \frac{a_{n-1}s^{n-1} + a_{n-2}s^{n-2} + \cdots + a_1 s + a_0}{b_n s^n + b_{n-1}s^{n-1} + \cdots + b_1 s + b_0} \qquad (8)$$

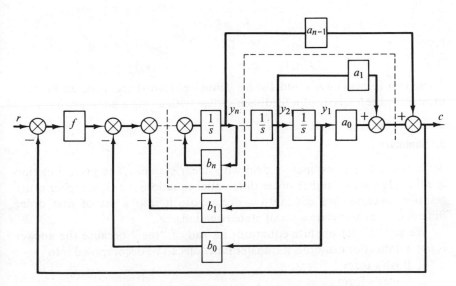

Fig. 6-22. The linear part is decomposed into Bush's form.

The diagram corresponding to Fig. 6-17 is shown as Fig. 6-22. Assign a name after each integrator and write

$$\dot{y}_1 = y_2$$
$$\dot{y}_2 = y_3$$
$$\vdots$$
$$\dot{y}_{n-1} = y_n$$
$$\dot{y}_n = r - f(a_0 y_1 + a_1 y_2 + \cdots + a_{n-1} y_n) - (b_0 y_1 + b_1 y_2 + \cdots + b_n y_n)$$

In matrix notation,

$$
\begin{bmatrix} \dot{y}_1 \\ \dot{y}_2 \\ \dot{y}_3 \\ \cdot \\ \cdot \\ \cdot \\ \dot{y}_{n-1} \\ \dot{y}_n \end{bmatrix}
=
\begin{bmatrix}
0 & 1 & \cdots\cdots\cdots\cdots \\
0 & 0 & 1 & \cdots\cdots\cdots \\
0 & 0 & 0 & 1 & \cdots\cdots \\
\cdot \\
\cdot \\
\cdot \\
-b_0 & -b_1 & \cdots\cdots & -b_n
\end{bmatrix}
\begin{bmatrix} y_1 \\ y_2 \\ y_3 \\ \cdot \\ \cdot \\ \cdot \\ y_n \end{bmatrix}
$$

$$
+
\begin{bmatrix} 0 \\ \cdot \\ \cdot \\ \cdot \\ \cdot \\ \cdot \\ 0 \\ -f(a_0 y_1 + a_1 y_2 + \cdots + a_{n-1} y_n) \end{bmatrix}
+
\begin{bmatrix} 0 \\ \cdot \\ \cdot \\ \cdot \\ \cdot \\ 0 \\ r \end{bmatrix}
$$

These two forms (Lure's and Bush's canonical forms) are basic and fundamental for the analysis of a feedback system with a single nonlinearity.

5. Summary

In this chapter, we mainly deal with linear systems. The given function is either (1) a set of higher order differential equations, or (2) a higher order transfer function. We can always decompose it into a set of first order differential equations or a set of state equations.

We say "a" set of state equations instead of "the," because the answer is not unique. For example, a transfer function can be decomposed into

1. Bush's form
2. Cauer's form
3. Foster's form
4. Guillemin's form

It depends on how we decompose the transfer function. After a state diagram is drawn, the corresponding state equations can be readily written.

The last section is on the writing of state equations for a class of nonlinear feedback systems. Lure's and Bush's canonical forms were explained.

The topics of this chapter are a necessary basis for a stability study done in the Liapunov method. They are also an important part of the foundation necessary in solving the design and the optimization problems in modern control.

PROBLEMS

6-1. A servomechanism is shown in Fig. P6-1. The difference between input θ_i and output θ_o is sensed and transformed into e by a summer to which the following relation holds: $e = k(\theta_i - \theta_o)$, where k is a constant. The error signal e so produced is fed to an amplifier with amplification μ. The output of the amplifier controls the field circuit of the generator which drives the motor. The moment of inertia of the load and the rotor of the motor is J. The axis of the load is directly connected to an arm of the summing device. The generator-motor set obeys the following laws:

$$K_g i_f = e_0$$

$$e_0 = i_a R_a + K_v \frac{d\theta_o}{dt}$$

$$K_t i_a = J \frac{d^2\theta_o}{dt^2}$$

Consider $\theta_o = x_1$, $\dot{\theta}_o = x_2$, $\ddot{\theta}_o = x_3$ as state variables, and write the state equations for the system.

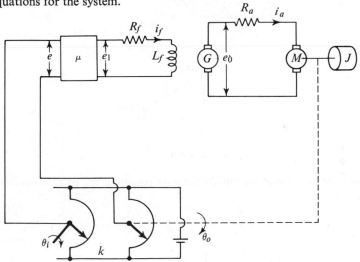

Fig. P6-1

6-2. Write the corresponding state equations for the system in Problem 6-1 if $i_f = y_1$, $\theta_o = y_2$, $\dot{\theta} = y_3$ are considered as a set of state variables.

6-3. For the given system:

$$M(s) = \frac{(s+2)(s+4)}{(s+1)(s+3)(s+6)}$$

$$= \frac{s^2 + 6s + 8}{s^3 + 10s^2 + 27s + 18}$$

write state variable equations by

(a) decomposing the transfer function into Bush's form;
(b) decomposing the transfer function into Guillemin's form;
(c) decomposing the transfer function into Foster's form;
(d) decomposing the transfer function into Cauer's form.

6-4. Use Hamilton's technique to formulate a set of p_i-q_i equations for each of the systems in Fig. P6-4.

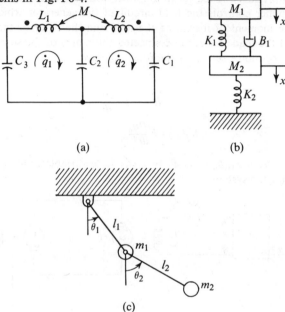

(a)   (b)

(c)

Fig. P6-4

6-5. Using the $2n$ method, formulate the state equations for each of the circuits in Fig. P6-5.

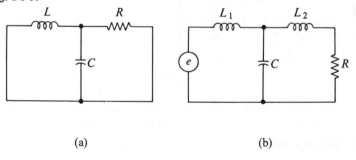

(a)   (b)

Fig. P6-5

6-6. Give state equations for the nonlinear system in Fig. P6-6 in terms of
(a) Bush's canonical form;
(b) Lure's canonical form.

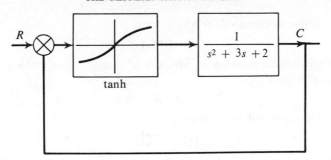

Fig. P6-6

6-7. For a nonconservative system, Hamilton's equation reads

$$\dot{p} = -\frac{\partial H}{\partial q_i} - \frac{\partial D}{\partial \dot{q}_i} + Q_i \tag{1}$$

This equation can be derived from Lagrange's equations in the following manner:

$$\frac{d}{dt}\left(\frac{\partial T}{\partial \dot{q}_i}\right) - \frac{\partial T}{\partial q_i} + \frac{\partial D}{\partial \dot{q}_i} + \frac{\partial V}{\partial q_i} = Q_i \tag{2}$$

or

$$\frac{d}{dt}\left(\frac{\partial L}{\partial \dot{q}_i}\right) - \frac{\partial L}{\partial q_i} + \frac{\partial D}{\partial \dot{q}_i} = Q_i \tag{2a}$$

where $L = T - V$, and V is a function of q only. The relationship between L and H is the Legendre transformation

$$L = \sum p_i \dot{q}_i - H \tag{3}$$

Let us partial-differentiate the equation with respect to $\dot{q}_i$:

$$\frac{\partial L}{\partial \dot{q}_i} = p_i - \frac{\partial H}{\partial \dot{q}_i} \tag{4}$$

Substituting (4) into (2a), we have

$$\frac{d}{dt}\left(p_i - \frac{\partial H}{\partial \dot{q}_i}\right) + \frac{\partial H}{\partial q_i} + \frac{\partial D}{\partial \dot{q}_i} = Q_i \tag{5}$$

or

$$\dot{p}_i = -\frac{\partial H}{\partial q_i} - \frac{\partial D}{\partial \dot{q}_i} + Q_i \tag{5a}$$

Use these equations to formulate the canonical equations for the circuit in Fig. P6-7.

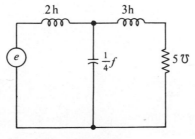

Fig. P6-7

6-8. It is possible to change an arbitrary form

$$[\dot{x}] = [A][x] + [b]r$$

(the characteristic equation of which is $\lambda^n + a_1\lambda^{n-1} + a_2\lambda^{n-2} + \cdots + a_n = 0$) into Bush's form:

$$[\dot{y}] = [A_0][y] + [b_0]r$$

by the following transformation:

$$[x] = [T][y]$$

where

$$[T] = [\mathbf{A}^{n-1}\mathbf{b},\ \mathbf{A}^{n-2}\mathbf{b},\ \cdots,\ \mathbf{A}\mathbf{b},\ \mathbf{b}] \begin{bmatrix} 1 & 0 & 0 & 0 & \cdot & \cdot & \cdot & 0 \\ a_1 & 1 & 0 & 0 & 0 & \cdot & \cdot & 0 \\ a_2 & a_1 & 1 & 0 & 0 & \cdot & \cdot & 0 \\ a_3 & a_2 & a_1 & 1 & 0 & \cdot & \cdot & 0 \\ \cdot & a_3 & a_2 & a_1 & 1 & \cdot & \cdot & 0 \\ \cdot & \cdot & a_3 & a_2 & a_1 & \cdot & \cdot & \cdot \\ \cdot & \cdot & \cdot & \cdot & \cdot & \cdot & \cdot & \cdot \\ a_{n-1} & \cdot & \cdot & \cdot & \cdot & \cdot & a_1 & 1 \end{bmatrix}$$

Verify this transformation by changing the following arbitrary form

$$\begin{bmatrix} \dot{x}_1 \\ \dot{x}_2 \\ \dot{x}_3 \end{bmatrix} = \begin{bmatrix} -3 & 0 & 0 \\ 0 & 0 & 1 \\ 0 & -5 & -2 \end{bmatrix} \begin{bmatrix} x_1 \\ x_2 \\ x_3 \end{bmatrix} + \begin{bmatrix} 1 \\ 0 \\ 1 \end{bmatrix} r$$

into the corresponding Bush form.

CHAPTER 7

The Solution Problem

The preceding chapter dealt with the various mathematical constructs used to represent a system. The present chapter is concerned with the development of methods for getting solutions.

Finding the solution of a differential equation means finding a relation between the variables involved, free of derivatives, which satisfies the equation identically. It might be said that the solution of a set of first order differential equations is a certain set of algebraic equations. In other words, the following set of differential equations

$$\dot{x}_1 = f_1(x_1, x_2, x_3, \ldots, t)$$
$$\dot{x}_2 = f_2(x_1, x_2, x_3, \ldots, t)$$
$$\dot{x}_3 = f_3(x_1, x_2, x_3, \ldots, t)$$
$$\cdots \cdots \cdots \cdots \cdots \cdots$$
$$\dot{x}_n = f_n(x_1, x_2, x_3, \ldots, t)$$

is said to have

$$x_1 = g_1(t)$$
$$x_2 = g_2(t)$$
$$\cdots \cdots \cdots$$
$$x_n = g_n(t)$$

as a solution if

$$\dot{g}_1 = f_1(g_1, g_2, g_3, \ldots, t)$$
$$\dot{g}_2 = f_2(g_1, g_2, g_3, \ldots, t)$$
$$\cdots \cdots \cdots \cdots \cdots \cdots$$
$$g_n = f_n(g_1, g_2, g_3, \ldots, t)$$

In the first part of this chapter, we will demonstrate solution techniques for linear equations with constant coefficients using the Laplace transfor-

319

mation method. In doing so, we will derive the "transition matrix" and explain its properties.

Then, a classical approach to the same equations will be presented, including (1) the matrizant method for evaluating the transition matrix; (2) the Lagrange variational parameter technique.

Next, in view of the fact that time-varying systems and nonlinear systems (with the exception of very special cases) can only be solved by some type of iteration scheme, we will develop these numerical methods with an introduction to digital computer programs.

I.   Laplace Transformation Method

1.   Free Motion or Zero Input Response

A set of homogeneous differential equations with constant coefficients is a mathematical model of a time-invariant system. To find the solution of the set of differential equations is to evaluate the "zero input response" of the system. A set of linear differential equations is given.

$$[\dot{x}] = [A][x] \tag{1}$$

where $[A]$ is a square constant matrix, $[x]$ is a column matrix.

Perform Laplace transformation on (1):

$$[sX(s) - x(0)] = [A][X(s)]$$

or

$$\{s[I] - [A]\}[X(s)] = [x(0)]$$

Rearrangement gives

$$[X(s)] = \{s[I] - [A]\}^{-1}[x(0)] \tag{2}$$

where $[X(s)]$ is the Laplace transform of the zero input response and $[x(0)]$ is the initial vector in the time domain.

Taking the inverse Laplace transformation, we have

$$[x(t)] = \mathscr{L}^{-1}\{s[I] - [A]^{-1}\}[x(0)] \tag{2a}$$

where $[x(t)]$ is the zero input response of the system.

2.   Example

A system described by the following differential equation is given.

$$\frac{d^3x}{dt^3} + 6\frac{d^2x}{dt^2} + 11\frac{dx}{dt} + 6x = 0$$

or

$$\dot{x}_1 = x_2$$
$$\dot{x}_2 = x_3$$
$$\dot{x}_3 = -6x_1 - 11x_2 - 6x_3$$

In the matrix form:

$$\begin{bmatrix} \dot{x}_1 \\ \dot{x}_2 \\ \dot{x}_3 \end{bmatrix} = \begin{bmatrix} 0 & 1 & 0 \\ 0 & 0 & 1 \\ -6 & -11 & -6 \end{bmatrix} \begin{bmatrix} x_1 \\ x_2 \\ x_3 \end{bmatrix}$$

After taking Laplace transforms,

$$\begin{bmatrix} sX_1(s) \\ sX_2(s) \\ sX_3(s) \end{bmatrix} - \begin{bmatrix} x_1(0) \\ x_2(0) \\ x_3(0) \end{bmatrix} = \begin{bmatrix} 0 & 1 & 0 \\ 0 & 0 & 1 \\ -6 & -11 & -6 \end{bmatrix} \begin{bmatrix} X_1(s) \\ X_2(s) \\ X_3(s) \end{bmatrix} \tag{3}$$

is obtained.

Equation (3) also can be written as

$$\left\{ s\begin{bmatrix} 1 & 0 & 0 \\ 0 & 1 & 0 \\ 0 & 0 & 1 \end{bmatrix} - \begin{bmatrix} 0 & 1 & 0 \\ 0 & 0 & 1 \\ -6 & -11 & -6 \end{bmatrix} \right\} \begin{bmatrix} X_1(s) \\ X_2(s) \\ X_3(s) \end{bmatrix} = \begin{bmatrix} x_1(0) \\ x_2(0) \\ x_3(0) \end{bmatrix}$$

$$\begin{bmatrix} X_1(s) \\ X_2(s) \\ X_3(s) \end{bmatrix} = \begin{bmatrix} s & -1 & 0 \\ 0 & s & -1 \\ 6 & 11 & s+6 \end{bmatrix}^{-1} \begin{bmatrix} x_1(0) \\ x_2(0) \\ x_3(0) \end{bmatrix}$$

Using the standard inversing matrix procedure, we have

$$\begin{bmatrix} X_1(s) \\ \\ X_2(s) \\ \\ X_3(s) \end{bmatrix} =$$

$$\begin{bmatrix} \dfrac{s^2 + 6s + 11}{s^3 + 6s^2 + 11s + 6} & \dfrac{s + 6}{s^3 + 6s^2 + 11s + 6} & \dfrac{1}{s^3 + 6s^2 + 11s + 6} \\ \dfrac{-6}{s^3 + 6s^2 + 11s + 6} & \dfrac{s^2 + 6s}{s^3 + 6s^2 + 11s + 6} & \dfrac{s}{s^3 + 6s^2 + 11s + 6} \\ \dfrac{-6s}{s^3 + 6s^2 + 11s + 6} & \dfrac{-11s - 6}{s^3 + 6s^2 + 11s + 6} & \dfrac{s^2}{s^3 + 6s^2 + 11s + 6} \end{bmatrix} \begin{bmatrix} x_1(0) \\ x_2(0) \\ x_3(0) \end{bmatrix}$$

$$\tag{4}$$

Inverse Laplace transformation gives

$$
\begin{bmatrix} x_1(t) \\ x_2(t) \\ x_3(t) \end{bmatrix} =
$$

$$
\begin{bmatrix}
3e^{-t} - 3e^{-2t} + e^{-3t} & \tfrac{5}{2}e^{-t} - 4e^{-2t} + \tfrac{3}{2}e^{-3t} & \tfrac{1}{2}e^{-t} - e^{-2t} + \tfrac{1}{2}e^{-3t} \\
-3e^{-t} + 6e^{-2t} - 3e^{-3t} & -\tfrac{5}{2}e^{-t} + 8e^{-2t} - \tfrac{9}{2}e^{-3t} & -\tfrac{1}{2}e^{-t} + 2e^{-2t} - \tfrac{3}{2}e^{-3t} \\
3e^{-t} - 12e^{-2t} + 9e^{-3t} & \tfrac{5}{2}e^{-t} - 16e^{-2t} + \tfrac{27}{2}e^{-3t} & \tfrac{1}{2}e^{-t} - 4e^{-2t} + \tfrac{9}{2}e^{-3t}
\end{bmatrix}
$$

$$
\cdot \begin{bmatrix} x_1(0) \\ x_2(0) \\ x_3(0) \end{bmatrix} \qquad (4\text{a})
$$

If (4a) is written in symbol form, we have the following

$$
\begin{bmatrix} x_1(t) \\ x_2(t) \\ x_3(t) \end{bmatrix} =
\begin{bmatrix} \phi_{11}(t) & \phi_{12}(t) & \phi_{13}(t) \\ \phi_{21}(t) & \phi_{22}(t) & \phi_{23}(t) \\ \phi_{31}(t) & \phi_{32}(t) & \phi_{33}(t) \end{bmatrix}
\begin{bmatrix} x_1(0) \\ x_2(0) \\ x_3(0) \end{bmatrix}
$$

The general form of the solution of (1) is usually written as follows

$$
[x(t)] = [\Phi(t)][x(0)] \qquad (5)
$$

where $[\Phi(t)]$ is called the transition matrix; $[x(0)]$, the initial state; $[x(t)]$, the state.

Geometrically interpreted, $[x(t)]$ is a point in an n-dimensional space which defines the state of the system at time t. $[x(0)]$ is the initial state of the system. Now it follows that if $[x(0)]$ is specified, the future behavior of $[x(t)]$ is completely determined. $[\Phi(t)]$ plays the role of an operator changing the system from $[x(0)]$ state to $[x(t)]$ state.

3. Generalization of Heaviside's Expansion

The process by which we obtain the direct inversion of (4) from (3) is lengthy, because the s variable is present for all steps preceding the next-to-last. After obtaining (4), one still has to perform n^2 partial fraction expansions. Because of the drudgery of doing all the above operations, we generalize Heaviside's expansion into a matrix coefficient expansion.

The transition matrix in the s domain can be written as an adjoint matrix and a determinant:

$$
[\hat{\Phi}(s)] = \{s[I] - [A]\}^{-1} \triangleq \frac{[J]}{\det [sI - A]} \qquad (6)
$$

where $[J]$ is the adjoint matrix of $[sI - A]$.

The roots of the characteristic equation are given as $s_1, s_2, \ldots, s_n$.

Step 1. Assume an undetermined constant matrix for each term of the partial fraction expansion, letting

$$[\hat{\Phi}(s)] = \frac{[K_1]}{s - s_1} + \frac{[K_2]}{s - s_2} + \cdots + \frac{[K_k]}{s - s_k} + \cdots + \frac{[K_n]}{s - s_n} \tag{6a}$$

Step 2. Multiply both sides by $(s - s_k)$ and then let $s = s_k$.

$$\{s[I] - [A]\}^{-1}(s - s_k)\Big|_{s \to s_k} = [K_k]\frac{1}{s - s_k}(s - s_k)\Big|_{s \to s_k} \tag{7}$$

Step 3. Solve for $[K_k]$ in terms of the adjoint matrix of $\{s[I] - [A]\}$.

$$[K_k] = \frac{[J(s)]|_{s \to s_k}}{(s_k - s_1)(s_k - s_2)\ldots(s_k - s_{k-1})(s_k - s_{k+1})\ldots(s_k - s_n)} \tag{8}$$

where $[J(s)]|_{s \to s_k}$ is the adjoint matrix of $\{s[I] - [A]\}$ after s is substituted for s_k.

Step 4. The evaluation of $[J]|_{s-s_k}$ can be carried out by any numerical method; for example, Cayley-Hamilton's or Faddeev's method (see Problem 7-3). It is noted that evaluating the adjoint matrix $[J(s)]|_{s \to s_k}$ can be done numerically, since the elements of the matrix do not involve the variable s any more.

Once $[K_1]$, $[K_2]$, $\ldots$, $[K_n]$ are obtained, the transition matrix is found by substituting them into (6a).

It is interesting to note that all four steps are comparable to our regular partial fraction expansion technique except the last, and the last step can be easily performed with a digital computer.

4. Sylvester's Expansion

Sylvester has an expansion as follows:

$$[\Phi(t)] = \sum_{k=1}^{n} \frac{[A - s_1 I] \cdots [A - s_{k-1}I][A - s_{k+1}I] \cdots [A - s_n I]}{(s_k - s_1) \cdots (s_k - s_{k-1})(s_k - s_{k+1}) \cdots (s_k - s_n)} e^{s_k t} \tag{9}$$

Equation (9) is simply the inverse Laplace transformation of Eq. (6) and the numerator of the coefficient is equal to the adjoint matrix after s variable has been replaced by s_k.

Equation (9) is usually called the Sylvester expansion formula, which is the generalized Heaviside expansion.

5. Example 1—System with Distinct Roots

The method can best be illustrated by the following example, which involves only distinct roots.

For the given system,

$$\begin{bmatrix} \dot{x}_1 \\ \dot{x}_2 \\ \dot{x}_3 \end{bmatrix} = \begin{bmatrix} 0 & 1 & 0 \\ 0 & 0 & 1 \\ -6 & -11 & -6 \end{bmatrix} \begin{bmatrix} x_1 \\ x_2 \\ x_3 \end{bmatrix} \tag{10}$$

Its transition matrix is desired.

Performing the Laplace transformation and collecting terms, we have

$$\begin{bmatrix} X_1(s) \\ X_2(s) \\ X_3(s) \end{bmatrix} = \begin{bmatrix} s & -1 & 0 \\ 0 & s & -1 \\ 6 & 11 & s+6 \end{bmatrix}^{-1} \begin{bmatrix} x_1(0) \\ x_2(0) \\ x_3(0) \end{bmatrix}$$

Step 1. Assume the undetermined matrix coefficients

$$\begin{bmatrix} s & -1 & 0 \\ 0 & s & -1 \\ 6 & 11 & s+6 \end{bmatrix}^{-1} = \frac{[K_1]}{s+1} + \frac{[K_2]}{s+2} + \frac{[K_3]}{s+3}$$

where $[K_1]$, $[K_2]$, and $[K_3]$ are 3×3 matrices whose elements are to be determined.

Step 2. Multiply each side by $(s + 1)$ and set $s = -1$.

$$(s+1)\begin{bmatrix} s & -1 & 0 \\ 0 & s & -1 \\ 6 & 11 & s+6 \end{bmatrix}^{-1} \Bigg|_{s \to -1} = [K_1]$$

Step 3. Evaluate $[K_1]$ in terms of the adjoint matrix and determinant,

$$[K_1] = \frac{[J(s)]|_{s \to -1}}{(-1+2)(-1+3)}$$

Step 4. Find $[J(s)]|_{s \to -1}$ by Faddeev's method.

$$[J_1] \triangleq [J(s)]|_{s \to -1} = \begin{bmatrix} -1 & -1 & 0 \\ 0 & -1 & -1 \\ 6 & 11 & 5 \end{bmatrix} \quad \text{and} \quad p_1 = \text{tr}\,[J_1] = 3$$

where "tr" stands for "trace" which means $\sum a_{ij} = (-1) + (-1) + 5 = 3$.

$$[B_1] \triangleq [J_1] - p_1[I] = \begin{bmatrix} -4 & -1 & 0 \\ 0 & -4 & -1 \\ 6 & 11 & 2 \end{bmatrix}$$

where $[I]$ is the identity matrix.
$[J_2]$ is then found as follows:

$$[J_2] = [J_1][B_2] = \begin{bmatrix} 4 & 5 & 1 \\ -6 & -7 & -1 \\ 6 & 5 & -1 \end{bmatrix} \quad \text{and} \quad p_2 = \tfrac{1}{2}\text{tr}\,[J_2] = -2$$

$$[B_2] = [J_2] - p_2[I] = \begin{bmatrix} 6 & 5 & 1 \\ -6 & -5 & -1 \\ 6 & 5 & 1 \end{bmatrix}$$

Therefore,

$$[K_1] = \frac{[J(s)]|_{s \to -1}}{(-1+2)(-1+3)} = \frac{B_2}{(-1+2)(-1+3)} = \begin{bmatrix} 3 & \frac{5}{2} & \frac{1}{2} \\ -3 & -\frac{5}{2} & -\frac{1}{2} \\ 3 & \frac{5}{2} & \frac{1}{2} \end{bmatrix}$$

Similarly, we can find $[K_2]$ and $[K_3]$.

The transition matrix is then found as follows:

$$[\Phi(t)] = [K_1]e^{s_1 t} + [K_2]e^{s_2 t} + [K_3]e^{s_3 t} =$$

$$\begin{bmatrix} 3 & \frac{5}{2} & \frac{1}{2} \\ -3 & -\frac{5}{2} & -\frac{1}{2} \\ 3 & \frac{5}{2} & \frac{1}{2} \end{bmatrix} e^{-t} + \begin{bmatrix} 3 & -4 & -1 \\ 6 & 8 & 2 \\ -12 & -16 & -4 \end{bmatrix} e^{-2t} + \begin{bmatrix} 1 & \frac{3}{2} & \frac{1}{2} \\ -3 & -\frac{9}{2} & -\frac{3}{2} \\ 9 & \frac{27}{2} & \frac{9}{2} \end{bmatrix} e^{-3t} \quad (11)$$

After combining the three matrices, we finally have the transition matrix $[\Phi(t)]$. The answer, of course, coincides with (4a), which was obtained from the direct inversion.

6. Example 2—System with Multiple Roots

With certain modifications being made, the generalized Heaviside expansion method is applicable to a system with multiple roots. We use an example to explain the procedure.

Consider the system

$$\begin{bmatrix} \dot{x}_1 \\ \dot{x}_2 \\ \dot{x}_3 \end{bmatrix} = \begin{bmatrix} 3 & -1 & 1 \\ 2 & 0 & 1 \\ 1 & -1 & 2 \end{bmatrix} \begin{bmatrix} x_1 \\ x_2 \\ x_3 \end{bmatrix} \quad (12)$$

Its transition matrix is required.

Step 1. A partial fraction form is assumed.

$$\begin{bmatrix} s-3 & 1 & -1 \\ -2 & s & -1 \\ -1 & 1 & s-2 \end{bmatrix}^{-1} = \frac{[K_1]}{s-1} + \frac{[K_2]}{s-2} + \frac{[K_3]}{(s-2)^2} \quad (13)$$

where $[K_1]$ is easily found by applying the procedures shown in the last section, or

$$[K_1] = \begin{bmatrix} 0 & 0 & 0 \\ -1 & 1 & 0 \\ -1 & 1 & 0 \end{bmatrix} \quad (14)$$

Step 2. $[K_2]$ and $[K_3]$ can be found simultaneously by substituting the s variable by $(s'+2)$, which means to shift the vertical axis by 2, or

$$[J(s')] = \begin{bmatrix} s'-1 & 1 & -1 \\ -2 & s'+2 & -1 \\ -1 & 1 & s' \end{bmatrix}$$

Following Faddeev's adjoint matrix evaluation procedure, we have

$$[J_1] = \begin{bmatrix} s' - 1 & 1 & -1 \\ -2 & s' + 2 & -1 \\ -1 & 1 & s' \end{bmatrix} \quad \text{and} \quad p_1 = \text{tr}\,[J_1] = 3s' + 1$$

Then

$$[B_1] = \begin{bmatrix} s' - 1 & 1 & -1 \\ -2 & s' + 2 & -1 \\ -1 & 1 & s' \end{bmatrix} - p_1[I] = \begin{bmatrix} 2s' - 2 & 1 & -1 \\ -2 & -2s' + 1 & -1 \\ -1 & 1 & -2s' - 1 \end{bmatrix}$$

$$[J_2] = [J_1][B_1] = \begin{bmatrix} -2s'^2 + 1 & -s' - 1 & s' + 1 \\ 2s' + 1 & -2s' - 3s' - 1 & s' + 1 \\ s' & s' & 2s'^2 - s' \end{bmatrix}$$

and $p_2 = \frac{1}{2}\,\text{tr}\,[J_2] = -3s' - 2s'$.

$$[B_2] = [J_2] - p[I] = \begin{bmatrix} s'^2 + 2s' + 1 & -s' - 1 & s' + 1 \\ 2s' + 1 & s'^2 - s' & s' + 1 \\ s' & -s' & s'^2 + s' \end{bmatrix}$$

Step 3. $[K_2]$ and $[K_3]$ are found by rewriting each element of $[B_2]$ in the reversing order and dividing $[B_2]$ by $(s' + 2 - 1)$. These correspond to a Taylor series expansion, and the differentiations which are needed in regular Heaviside expansions are avoided here.

$$\frac{[B_2]}{1 + s'} = \frac{1}{1 + s'} \begin{bmatrix} 1 + 2s' + s'^2 & -1 - s' & 1 + s' \\ 1 + 2s' & -1 - s' + s'^2 & 1 + s' \\ s' & -s' & s' + s'^2 \end{bmatrix}$$

$$= \begin{bmatrix} 1 + s' & -1 & 1 \\ 1 + s' - s'^2 & -1 + 0 + s'^2 & 1 \\ 0 + s^1 - s'^2 + \cdots & 0 - s' + s'^2 - \cdots & 0 + s' \end{bmatrix} \quad (15)$$

$$= \begin{bmatrix} 1 & -1 & 1 \\ 1 & -1 & 1 \\ 0 & 0 & 0 \end{bmatrix} + \begin{bmatrix} 1 & 0 & 0 \\ 1 & 0 & 0 \\ 1 & -1 & 1 \end{bmatrix} s' + \cdots$$

The matrix coefficient of s' is equal to $[K_2]$ and the first matrix is equal to $[K_3]$.

Substituting (13) and (14) into (12), we obtain

$$[\Phi(s)] = \begin{bmatrix} 0 & 0 & 0 \\ -1 & 1 & 0 \\ -1 & 1 & 0 \end{bmatrix} \frac{1}{s - 1} + \begin{bmatrix} 1 & 0 & 0 \\ 1 & 0 & 0 \\ 1 & -1 & 1 \end{bmatrix} \frac{1}{s - 2} + \begin{bmatrix} 1 & -1 & 1 \\ 1 & -1 & 1 \\ 0 & 0 & 0 \end{bmatrix} \frac{1}{(s - 2)^2}$$

$$(16)$$

Inverse Laplace transformation gives

$$[\Phi(t)] = \begin{bmatrix} 0 & 0 & 0 \\ -1 & 1 & 0 \\ -1 & 1 & 0 \end{bmatrix} e^t + \begin{bmatrix} 1 & 0 & 0 \\ 1 & 0 & 0 \\ 1 & -1 & 1 \end{bmatrix} e^{2t} + \begin{bmatrix} 1 & -1 & 1 \\ 1 & -1 & 1 \\ 0 & 0 & 0 \end{bmatrix} t e^{2t} \qquad (16a)$$

which is the required transition matrix of (12).

7. Properties of the Transition Matrix

(a) $[\Phi(t)]$ has the multiplicative property

$$[\Phi(t)][\Phi(\tau)] = [\Phi(t+\tau)] \qquad (17)$$

We will use an example, shown in Fig. 7-1, to demonstrate this property.

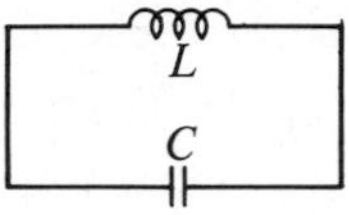

Fig. 7-1. A circuit exemplifying the properties of the transition matrix.

The differential equations for describing the circuit are

$$L \frac{di}{dt} = v$$

$$C \frac{dv}{dt} = -i$$

(assume $L = 1$ and $C = 1$).

Using new notations, let

$$x_1 = i$$
$$x_2 = v$$

Equations (8) and (9) become

$$\frac{dx_1}{dt} = x_2$$

$$\frac{dx_2}{dt} = -x_1$$

or

$$\begin{bmatrix} \dot{x}_1 \\ \dot{x}_2 \end{bmatrix} = \begin{bmatrix} 0 & 1 \\ -1 & 0 \end{bmatrix} \begin{bmatrix} x_1 \\ x_2 \end{bmatrix}$$

The solution in the Laplace transformation domain is

$$
\begin{bmatrix} X_1(s) \\ X_2(s) \end{bmatrix} = \begin{bmatrix} \dfrac{s}{s^2+1} & \dfrac{s}{s^2+1} \\ \dfrac{-1}{s^2+1} & \dfrac{s}{s^2+1} \end{bmatrix} \begin{bmatrix} x_1(0) \\ x_2(0) \end{bmatrix}
$$

The solution in the time domain is

$$
\begin{bmatrix} x_1(t) \\ x_2(t) \end{bmatrix} = \begin{bmatrix} \cos t & \sin t \\ -\sin t & \cos t \end{bmatrix} \begin{bmatrix} x_1(0) \\ x_2(0) \end{bmatrix}
$$

In the problem, $[\Phi(t)]$ is

$$
\begin{bmatrix} \cos t & \sin t \\ -\sin t & \cos t \end{bmatrix}
$$

It is obvious that the following identity is true:

$$
[\Phi(t+\tau)] = \begin{bmatrix} \cos(t+\tau) & \sin(t+\tau) \\ -\sin(t+\tau) & \cos(t+\tau) \end{bmatrix}
$$

$$
= \begin{bmatrix} \cos t & \sin t \\ -\sin t & \cos t \end{bmatrix} \begin{bmatrix} \cos \tau & \sin \tau \\ -\sin \tau & \cos \tau \end{bmatrix}
$$

$$
= [\Phi(t)][\Phi(\tau)]
$$

The proof of (17) is easily established if the basic definition is considered.

$$
[x(t+\tau)] = [\Phi(t+\tau)][x(0)]
$$
$$
[x(t+\tau)] = [\Phi(t)][x(\tau)]
$$
$$
= [\Phi(t)][\Phi(\tau)][x(0)]
$$

By comparison,

$$
[\Phi(t+\tau)] = [\Phi(t)][\Phi(\tau)]
$$

which is called the multiplicative property.

(b) $[\Phi(t)]$ has the commutative property.
An interchange of the positions of t and τ of (7) gives

$$
[\Phi(\tau+t)] = [\Phi(\tau)][\Phi(t)]
$$

or, rewritten,

$$
[\Phi(t+\tau)] = [\Phi(\tau)][\Phi(t)]
$$

Thus, we can write

$$
[\Phi(\tau)][\Phi(t)] = [\Phi(t)][\Phi(\tau)] \tag{18}
$$

which is called the commutative property.

(c) $[\Phi(t)]$ is the solution of the following differential equation:

$$
[\dot{\Phi}(t)] = [A][\Phi(t)], \qquad [\Phi(0)] = [I] \tag{19}
$$

where $[A]$ is the constant square matrix.

To prove this property, take the Laplace transform of (12):

$$s[\hat{\Phi}(s)] - [\Phi(0)] = [A][\hat{\Phi}(s)]$$

Substitution of the initial condition gives

$$s[\hat{\Phi}(s)] - [I] = [A][\hat{\Phi}(s)]$$

or

$$[\hat{\Phi}(s)] = \{s[I] - [A]\}^{-1}[I]$$

Then take the inverse Laplace transform:

$$[\Phi(t)] = \mathscr{L}^{-1}\{s[I] - [A]\}^{-1}$$

which is the definition of the transition matrix.

Equation (19) is also referred to as the normalized state equations of the original state equations. Usually it is by solving (19) that we obtain $[\Phi(t)]$.

8. Forced Motion: Zero State Response

If a forcing function is applied to a system with certain initial conditions given, the response can be considered in two parts: (1) the zero input response and (2) the zero state response. The former is the effect of the initial conditions only, and the latter results from the forcing function with zero initial conditions.

The zero state response can be found by Laplace transformation in a straightforward manner. A single example can make this clear.

Consider the following

$$[\dot{x}] = [A][x] + [b] \tag{20}$$

Performing Laplace transformation gives

$$\{s[I] - [A]\}[X(s)] = [x(0)] + [B(s)]$$
$$[X(s)] = \{s[I] - [A]\}^{-1}[x(0)] + \{s[I] - [A]\}^{-1}[B(s)]$$

or

$$[X(s)] = [\hat{\Phi}(s)][x(0)] + [\hat{\Phi}(s)][B(s)] \tag{21}$$

The complete solution $[x(t)]$ is obtained by the inverse transformation:

$$[x(t)] = [\Phi(t)][x(0)] + \mathscr{L}^{-1}[\hat{\Phi}(s)][B(s)] \tag{22}$$

The second term on the right-hand side is the zero state response. This means that when $[x(0)] = [0]$, $[x(t)]$ is determined by only the last term of (22), $\mathscr{L}^{-1}[\hat{\Phi}(s)][B(s)]$.

9. An Example

A circuit is shown in Fig. 7-2; the initial conditions are zero. Find the zero state response of the circuit.

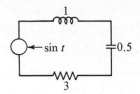

Fig. 7-2. An *RLC* circuit.

The state equation is

$$\begin{bmatrix} \dot{x}_1 \\ \dot{x}_2 \end{bmatrix} = \begin{bmatrix} 0 & 1 \\ -2 & -3 \end{bmatrix} \begin{bmatrix} x_1 \\ x_2 \end{bmatrix} + \begin{bmatrix} 0 \\ \sin t \end{bmatrix}$$

Its corresponding solution is as follows:

$$\underbrace{\begin{bmatrix} X_1(s) \\ \\ X_2(s) \end{bmatrix}}_{\text{Total response}} = \underbrace{\begin{bmatrix} \dfrac{s+3}{s^2+3s+2} & \dfrac{1}{s^2+3s+2} \\ \dfrac{-2}{s^2+3s+2} & \dfrac{s}{s^2+3s+2} \end{bmatrix} \begin{bmatrix} x_1(0) \\ \\ x_2(0) \end{bmatrix}}_{\text{Zero input response}}$$

$$+ \underbrace{\begin{bmatrix} \dfrac{s+3}{s^2+3s+2} & \dfrac{1}{s^2+3s+2} \\ \dfrac{-2}{s^2+3s+2} & \dfrac{s}{s^2+3s+2} \end{bmatrix} \begin{bmatrix} 0 \\ \\ \dfrac{1}{s^2+1} \end{bmatrix}}_{\text{Zero state response}}$$

Thus, the zero state response in the Laplace domain is

$$\begin{bmatrix} X_1(s)' \\ \\ X_2(s)' \end{bmatrix} = \begin{bmatrix} \dfrac{s+3}{s^2+3s+2} & \dfrac{1}{s^2+3s+2} \\ \dfrac{-2}{s^2+3s+2} & \dfrac{s}{s^2+3s+2} \end{bmatrix} \begin{bmatrix} 0 \\ \\ \dfrac{1}{s^2+1} \end{bmatrix}$$

II. Classical Method

1. The Matrizant

The classical term for transition matrix is matrizant. Deriving a matrizant from a set of first order differential equations is the core of the solution problem in the classical approach. We will first explain the idea by considering some operations on a scalar differential equation:

$$\frac{dx}{dt} = ax, \qquad x = x_0 \qquad \text{when } t = 0$$

a is a constant; x_0, the initial condition for the differential equation.

Direct integration with respect to t gives

$$x(t) = x_0 + \int_0^t ax(t)\, dt \tag{1}$$

We also can write (1) as follows:

$$x(t) = x_0 + \int_0^t ax(\tau_1)\, d\tau_1 \tag{2}$$

in which τ_1 is a subsidiary variable. For a particular value $t = \tau_1$, (2) becomes

$$x(\tau_1) = x_0 + \int_0^{\tau_1} ax(\tau_1)\, d\tau_1 \tag{3}$$

We can change the subsidiary variable again;

$$x(\tau_1) = x_0 + \int_0^{\tau_1} ax(\tau_2)\, d\tau_2 \tag{4}$$

Substituting (4) into (2), we have

$$x(t) = x_0 + \int_0^t ax(\tau_1)\, d\tau_1 = x_0 + \int_0^t a\left[x_0 + \int_0^{\tau_1} ax(\tau_2)\, d\tau_2 \right] d\tau_1 \tag{5}$$

$$= x_0 + x_0 \int_0^t a\, d\tau_1 + \int_0^t a \int_0^{\tau_1} ax(\tau_2)\, d\tau_2\, d\tau_1 \tag{6}$$

But

$$x(\tau_2) = x_0 + \int_0^{\tau_2} ax(\tau_3)\, d\tau_3 \tag{7}$$

Substituting (7) into (6) gives

$$x(t) = x_0 + x_0 \int_0^t a\, d\tau_1 + \int_0^t a \int_0^{\tau_1} a\left[x_0 + \int_0^{\tau_2} ax(\tau_3)\, d\tau_3 \right] d\tau_2\, d\tau_1$$

$$= x_0 + x_0 \int_0^t a\, d\tau_1 + \int_0^t a \int_0^{\tau_1} ax_0\, d\tau_2\, d\tau_1$$

$$+ \int_0^t a \int_0^{\tau_1} a \int_0^{\tau_2} ax(\tau_3)\, d\tau_3\, d\tau_2\, d\tau_1 \tag{8}$$

If a new operator Q is introduced,

$$Q = \int_0^t (\ )\, d\tau$$

then (2) may be written in the following form:

$$x = x_0 + Q(x)$$

and (8) becomes

$$x(t) = [1 + Q(x) + Q(x)Q(x) + Q(x)Q(x)Q(x) + \cdots]x_0$$

Example 1. Solve the differential equation

$$\frac{dx}{dt} = 2x, \qquad x = x_0 \text{ when } t = 0.$$

Solution:

$$dx = 2x \, dt$$

$$x = x_0 + \int_0^t 2x \, dt$$

$$= x_0 + x_0 \int_0^t 2 \, d\tau_1 + x_0 \int_0^t 2 \int_0^{\tau_1} 2 \, d\tau_2 \, d\tau_1 + \cdots$$

$$= \left(1 + 2t + \frac{4}{2!} + 2 + \frac{8}{3!}t^3 + \cdots\right)x_0 = x_0 e^{2t}$$

The extension of this technique to a matrix differential equation is straightforward:

$$\frac{d}{dx}[x] = [A][x], \qquad [x(0)] = [x_0] \tag{9}$$

Direct integration gives

$$[x] = [x_0] + \int_0^t [A][x] \, d\tau$$

We introduce the Q operator again

$$Q = \int_0^t (\ ) \, d\tau$$

and then develop a formula for solving (9).

$$[x] = ([I] + Q[A] + Q[A]Q[A] + Q[A]Q[A]Q[A] + \cdots) [x_0]$$

We usually can use the expression,

$$[\Phi(t)] = [I] + Q[A] + Q[A]Q[A] + \cdots \tag{10}$$

the matrizant of $[A]$, which is the transition matrix.

Example 2. The state equation of the system of Fig. 7-3 is

$$\begin{bmatrix} \dot{x}_1 \\ \dot{x}_2 \end{bmatrix} = \begin{bmatrix} 0 & 1 \\ -1 & 0 \end{bmatrix} \begin{bmatrix} x_1 \\ x_2 \end{bmatrix}$$

Then

$$[A] = \begin{bmatrix} 0 & 1 \\ -1 & 0 \end{bmatrix}$$

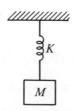

Fig. 7-3. Mechanical example corresponding to Eq. (10).

and

$$Q[A] = \begin{bmatrix} 0 & t \\ -t & 0 \end{bmatrix}$$

$$[A]Q[A] = \begin{bmatrix} 0 & 1 \\ -1 & 0 \end{bmatrix}\begin{bmatrix} 0 & t \\ -t & 0 \end{bmatrix} = \begin{bmatrix} -t & 0 \\ 0 & -t \end{bmatrix}$$

$$Q[A]Q[A] = \begin{bmatrix} \dfrac{-t^2}{2} & 0 \\ 0 & \dfrac{-t^2}{2} \end{bmatrix}$$

$$[A]Q[A]Q[A] = \begin{bmatrix} 0 & 1 \\ -1 & 0 \end{bmatrix}\begin{bmatrix} \dfrac{-t^2}{2} & 0 \\ 0 & \dfrac{-t^2}{2} \end{bmatrix} = \begin{bmatrix} 0 & \dfrac{-t^2}{2} \\ \dfrac{t^2}{2} & 0 \end{bmatrix}$$

$$Q[A]Q[A]Q[A] = \begin{bmatrix} 0 & \dfrac{1}{2\times 3}t^3 \\ \dfrac{1}{2\times 3}t^3 & 0 \end{bmatrix}$$

$$[\Phi(t)] = [I] + \begin{bmatrix} 0 & t \\ -t & 0 \end{bmatrix} + \begin{bmatrix} \dfrac{-t^2}{2} & 0 \\ 0 & \dfrac{-t^2}{2} \end{bmatrix} + \begin{bmatrix} 0 & \dfrac{t^3}{2\times 3} \\ \dfrac{t^2}{2\times 3} & 0 \end{bmatrix} + \cdots$$

$$= \begin{bmatrix} 1 & 0 \\ 0 & 1 \end{bmatrix} + \begin{bmatrix} 0 & t \\ -t & 0 \end{bmatrix} + \begin{bmatrix} \dfrac{-t^2}{2} & 0 \\ 0 & \dfrac{-t^2}{2} \end{bmatrix} + \begin{bmatrix} 0 & \dfrac{t^3}{2\times 3} \\ \dfrac{t^2}{2\times 3} & 0 \end{bmatrix} + \cdots$$

$$= \begin{bmatrix} 1 + 0 - \dfrac{t^2}{2} + 0 + \cdots & t - \dfrac{t^3}{2\times 3} + \cdots \\ -t + \dfrac{t^3}{2\times 3} + \cdots & 1 - \dfrac{t^2}{2} + \cdots \end{bmatrix}$$

$$= \begin{bmatrix} 1 - \dfrac{t^2}{2!} + \cdots & t - \dfrac{t^3}{3!} + \cdots \\ -t + \dfrac{t^3}{3!} + \cdots & 1 - \dfrac{t^2}{2!} + \cdots \end{bmatrix}$$

$$= \begin{bmatrix} \cos t & \sin t \\ -\sin t & \cos t \end{bmatrix}$$

2. Another Point of View

Let us consider the scalar equation (1) again.

$$\dot{x} = ax \tag{1}$$

The solution of (1) is the type e^{at}. Then what is the solution of the following?

$$[\dot{\Phi}(t)] = [A][\Phi(t)], \qquad [\Phi(0)] = [I] \tag{11}$$

Once we have the solution of (11), we have $[\Phi(t)]$; and when we know $[\Phi(t)]$, (9) is solved. This is because the solution of (9) is $[\Phi(t)][x(0)]$.

From the solution of (1), we naturally think that the solution of (11) may be assumed as

$$e^{[A]t} \text{ type} \tag{12}$$

In the special case in which the elements of the matrix $[A]$ are constants,

$$[\Phi(t)] = [I] + [A]\frac{t}{1!} + [A]^2\frac{t^2}{2!} + [A]^3\frac{t^3}{3!} + \cdots$$

$$= e^{[A]t} \tag{13}$$

$$= \sum_{n=0}^{\infty} \frac{1}{n!}[A]^n t^n$$

Therefore, the solution of (9) may be expressed in the convenient form,

$$[x] = e^{[A]t}[x_0] \tag{14}$$

3. Variation of Parameters

The method shown in last section is for zero input response. We naturally consider next how to find the zero state response by a classical method. This method is due to Lagrange.

Consider the following matrix equation.

$$\frac{d}{dt}[x] = [A(t)][x] + [b(t)], \qquad [x(0)] = [x_0] \tag{15}$$

The corresponding homogeneous equation or the reduced equation of (15) is

$$\frac{d}{dt}[x] = [A(t)][x] \tag{16}$$

where $[x(0)] = [x_0]$.

Prior to solving (16), the following equation will be studied first.

$$\frac{d}{dt}[\Phi(t)] = [A(t)][\Phi(t)], \qquad [\Phi(0)] = [I] \tag{17}$$

Let

$$[x] = [\Phi(t)][u] \tag{18}$$

where $[u]$ is a variable matrix. (If we recall Lagrange's variation method in Chapter 2, this $[u]$ corresponds to u in that approach.)

Take the derivative of both sides.

$$\frac{d[x]}{dt} = [\Phi(t)]\frac{d[u]}{dt} + \frac{d[\Phi(t)]}{dt}[u]$$

$$= [\Phi(t)]\frac{d[u]}{dt} + [A(t)][\Phi(t)][u] \tag{19}$$

Then establish the following relationship:

$$[\Phi(t)]\frac{d[u]}{dt} = [b(t)]$$

$$\frac{d[u]}{dt} = [\Phi(t)]^{-1}[b(t)] \tag{20}$$

$$[u] = \int_0^t [\Phi(t)]^{-1}[b(t)] \, dt + [u(0)]$$

But

$$[u(0)] = [\Phi(0)][u(0)] = [x(0)] = [x_0] \tag{21}$$

Therefore,

$$[u] = \int_0^t [\Phi(\tau)]^{-1}[b(\tau)] \, d\tau + [x_0]$$

Substitution of the above formula into (18) gives the solution

$$[x(t)] = [\Phi(t)][x_0] + \int_0^t [\Phi(t)][\Phi(\tau)]^{-1}[b(\tau)] \, d\tau \tag{22}$$

Thus, we see that the last term of (22) is the matrix form of a convolution integral.

Example. The state equation for the system shown in Fig. 7-4 is

$$\begin{bmatrix} \dot{x}_1 \\ \dot{x}_2 \end{bmatrix} = \begin{bmatrix} 0 & 1 \\ -1 & 0 \end{bmatrix}\begin{bmatrix} x_1 \\ x_2 \end{bmatrix} + \begin{bmatrix} 0 \\ 10 \end{bmatrix} \tag{23}$$

$$\begin{bmatrix} x_1(0) \\ x_2(0) \end{bmatrix} = \begin{bmatrix} 0 \\ 0 \end{bmatrix} \tag{24}$$

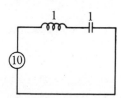

Fig. 7-4. Electrical example corresponding to Eq. (26).

Before trying to solve (23), consider the following equation.

$$\frac{d}{dt}[\Phi(t)] = [A(t)][\Phi(t)], \qquad [\Phi(0)] = [I]$$

The transition matrix can be formed by any method shown in the last section.

$$[\Phi(t)] = \begin{bmatrix} \cos t & \sin t \\ -\sin t & \cos t \end{bmatrix} \tag{25}$$

Substituting (24) and (25) into (22) gives

$$[x(t)] = [\Phi(t)][x_0] + \int_0^t [\Phi(t)][\Phi(\tau)]^{-1}[b(\tau)]\, d\tau$$

$$= \begin{bmatrix} \cos t & \sin t \\ -\sin t & \cos t \end{bmatrix}\begin{bmatrix} 0 \\ 0 \end{bmatrix} + \int_0^t \begin{bmatrix} \cos t & \sin t \\ -\sin t & \cos t \end{bmatrix}\begin{bmatrix} \cos \tau & \sin \tau \\ -\sin \tau & \cos \tau \end{bmatrix}\begin{bmatrix} 0 \\ 10 \end{bmatrix} d\tau$$

$$= \begin{bmatrix} 0 \\ 0 \end{bmatrix} + \int_0^t \begin{bmatrix} \cos (t - \tau) & \sin (t - \tau) \\ -\sin (t - \tau) & \cos (t - \tau) \end{bmatrix}\begin{bmatrix} 0 \\ 10 \end{bmatrix} d\tau$$

$$\begin{bmatrix} x_1(t) \\ x_2(t) \end{bmatrix} = \begin{bmatrix} 10 - 10 \cos t \\ 10 \sin t \end{bmatrix}$$

which is our expected solution, or the zero state response of the system.

III.   Linear Transformation Method

1. Primer System

Given three independent circuits, as shown in Fig. 7-5, how does one describe them together using a matrix differential equation? One has the following equations.

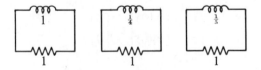

Fig. 7-5. Three independent subcircuits.

$$\dot{x}_1 = -x_1$$
$$\dot{x}_2 = -4x_2 \tag{1}$$
$$\dot{x}_3 = -5x_3$$

where x_1, x_2, and x_3 express i_1, i_2, and i_3, respectively.

The matrix form of (1) is

$$\begin{bmatrix} \dot{x}_1 \\ \dot{x}_2 \\ \dot{x}_3 \end{bmatrix} = \begin{bmatrix} -1 & 0 & 0 \\ 0 & -4 & 0 \\ 0 & 0 & -5 \end{bmatrix}\begin{bmatrix} x_1 \\ x_2 \\ x_3 \end{bmatrix} \tag{1a}$$

The constant matrix

$$[A] = \begin{bmatrix} -1 & & \\ & -4 & \\ & & -5 \end{bmatrix}$$

has values in the diagonal only, and the solution is directly obtained by inspection.

$$\begin{bmatrix} x_1 \\ x_2 \\ x_3 \end{bmatrix} = \begin{bmatrix} e^{-t} & & \\ & e^{-4t} & \\ & & e^{-5t} \end{bmatrix} \begin{bmatrix} x_1(0) \\ x_2(0) \\ x_3(0) \end{bmatrix} \tag{2}$$

The transition matrix is also a diagonal one. From this fact we can see that a system formed by three independent circuits is much easier to solve and gives a closed form transition matrix immediately.

A system which is merely a collection of independent subsystems in which the subsystems do not influence each other is called a primer system.

We then want to know if we can change the general system into an equivalent primer system. If so, this will save time and trouble.

2. Diagonalization

Changing a general system into a primer system is an engineer's technique. Mathematicians have a corresponding procedure called diagonalization of a matrix.

Consider the following system:

$$\begin{bmatrix} \dot{x}_1 \\ \dot{x}_2 \\ \dot{x}_3 \end{bmatrix} = \begin{bmatrix} 0 & 1 & 0 \\ 0 & 0 & 1 \\ -6 & -11 & -6 \end{bmatrix} \begin{bmatrix} x_1 \\ x_2 \\ x_3 \end{bmatrix} \tag{3}$$

The square matrix of (3) is not a diagonalized form. We shall obtain the primer system by the following procedure. Let

$$\begin{bmatrix} x_1 \\ x_2 \\ x_3 \end{bmatrix} = \begin{bmatrix} 1 & 1 & 1 \\ \lambda_1 & \lambda_2 & \lambda_3 \\ \lambda_1^2 & \lambda_2^2 & \lambda_3^2 \end{bmatrix} \begin{bmatrix} y_1 \\ y_2 \\ y_3 \end{bmatrix} \tag{4}$$

where λ_1, λ_2, λ_3 are the characteristic roots of the system. It is of no concern at this time why we choose to use (4). Suffice it to say that it comes from the inspiration of Vandermonde. Substituting (4) into (3), we have

$$\begin{bmatrix} 1 & 1 & 1 \\ -1 & -2 & -3 \\ 1 & 4 & 9 \end{bmatrix} \begin{bmatrix} \dot{y}_1 \\ \dot{y}_2 \\ \dot{y}_3 \end{bmatrix} = \begin{bmatrix} 0 & 1 & 0 \\ 0 & 0 & 1 \\ -6 & -11 & -6 \end{bmatrix} \begin{bmatrix} 1 & 1 & 1 \\ -1 & -2 & -3 \\ 1 & 4 & 9 \end{bmatrix} \begin{bmatrix} y_1 \\ y_2 \\ y_3 \end{bmatrix}$$

or

$$\begin{bmatrix} \dot{y}_1 \\ \dot{y}_2 \\ \dot{y}_3 \end{bmatrix} = \begin{bmatrix} 1 & 1 & 1 \\ -1 & -2 & -3 \\ 1 & 4 & 9 \end{bmatrix}^{-1} \begin{bmatrix} 0 & 1 & 0 \\ 0 & 0 & 1 \\ -6 & -11 & -6 \end{bmatrix} \begin{bmatrix} 1 & 1 & 1 \\ -1 & -2 & -3 \\ 1 & 4 & 9 \end{bmatrix} \begin{bmatrix} y_1 \\ y_2 \\ y_3 \end{bmatrix}$$

or

$$\begin{bmatrix} \dot{y}_1 \\ \dot{y}_2 \\ \dot{y}_3 \end{bmatrix} = \begin{bmatrix} -3 & & \\ & -2 & \\ & & -1 \end{bmatrix} \begin{bmatrix} y_1 \\ y_2 \\ y_3 \end{bmatrix} \tag{5}$$

Equation (5) can describe the system well as (3) can, but (5) is much easier to solve.

Solving (5) by inspection, we have

$$\begin{bmatrix} y_1 \\ y_2 \\ y_3 \end{bmatrix} = \begin{bmatrix} e^{-3t} & & \\ & e^{-2t} & \\ & & e^{-t} \end{bmatrix} \begin{bmatrix} y_1(0) \\ y_2(0) \\ y_3(0) \end{bmatrix}$$

Then, substituting the result into (4), we obtain

$$\begin{bmatrix} y_1(0) \\ y_2(0) \\ y_3(0) \end{bmatrix} = \begin{bmatrix} 1 & 1 & 1 \\ -1 & -2 & -3 \\ 1 & 4 & 9 \end{bmatrix}^{-1} \begin{bmatrix} x_1(0) \\ x_2(0) \\ x_3(0) \end{bmatrix}$$

and

$$\begin{bmatrix} x_1 \\ x_2 \\ x_3 \end{bmatrix} = \begin{bmatrix} 1 & 1 & 1 \\ -1 & -2 & -3 \\ 1 & 4 & 9 \end{bmatrix} \begin{bmatrix} e^{-3t} & & \\ & e^{-2t} & \\ & & e^{-t} \end{bmatrix} \begin{bmatrix} 1 & 1 & 1 \\ -1 & -2 & -3 \\ 1 & 4 & 9 \end{bmatrix}^{-1} \begin{bmatrix} x_1(0) \\ x_2(0) \\ x_3(0) \end{bmatrix}$$

This is the solution.

In general, a given system

$$[\dot{x}] = [A][x] + [b] \tag{6}$$

can be transformed into

$$[\dot{y}] = [P]^{-1}[A][P][y] + [P]^{-1}[b] \tag{7}$$

where $[P]^{-1}[A][P]$ is the diagonalized matrix, $[P]$ is Vandermonde's matrix, and $[A]$ is Bush's form. Denote

$$[P]^{-1}[A][P] = [\Lambda] \tag{8}$$

$[A]$ and $[\Lambda]$ are said to be similar.

Two similar matrices have the following properties:
1. Their characteristic equations are the same.
2. Their traces are the same.
3. Their determinants, formed by the elements of the matrices, are the same.

3. Eigenvalues

A more general method for diagonalizing a matrix is by the use of eigenvectors. Before one can learn about them, it is necessary to know the eigenvalue problem. It is known that the premultiplication of the vector $[x]$ by the matrix $[A]$ generates a new vector $[z]$ so that

$$[A][x] = [z] \tag{9}$$

The vector $[z]$ can be considered a transformation of the original vector $[x]$. If $[x]$ and $[z]$ have the same direction (which is possible, of course), then

$$[A][x] = \lambda[x] \tag{10}$$

The scalar multiplier λ must now be determined.

Equation (10) is rewritten into the regular algebraic form,

$$
\begin{aligned}
a_{11}x_1 + a_{12}x_2 + \cdots + a_{1n}x_n &= \lambda x_1 \\
a_{21}x_1 + a_{22}x_2 + \cdots + a_{2n}x_n &= \lambda x_2 \\
& \vdots \\
a_{n1}x_1 + a_{n2}x_2 + \cdots + a_{nn}x_n &= \lambda x_n
\end{aligned}
\tag{10a}
$$

or

$$
\begin{aligned}
(a_{11} - \lambda)x_1 + a_{12}x_2 + \cdots + a_{1n}x_n &= 0 \\
a_{21}x_1 + (a_{22} - \lambda)x_2 + \cdots + a_{2n}x_n &= 0 \\
& \vdots \\
a_{n1}x_1 + a_{n2}x_2 + \cdots + (a_{nn} - \lambda)x_n &= 0
\end{aligned}
\tag{10b}
$$

or

$$\{[A] - \lambda[I]\}[x] = [0] \tag{10c}$$

Equation (10b) has no solution unless the determinant of the system is zero, so

$$\det([A] - \lambda[I]) = 0 \tag{11}$$

Equation (11) is a polynomial of λ, and is called the characteristic equation of the system. We express it by $C(\lambda) = 0$.

The roots $\lambda_1, \lambda_2, \lambda_3, \ldots, \lambda_n$ of the characteristic equation $C(\lambda) = 0$ are called eigenvalues of the matrix $[A]$.

4. Eigenvectors

For every possible value of $\lambda = \lambda_i$, $i = 1, 2, \ldots, n$, a solution of (10b) can be found. Let $[x] = [x]_i$ be the vector associated with $\lambda = \lambda_i$; one may then write

$$([A] - \lambda_i[I])[x]_i = [0] \qquad i = 1, 2, \ldots \quad (12)$$

The vectors $[x]_i$ are called the eigenvectors of $[A]$.

Let a square matrix $[M]$ be constructed from the eigenvector columns $[M]_i$ in the following manner:

$$[M] = ([M]_1, [M]_2, [M]_3, \ldots, [M]_n) \qquad (13)$$

$[M]$ is called the modal matrix of $[A]$. It is easy to show that (12) may be written as one equation, or

$$[M]^{-1}[A][M] = [\Lambda] \qquad (14)$$

Using the eigenvector method to diagonalize a square matrix is illustrated by the following example.

5. An Example

Diagonalize the following matrix:

$$[A] = \begin{bmatrix} 0 & 1 & 0 \\ 0 & 0 & 1 \\ -6 & -11 & -6 \end{bmatrix} \qquad (15)$$

The first step is to find the eigenvalues

$$\det ([A] - \lambda[I]) = 0$$

or

$$\lambda^3 + 6\lambda^2 + 11\lambda + 6 = 0$$

$$\lambda = -1, -2, \text{ and } -3$$

The eigenvector associated with $\lambda = -1$ can be obtained from (12).

$$\left\{ \begin{bmatrix} 0 & 1 & 0 \\ 0 & 0 & 1 \\ -6 & -11 & -6 \end{bmatrix} - (-1) \begin{bmatrix} 1 & & \\ & 1 & \\ & & 1 \end{bmatrix} \right\} \begin{bmatrix} m_{11} \\ m_{21} \\ m_{31} \end{bmatrix} = \begin{bmatrix} 0 \\ 0 \\ 0 \end{bmatrix}$$

or

$$\begin{bmatrix} 1 & 1 & 0 \\ 0 & 1 & 1 \\ -6 & -11 & -5 \end{bmatrix} \begin{bmatrix} m_{11} \\ m_{21} \\ m_{31} \end{bmatrix} = \begin{bmatrix} 0 \\ 0 \\ 0 \end{bmatrix} \qquad (16)$$

where m_{i1} is the vector to be evaluated.

Equation (16) can be written into three equations:

$$m_{11} + m_{21} = 0$$

$$m_{21} + m_{31} = 0 \qquad (17)$$

$$-6m_{11} - 11m_{21} - 5m_{31} = 0$$

If $[M]_i$ is regarded as a vector, its direction, but not its magnitude, is uniquely determined. If we assume

$$m_{11} = 1$$

then

$$m_{21} = -1$$

and

$$m_{31} = 1$$

The first eigenvector is

$$[M]_1 = \begin{bmatrix} 1 \\ -1 \\ 1 \end{bmatrix} \qquad (18)$$

The second eigenvector, which is associated with $\lambda = -2$, is obtained by

$$\begin{bmatrix} 2 & 1 & 0 \\ 0 & 2 & 1 \\ -6 & -11 & -4 \end{bmatrix} \begin{bmatrix} m_{12} \\ m_{22} \\ m_{32} \end{bmatrix} = \begin{bmatrix} 0 \\ 0 \\ 0 \end{bmatrix}$$

$$2m_{12} + m_{22} = 0$$

$$2m_{22} + m_{32} = 0$$

$$-6m_{12} - 11m_{22} - 4m_{32} = 0$$

If we assume

$$m_{12} = 1$$

then

$$m_{22} = -2$$

and

$$m_{32} = +4$$

The second eigenvector is then

$$[M]_2 = \begin{bmatrix} 1 \\ -2 \\ 4 \end{bmatrix} \qquad (19)$$

Similarly, the third one is

$$[M]_3 = \begin{bmatrix} m_{13} \\ m_{23} \\ m_{33} \end{bmatrix} = \begin{bmatrix} 1 \\ -3 \\ 9 \end{bmatrix} \qquad (20)$$

Collecting (18), (19), and (20) together, we have

$$[M] = \begin{bmatrix} 1 & 1 & 1 \\ -1 & -2 & -3 \\ 1 & 4 & 9 \end{bmatrix} \qquad (21)$$

$$[M]^{-1}[A][M] = \begin{bmatrix} 3 & \frac{5}{2} & \frac{1}{2} \\ -3 & 4 & -1 \\ 1 & \frac{3}{2} & \frac{1}{2} \end{bmatrix} \begin{bmatrix} 0 & 1 & 0 \\ 0 & 0 & 1 \\ -6 & -11 & -6 \end{bmatrix} \begin{bmatrix} 1 & 1 & 1 \\ -1 & -2 & -3 \\ 1 & 4 & 9 \end{bmatrix}$$

$$[\Lambda] = \begin{bmatrix} -3 & & \\ & -2 & \\ & & -1 \end{bmatrix}$$

It is noted that the Vandermonde matrix can be derived from the eigenvector approach. However, the latter is a general method.

The Vandermonde matrix for n distinct roots can be derived from eigenvectors and is written as follows:

$$[M] = \begin{bmatrix} 1 & 1 & \cdot & \cdot & 1 \\ \lambda_1 & \lambda_2 & \cdot & \cdot & \lambda_n \\ \lambda_1^2 & \lambda_2^2 & \cdot & \cdot & \lambda_n^2 \\ \cdot & \cdot & \cdot & \cdot & \cdot \\ \cdot & \cdot & \cdot & \cdot & \cdot \\ \lambda_1^{n-1} & \lambda_2^{n-1} & \cdot & \cdot & \lambda_n^{n-1} \end{bmatrix} \tag{22}$$

6. Example with Repeated Roots

If the roots of the characteristic equation of a Bush's matrix are not distinct, it cannot be diagonalized. But a Jordan matrix can be obtained by using a modified Vandermonde matrix:

$$[J] = [W]^{-1}[A][W] \tag{23}$$

in which $[J]$ is the Jordan form, $[A]$ is the Bush form, and $[W]$ is the modified Vandermonde matrix for repeated roots.

If the characteristic equation of $[A]$ has an m-multiple root λ_1 and a single root λ_2, the modified Vandermonde matrix reads:

$$[W] =$$

$$\begin{bmatrix} 1 & 0 & 0 & 0 & 0 & 1 \\ \lambda\big|_{\lambda=\lambda_1} & \frac{\partial\lambda}{\partial\lambda}\big|_{\lambda=\lambda_1} & 0 & 0 & 0 & \lambda_2 \\ \lambda^2\big|_{\lambda=\lambda_1} & \frac{\partial\lambda^2}{\partial\lambda}\big|_{\lambda=\lambda_1} & \frac{1}{2!}\frac{\partial^2(\lambda^2)}{\partial\lambda^2}\big|_{\lambda=\lambda_1} & \cdot & \cdot & \lambda_2^2 \\ \cdot & \cdot & \cdot & & & \cdot \\ \cdot & \cdot & \cdot & & & \cdot \\ \cdot & \cdot & \cdot & & & \cdot \\ \lambda^{n-1}\big|_{\lambda=\lambda_1} & \frac{\partial\lambda^{n-1}}{\partial\lambda}\big|_{\lambda=\lambda_1} & \frac{1}{2!}\frac{\partial^2(\lambda^{n-1})}{\partial\lambda^2}\big|_{\lambda=\lambda_1} & \cdot & \frac{1}{(m-1)!}\frac{\partial^{m-1}(\lambda^{n-1})}{\partial\lambda^{m-1}}\big|_{\lambda=\lambda_1} & \lambda_2^{n-1} \end{bmatrix}$$

$$\tag{24}$$

The transform procedure can best be illustrated by an example. For the following matrix $[A]$, find its Jordan form.

$$[A] = \begin{bmatrix} 0 & 1 & 0 & 0 \\ 0 & 0 & 1 & 0 \\ 0 & 0 & 0 & 1 \\ -2 & -7 & -9 & -5 \end{bmatrix} \tag{25}$$

The characteristic equation for the matrix is found by

$$\det |[A] - \lambda[I]| = 0$$
$$\lambda^4 + 5\lambda^3 + 9\lambda^2 + 7\lambda + 2 = 0$$

or

$$(\lambda + 1)^3(\lambda + 2) = 0 \tag{26}$$

We have a triple root, $\lambda_1 = -1$, and a single root, $\lambda_2 = -2$.

Then the Vandermonde matrix is formed by substituting (26) into (24).

$$[W] = \begin{bmatrix} 1 & 0 & 0 & 1 \\ -1 & 1 & 0 & -2 \\ 1 & -2 & 1 & 4 \\ -1 & 3 & -3 & -8 \end{bmatrix} \tag{27}$$

The Jordan form is obtained by substituting (27) into (23).

$$[J] = \begin{bmatrix} 1 & 0 & 0 & 1 \\ -1 & 1 & 0 & -2 \\ 1 & -2 & 1 & 4 \\ -1 & 3 & -3 & -8 \end{bmatrix}^{-1} \begin{bmatrix} 0 & 1 & 0 & 0 \\ 0 & 0 & 1 & 0 \\ 0 & 0 & 0 & 1 \\ -2 & -7 & -9 & -5 \end{bmatrix} \begin{bmatrix} 1 & 0 & 0 & 1 \\ -1 & 1 & 0 & -2 \\ 1 & -2 & 1 & 4 \\ -1 & 3 & -3 & -8 \end{bmatrix}$$

$$= \begin{bmatrix} -1 & 1 & 0 & 0 \\ 0 & -1 & 1 & 0 \\ 0 & 0 & -1 & 0 \\ 0 & 0 & 0 & -2 \end{bmatrix} \tag{28}$$

An arbitrary constant matrix can always be reduced to a Bush's form through finding its characteristic equation. Therefore, the Vandermonde matrices for a system with distinct roots and for a system with multiple roots can always be applied in the linear transformation. Once the system in question is diagonalized or is changed into a Jordan form, the solution of a differential equation can be obtained almost by inspection.

7. Inverse Laplace Transformation of a Transfer Function

The power of the linear transformation method can best be illustrated by obtaining the inverse Laplace transformation of a transfer function.

Consider the following transfer function.

$$X(s) = \frac{b_1 s^{n-1} + b_2 s^{n-2} + \cdots + b_n}{s^n + a_1 s^{n-1} + a_2 s^{n-2} + \cdots + a_n} \tag{29}$$

Its inverse Laplace transformation is desired.

Cross multiplication gives

$$\{s^n + a_1 s^{n-1} + a_2 s^{n-2} + \cdots + a_n\} X(s) = b_1 s^{n-1} + b_2 s^{n-2} + \cdots + b_n \tag{30}$$

Writing the right-hand side of Eq. 30 into the inner product form, we have

$$\{s^n + a_1 s^{n-1} + a_2 s^{n-2} + \cdots + a_n\} X(s) = [s^{n-1}, s^{n-2}, \ldots, s, 1] \begin{bmatrix} b_1 \\ b_2 \\ \cdot \\ \cdot \\ \cdot \\ b_n \end{bmatrix} \tag{31}$$

From the differential equation point of view, we can consider the transfer function (29) as being generated from a set of initial conditions

$$\begin{bmatrix} x(0) \\ x'(0) \\ x''(0) \\ \cdot \\ \cdot \\ \cdot \\ x^{(n-1)}(0) \end{bmatrix} \tag{32}$$

and the following differential equation,

$$x^{(n)} + a_1 x^{(n-1)} + a_2 x^{(n-2)} + \cdots + a_{n-1} x' + a_n x = 0 \tag{33}$$

Performing the Laplace transformation, we have

$$\{s^n + a_1 s^{n-1} + a_2 s^{n-2} + \cdots + a_n\} X(s)$$
$$= \{s^{n-1} x(0) + s^{n-2} x'(0) + \cdots + x^{(n-1)}(0)\}$$
$$+ a_1 \{s^{n-2} x(0) + \cdots + x^{(n-2)}(0)\}$$
$$+ a_2 \{s^{n-3} x(0) + \cdots + x^{(n-3)}(0)\}$$
$$+ \cdots$$
$$+ a_{n-1} \{x(0)\} \tag{34}$$

The right-hand side of (34) can be written as a matrix product and becomes

$$\{s^n + a_1 s^{n-1} + a_2 s^{n-2} + \cdots + a_n\} X(s)$$
$$= [s^{n-1}, s^{n-2}, \ldots, s, 1] \begin{bmatrix} 1 & 0 & 0 & \cdot & \cdot & 0 \\ a_1 & 1 & 0 & \cdot & \cdot & 0 \\ a_2 & a_1 & 1 & 0 & \cdot & 0 \\ \cdot & \cdot & \cdot & \cdot & \cdot & \cdot \\ \cdot & & \cdot & a_1 & 1 & \cdot \\ a_{n-1} & a_{n-2} & \cdot & \cdot & a_1 & 1 \end{bmatrix} \begin{bmatrix} x(0) \\ x'(0) \\ x''(0) \\ \cdot \\ \cdot \\ x^{(n-1)}(0) \end{bmatrix} \tag{35}$$

The left-hand sides of (31) and (35) are identical; therefore, the right-hand side of both equations must be equal. Thus, we obtain

$$
\begin{bmatrix} b_1 \\ b_2 \\ \cdot \\ \cdot \\ \cdot \\ b_n \end{bmatrix} =
\begin{bmatrix}
1 & 0 & 0 & \cdot & \cdot & 0 \\
a_1 & 1 & 0 & \cdot & \cdot & 0 \\
a_2 & a_1 & 1 & \cdot & \cdot & 0 \\
\cdot & \cdot & \cdot & \cdot & \cdot & \cdot \\
\cdot & \cdot & \cdot & \cdot & \cdot & \cdot \\
a_{n-1} & a_{n-2} & \cdot & \cdot & a_1 & 1
\end{bmatrix}
\begin{bmatrix} x(0) \\ x'(0) \\ x''(0) \\ \cdot \\ \cdot \\ x^{(n-1)}(0) \end{bmatrix}
\tag{36}
$$

or, alternatively,

$$
\begin{bmatrix} x(0) \\ x'(0) \\ x''(0) \\ \cdot \\ \cdot \\ x^{n-1}(0) \end{bmatrix} =
\begin{bmatrix}
1 & 0 & 0 & \cdot & \cdot & 0 \\
a_1 & 1 & 0 & \cdot & \cdot & 0 \\
a_2 & a_1 & 1 & \cdot & \cdot & 0 \\
\cdot & \cdot & \cdot & \cdot & \cdot & \cdot \\
\cdot & \cdot & \cdot & \cdot & \cdot & \cdot \\
a_{n-1} & a_{n-2} & \cdot & \cdot & a_1 & 1
\end{bmatrix}^{-1}
\begin{bmatrix} b_1 \\ b_2 \\ \cdot \\ \cdot \\ \cdot \\ b_n \end{bmatrix}
\tag{37}
$$

When the denominator coefficients (a's) and the numerator coefficients (b's) are given, the set of initial conditions $x(0)$, $x'(0)$, ... is uniquely defined. The a and b matrices can be called the denominator matrix and the numerator matrix, respectively. The set of initial conditions is usually called the initial vector.

The transfer function [Eq. (29)] could be interpreted in many ways, one of which is to consider it as being generated from a differential equation

$$
x^{(n)} + a_1 x^{(n-1)} + a_2 x^{(n-2)} + \cdots + a_{n-1} x' + a_n x = 0 \tag{38}
$$

with a set of initial conditions.

It is well known that Eq. (38) can be written as

$$
[\dot{X}] = [A][X] \tag{39}
$$

where

$$
[A] =
\begin{bmatrix}
0 & 1 & 0 & \cdot & \cdot & 0 \\
0 & 0 & 1 & \cdot & \cdot & 0 \\
0 & 0 & 0 & \cdot & \cdot & 0 \\
\cdot & \cdot & \cdot & \cdot & \cdot & \cdot \\
\cdot & \cdot & \cdot & \cdot & \cdot & 1 \\
-a_n & -a_{n-1} & \cdot & \cdot & -a_2 & -a_1
\end{bmatrix}
$$

Laplace transformation of Eq. (39) gives

$$
[X(s)] = [sI - A]^{-1}[x(0)] \tag{40}
$$

Let

$$[X] = [V][Y] \tag{41}$$

where

$$[V] = \begin{bmatrix} 1 & 1 & \cdot & \cdot & \cdot & 1 \\ \lambda_1 & \lambda_2 & \cdot & \cdot & \cdot & \lambda_n \\ \lambda_1^2 & \lambda_2^2 & \cdot & \cdot & \cdot & \lambda_n^2 \\ \cdot & \cdot & \cdot & \cdot & & \cdot \\ \cdot & \cdot & \cdot & \cdot & & \cdot \\ \lambda_1^{n-1} & \lambda_2^{n-1} & \cdot & \cdot & \cdot & \lambda_n^{n-1} \end{bmatrix} \tag{42}$$

in which $\lambda_1, \lambda_2, \ldots, \lambda_n$ are the roots of the characteristic equation of Eq. (38). Equation (42) is the Vandermonde matrix and Y variables are canonical variables. Substituting Eq. (41) into Eq. (39), we have

$$[V][\dot{Y}] = [A][V][Y]$$

or

$$[\dot{Y}] = [V]^{-1}[A][V][Y] \tag{43}$$

The solution of Eq. (43) would be

$$[Y(s)] = [sI - V^{-1}AV]^{-1}[y(0)] \tag{44}$$

in which

$$V^{-1}AV = \Lambda$$

is a diagonal matrix or

$$\Lambda = \begin{bmatrix} \lambda_1 & & & & \\ & \lambda_2 & & & \\ & & \cdot & & \\ & & & \cdot & \\ & & & & \cdot \\ & & & & & \lambda_n \end{bmatrix}$$

The transition matrix $[sI - \Lambda]^{-1}$ may be directly written by inspection, and its inverse is obvious. Instead of using Y coordinates, we return to X coordinates. Equation (44) then becomes

$$[X(s)] = [V][sI - \Lambda]^{-1}[V]^{-1}[x(0)] \tag{45}$$

where $[x(0)]$ is our initial vector and $[X(s)]$ is our Laplace transform of the phase variables. Its first component corresponds to the given transfer function.

If Eq. (45) is expanded, the first row equation would be the Heaviside expansion expression of the given transfer function.

Combining Eqs. (37) and (45), we obtain the general formula

$$
\begin{bmatrix} X_1(s) \\ X_2(s) \\ X_3(s) \\ \cdot \\ X_n(s) \end{bmatrix} =
\begin{bmatrix}
1 & 1 & \cdot & \cdot & 1 \\
\lambda_1 & \lambda_2 & \cdot & \cdot & \lambda_n \\
\lambda_1^2 & \lambda_2^2 & \cdot & \cdot & \lambda_n^2 \\
\cdot & \cdot & \cdot & \cdot & \cdot \\
\lambda_1^{n-1} & \lambda_2^{n-1} & \cdot & \cdot & \lambda_n^{n-1}
\end{bmatrix}
\begin{bmatrix}
s - \lambda_1 & 0 & \cdot & \cdot & 0 \\
0 & s - \lambda_2 & 0 & \cdot & 0 \\
0 & & \cdot & \cdot & \cdot \\
& & & \cdot & 0 \\
0 & & \cdot & 0 & s - \lambda_n
\end{bmatrix}^{-1}
$$

$$
\cdot
\begin{bmatrix}
1 & 1 & \cdot & \cdot & 1 \\
\lambda_1 & \lambda_2 & \cdot & \cdot & \lambda_n \\
\lambda_1^2 & \lambda_2^2 & \cdot & \cdot & \lambda_n^2 \\
\cdot & \cdot & \cdot & \cdot & \cdot \\
\lambda_1^{n-1} & \lambda_2^{n-1} & \cdot & \cdot & \lambda_n^{n-1}
\end{bmatrix}^{-1}
\begin{bmatrix}
1 & 0 & \cdot & 0 & 0 \\
a_1 & 1 & \cdot & 0 & 0 \\
a_2 & a_1 & : & 0 & 0 \\
\cdot & a_2 & \cdot & 1 & 0 \\
a_{n-1} & \cdot & \cdot & a_1 & 1
\end{bmatrix}^{-1}
\begin{bmatrix} b_1 \\ b_2 \\ \cdot \\ \cdot \\ b_n \end{bmatrix}
\tag{46}
$$

If the roots of the characteristic equation are not distinct, the general formula is easily modified as follows:

$$
[X(s)] = [W]\{sI - J\}^{-1}[W]^{-1}[x(0)]
\tag{47}
$$

where $[W]$ is the Vandermonde matrix for the general case. For example, if the characteristic equation in question has a multiple root λ_1 of multiplicity m and a root λ_2 of multiplicity 1, the Vandermonde matrix would be (24). And the (J) matrix is the Jordan canonical form.

8. Examples of the Inverse Laplace Transformation by Using the General Matrix Formula

A transfer function is given.

$$
X_1(s) = \frac{s + \beta}{(s + \alpha)(s + \gamma)} = \frac{s + \beta}{s^2 + (\alpha + \gamma)s + \alpha\gamma}
\tag{48}
$$

Its inverse Laplace transform is required. For this problem

$$
[V] = \begin{bmatrix} 1 & 1 \\ -\alpha & -\gamma \end{bmatrix}, \quad
[\Lambda] = \begin{bmatrix} -\alpha & 0 \\ 0 & -\gamma \end{bmatrix}
$$

$$
[x(0)] = \begin{bmatrix} 1 & 0 \\ \alpha + \gamma & 1 \end{bmatrix}^{-1} \begin{bmatrix} 1 \\ \beta \end{bmatrix}
$$

Substituting these matrices into Eq. (45), we obtain

$$
\begin{bmatrix} X_1(s) \\ X_2(s) \end{bmatrix} =
\begin{bmatrix} 1 & 1 \\ -\alpha & -\gamma \end{bmatrix}
\begin{bmatrix} \dfrac{1}{(s + \alpha)} & 0 \\ 0 & \dfrac{1}{(s + \gamma)} \end{bmatrix}
\begin{bmatrix} 1 & 1 \\ -\alpha & -\gamma \end{bmatrix}^{-1}
\begin{bmatrix} 1 & 0 \\ \alpha + \gamma & 0 \end{bmatrix}^{-1}
\begin{bmatrix} 1 \\ \beta \end{bmatrix}
$$

$$\begin{bmatrix} X_1(s) \\ X_2(s) \end{bmatrix} =$$

$$\begin{bmatrix} -\dfrac{\gamma}{\alpha-\gamma}\left(\dfrac{1}{s+\alpha}\right) + \dfrac{\alpha}{\alpha-\gamma}\left(\dfrac{1}{s+\gamma}\right), & \dfrac{1}{s+\alpha}\left(\dfrac{-1}{\alpha-\gamma}\right) + \dfrac{1}{s+\gamma}\left(\dfrac{1}{\alpha-\gamma}\right) \\[2ex] \dfrac{-\gamma}{\alpha-\gamma}\left(\dfrac{-\alpha}{s+\alpha}\right) + \dfrac{\alpha}{\alpha-\gamma}\left(\dfrac{-\gamma}{s+\gamma}\right), & \dfrac{-\alpha}{s+\alpha}\left(\dfrac{-1}{\alpha-\gamma}\right) + \dfrac{-\gamma}{s+\gamma}\left(\dfrac{1}{\alpha-\gamma}\right) \end{bmatrix}$$

$$\cdot \begin{bmatrix} 1 \\ -\alpha-\gamma+\beta \end{bmatrix} \qquad (49)$$

If the matrix equation is expanded into its two scalar equations, the first one would be the Heaviside expansion of Eq. (48), from which the inverse Laplace transform can be written directly.

Then we consider another example, which has multiple roots. A transfer function is given.

$$X_1(s) = \frac{s^3 + 12s^2 + 47s + 60}{(s+1)^3(s+2)}$$

$$= \frac{s^3 + 12s^2 + 47s + 60}{s^4 + 5s^3 + 9s^2 + 7s + 2} \qquad (50)$$

Its inverse Laplace transformation is required. Substituting the eigenvalues into Eq. (24), one obtains

$$[W] = \begin{bmatrix} 1 & 0 & 0 & 1 \\ \lambda\big|_{\lambda=-1} & 1 & 0 & \lambda\big|_{\lambda=-2} \\ \lambda^2\big|_{\lambda=-1} & 2\lambda\big|_{\lambda_1=-1} & 1 & \lambda^2\big|_{\lambda=-2} \\ \lambda^3\big|_{\lambda=-1} & 3\lambda^2\big|_{\lambda_1=-1} & \dfrac{1}{2!}6\lambda\big|_{\lambda_1=-1} & \lambda^3\big|_{\lambda=-2} \end{bmatrix}$$

$$= \begin{bmatrix} 1 & 0 & 0 & 1 \\ -1 & 1 & 0 & -2 \\ 1 & -2 & 1 & 4 \\ -1 & 3 & -3 & -8 \end{bmatrix} \qquad (51)$$

The Jordan matrix for this problem would be

$$[J] = \begin{bmatrix} \lambda_1 & 1 & 0 & 0 \\ 0 & \lambda_1 & 1 & 0 \\ 0 & 0 & \lambda_1 & 0 \\ 0 & 0 & 0 & \lambda_2 \end{bmatrix} = \begin{bmatrix} -1 & 1 & 0 & 0 \\ 0 & -1 & 1 & 0 \\ 0 & 0 & -1 & 0 \\ 0 & 0 & 0 & -2 \end{bmatrix} \qquad (52)$$

$$[x(0)] = \begin{bmatrix} 1 & 0 & 0 & 0 \\ a_1 & 1 & 0 & 0 \\ a_2 & a_1 & 1 & 0 \\ a_3 & a_2 & a_1 & 1 \end{bmatrix}^{-1} \begin{bmatrix} b_1 \\ b_2 \\ b_3 \\ b_4 \end{bmatrix}$$

$$= \begin{bmatrix} 1 & 0 & 0 & 0 \\ 5 & 1 & 0 & 0 \\ 9 & 5 & 1 & 0 \\ 7 & 9 & 5 & 1 \end{bmatrix}^{-1} \begin{bmatrix} 1 \\ 12 \\ 47 \\ 60 \end{bmatrix} \tag{53}$$

Substitution of Eqs. (51), (52), and (53) into Eq. (47) gives

$$[X(s)] = \begin{bmatrix} 1 & 0 & 0 & 1 \\ -1 & 1 & 0 & -2 \\ 1 & -2 & 1 & 4 \\ -1 & 3 & -3 & -8 \end{bmatrix} \begin{bmatrix} s+1 & -1 & 0 & 0 \\ 0 & s+1 & -1 & 0 \\ 0 & 0 & s+1 & 0 \\ 0 & 0 & 0 & s+2 \end{bmatrix}^{-1}$$

$$\cdot \begin{bmatrix} 1 & 0 & 0 & 1 \\ -1 & 1 & 0 & -2 \\ 1 & -2 & 1 & 4 \\ -1 & 3 & -3 & -8 \end{bmatrix}^{-1} \begin{bmatrix} 1 & 0 & 0 & 0 \\ 5 & 1 & 0 & 0 \\ 9 & 5 & 1 & 0 \\ 7 & 9 & 5 & 1 \end{bmatrix}^{-1} \begin{bmatrix} 1 \\ 12 \\ 47 \\ 60 \end{bmatrix}$$

$$= \begin{bmatrix} 1 & 0 & 0 & 1 \\ -1 & 1 & 0 & -2 \\ 1 & -2 & 1 & 4 \\ -1 & 3 & -3 & -8 \end{bmatrix} \begin{bmatrix} \dfrac{1}{s+1} & \dfrac{1}{(s+1)^2} & \dfrac{1}{(s+1)^3} & 0 \\ 0 & \dfrac{1}{s+1} & \dfrac{1}{(s+1)^2} & 0 \\ 0 & 0 & \dfrac{1}{s+1} & 0 \\ 0 & 0 & 0 & \dfrac{1}{s+2} \end{bmatrix} \begin{bmatrix} 7 \\ 2 \\ 24 \\ -6 \end{bmatrix} \tag{54}$$

If the matrix is expanded into four algebraic equations, the first one is

$$X_1(s) = \frac{7}{s+1} + \frac{2}{(s+1)^2} + \frac{24}{(s+1)^3} + \frac{-6}{(s+2)}$$

Inverse Laplace transformation yields

$$x_1(t) = 7e^{-t} + 2te^{-t} + 12t^2 e^{-t} - 6e^{-2t}$$

It is interesting to note that the complicated multiple differentiations of the s polynomial involved in the classical inverse Laplace transformation by the Heaviside expansion become a simple series differentiation of λ's. In very high order systems with repeated and complex roots, this new general formula is more straightforward and more easily adapted to digital computation techniques.

9. A Digital Computer Program

Consider (45) with (42). The elements of (42) can be written as follows:

$$\sum_{K=1}^{N} \sum_{L=1}^{N} \lambda_K^{L-1} \tag{55}$$

We call it $CAW(L, K)$, which means the element at Lth row and Kth column; N is the order of the characteristic equation in question.

When repeated roots appear in a transfer function, each element of (24) is obtained from the following:

(1) $\displaystyle\sum_{K=1}^{1} \sum_{L=1}^{N} CAW(L, K) = \lambda^{L-1}$

(2) $\displaystyle\sum_{L=1}^{1} \sum_{K=2}^{MGK} CAW(L, K) = (0., 0.)$ (56)

(3) $\displaystyle\sum_{K=2}^{MGK} \sum_{L=2}^{N} CAW(L, K) = CAW(L-1, K-1) * \dfrac{L-1}{K-1}$

where MGK is the order of each repeated root. In other words, (56) is an algorithm for calculating the repeated roots of the modified Vandermonde matrix or (24).

The digital computer program (Program 9) and its flow chart are shown on the following pages.

Program 9. GENERAL MATRIX METHOD FOR INVERSE LAPLACE TRANSFOR-
MATION

```
THE GENERALIZED MATRIX FORMULA FOR THE INVERSE LAPLACE
TRANSFORMATION PROGRAM....
          GIVEN    TRANSFER    FUNCTION  ...
X(S)=(B(1)*S**(N-1) + B(2)*S**(N-2) +...+ B(N)) / (A0*S**N +
      A(1)*S**(N-1) + A(2)*S**(N-2) +...+ A(N))
THE  CHARACTERISTIC EQUATION IS  ...
A0*S**N + A(1)*S**(N-1) +...+ A(N) =
     A0*(S-RR(1)-J*RI(1))**MG(1) * (S-RR(2)-J*RI(2))**MG(2) *...*
     (S-RR(M)-J*RI(M))**MG(M)=0.
WHERE    MG(L)=POWERS OF EACH DIFFERENT ROOTS
          (FOR REAL AND COMPLEX ROOTS)
MG(1)=MG(2)=...=MG(M)=1   .AND.   N=M
          (FOR REPEAT , REAL AND COMPLEX ROOTS)
N=MG(1) + MG(2) +...+ MG(M) .AND.   M=NO. OF DIFFERENT ROOTS
N..ORDER OF FUNCTION , M..NO. OF DIFFERENT ROOTS , NN..REQUIRED
TRANSIENT POINTS , T..STARTING TIME , DELT..TIME/INCREMENT

                MAIN    PROGRAM    BEGINS    HERE
500 FORMAT(3I4,2F16.6/(20I4))
501 FORMAT((5F16.6))
600 FORMAT(////1X,113HN..ORDER OF FUNCTION , M..NO. OF DIFFERENT ROOTS
   1 , NN..TRANSIENT POINTS , T..STARTING TIME , DELT..TIME INCREMENT/
   23X,1HN,3X,1HM,2X,2HNN,8X,1HT,14X,4HDELT)
601 FORMAT(3(1X,I3),2F16.6)
602 FORMAT(2X,45HMG(1)...MG(M)   POWERS OF EACH DIFFERENT ROOTS)
603 FORMAT((1X,25I4))
604 FORMAT(1X,34HDENOMINATOR  COEFF. A0,A(1)...A(N))
605 FORMAT((1X,6F16.6))
606 FORMAT(1X,32HNUMERATOR  COEFF.    B(1)...B(N))
607 FORMAT(1X,45HGIVEN   ROOTS.....REAL   PARTS    RR(1)...RR(N))
608 FORMAT(1X,46HGIVEN   ROOTS.....IMAGE  PARTS    RI(1)...RI(N))
609 FORMAT(////1X,34HTHE REQUIRED TRANSIENT EQUATION IS/3X,5HX(T)=)
610 FORMAT(9X,2H+(,E14.6,3H+J(,E14.6,11H)) * EXP ((,E14.6,3H+J(,E14.6,
   17H)) * T))
611 FORMAT(9X,2H+(,E14.6,3H+J(,F14.6,10H)) * T **(,I2,10H) * EXP ((,
   1E14.6,3H+J(,F14.6,7H)) * T))
612 FORMAT(////1X,33HTHE REQUIRED TRANSIENT VALUES ARE//3X,8HT...TIME,
   15X,22HX(T)...TRANSIENT VALUE)
613 FORMAT(1X,F10.6,13X,E14.6)

    DIMENSION A(100),B(100),RR(100),RI(100),MG(100),MH(100),
   1IPVOT(100),INDEX(100,2),X(500),Y(500)
    COMPLEX TT,SS,DET,CR(100),AB(100),WAB(100),PIVOT(100),CAW(100,100)
    EQUIVALENCE (IROW,JROW),(ICOL,JCOL),(A(1),MH(1),IPVOT(1)),
   1(WAB(1),PIVOT(1)),(RR(1),X(1)),(RI(1),Y(1))
    *** READ INPUT DATA AND ECHO CHECK ***
000 READ(5,500) N,M,NN,T,DELT,(MG(I),I=1,M)
    READ(5,501) A0,(A(I),I=1,N)
    READ(5,501) (B(I),I=1,N)
    READ(5,501) (RR(I),I=1,N)
    READ(5,501) (RI(I),I=1,N)
    WRITE(6,600)
    WRITE(6,601) N,M,NN,T,DELT
    WRITE(6,602)
    WRITE(6,603) (MG(I),I=1,M)
    WRITE(6,604)
    WRITE(6,605) A0, (A(I),I=1,N)
```

Program 9 (CONT.)

```
      WRITE(6,606)
      WRITE(6,605)  (B(I),I=1,N)
      WRITE(6,607)
      WRITE(6,605)  (RR(I),I=1,N)
      WRITE(6,608)
      WRITE(6,605)  (RI(I),I=1,N)
      KS=1
C     *** NORMALIZE TRANSFER FUNCTION AND REARRANGE COMPLEX VECTER **
      DO 10 I=1,N
      B(I)=B(I)/A0
      AB(I)=A(I)/A0
      AB(I)=CMPLX(AB,0.)
   10 CR(I)=CMPLX(RR(I),RI(I))
C     *** EVALUATE DENOMINATOR MATRIX ***
      DO 20 I=1,N
      DO 19 J=1,N
      IF(J-I) 4,5,6
    5 CAW(I,J)=(1.,0.)
      GO TO 19
    6 CAW(I,J)=(0.,0.)
      GO TO 19
    4 IJ=I-J
      CAW(I,J)=AB(IJ)
   19 CONTINUE
   20 CONTINUE
C     ***USING THE GAUSS-JORDAN METHOD TO FIND THE INVERSE OF
C         THE DENOMINATOR OR THE VANDERMONDE MATRIX ***
  200 DET=(1.,0.)
      DO 30 J=1,N
   30 IPVOT(J)=0
      DO 40 I=1,N
      TT=(0.,0.)
      DO 50 J=1,N
      IF(IPVOT(J)-1) 7,50,7
    7 DO 49 K=1,N
      IF(IPVOT(K)-1) 8,49,9
    8 IF(CABS(TT)-CABS(CAW(J,K))) 11,49,49
   11 IROW=J
      ICOL=K
      TT=CAW(J,K)
   49 CONTINUE
   50 CONTINUE
      IPVOT(ICOL)=IPVOT(ICOL)+1
      IF(IROW-ICOL) 12,13,12
   12 DET=DET*(-1.)
      DO 60 L=1,N
      TT=CAW(IROW,L)
      CAW(IROW,L)=CAW(ICOL,L)
   60 CAW(ICOL,L)=TT
   13 INDEX(I,1)=IROW
      INDEX(I,2)=ICOL
      PIVOT(I)=CAW(ICOL,ICOL)
      DET=DET*PIVOT(I)
      CAW(ICOL,ICOL)=(1.,0.)
      DO 70 L=1,N
   70 CAW(ICOL,L)=CAW(ICOL,L)/PIVOT(I)
      DO 39 LI=1,N
```

Program 9 (CONT.)

```
      IF(LI-ICOL) 14,39,14
   14 TT=CAW(LI,ICOL)
      CAW(LI,ICOL)=(0.,0.)
      DO 38 L=1,N
   38 CAW(LI,L)=CAW(LI,L)-CAW(ICOL,L)*TT
   39 CONTINUE
   40 CONTINUE
      DO 80 I=1,N
      L=N-I+1
      IF(INDEX(L,1)-INDEX(L,2)) 15,80,15
   15 JROW=INDEX(L,1)
      JCOL=INDEX(L,2)
      DO 79 K=1,N
      TT=CAW(K,JROW)
      CAW(K,JROW)=CAW(K,JCOL)
   79 CAW(K,JCOL)=TT
   80 CONTINUE
    9 IF(KS.EQ.2) GO TO 22
C        *** THE PRODUCT OF THE NUMERATOR AND INVERSE DENOMINATOR MATRIX**
      DO 100 I=1,N
      AB(I)=(0.,0.)
      DO 100 J=1,N
  100 AB(I)=CAW(I,J)*B(J)+AB(I)
C        *** MANIPULATE THE VANDERMONDE MATRIX ***
      LK=0
      DO 110 K=1,M
      MGK=MG(K)
      DO 109 J=1,MGK
      LK=LK+1
      IF(J-1) 16,16,17
   16 SS=(1.,0.)
      CAW(1,LK)=(1.,0.)
      DO 120 I=2,N
      SS=CR(LK)*SS
  120 CAW(I,LK)=SS
      GO TO 109
   17 DO 108 I=1,N
      IF(I-J) 18,26,21
   18 CAW(I,LK)=(0.,0.)
      GO TO 108
   26 CAW(I,LK)=(1.,0.)
      GO TO 108
   21 I1=I-1
      LK1=LK-1
      CAW(I,LK)=CAW(I1,LK1)*FLOAT(I1)/FLOAT(J-1)
  108 CONTINUE
  109 CONTINUE
  110 CONTINUE
      KS=2
      GO TO 200
C        *** DETERMINE RESIDUES ***
   22 DO 140 I=1,N
      WAB(I)=(0.,0.)
      DO 140 J=1,N
  140 WAB(I)=CAW(I,J)*AB(J)+WAB(I)
      WRITE(6,609)
      IF(N.EQ.M) GO TO 23
```

Program 9 (CONCL.)

```
C        *** FIND FACTORIALS ***
         LK=0
         DO 150 K=1,M
         MGK=MG(K)
         S=1.
         DO 149 J=1,MGK
         LK=LK+1
         IF(J-1) 24,24,25
   24    MH(LK)=J-1
         GO TO 149
   25    S=FLOAT(J-1)*S
C        *** RESIDUES FOR REPEAT ROOTS ***
         WAB(LK)=WAB(LK)/S
         MH(LK)=J-1
  149    CONTINUE
  150    CONTINUE
C        *** TRANSIENT EQ. FOR REPEAT ROOTS ***
         WRITE(6,611) (WAB(I),MH(I),CR(I),I=1,N)
         GO TO 300
C        *** TRANSIENT EQ. FOR SIMPLE ROOTS ***
   23    WRITE(6,610) (WAB(I),CR(I),I=1,N)
C        *** DETERMINE TRANSIENT VALUES ***
  300    DO 160 I=1,NN
         SS=(0.,0.)
         DO 170 J=1,N
         IF(N.EQ.M) MH(J)=0
         IF(MH(J).EQ.0) GO TO 27
         TL=T
         GO TO 28
   27    TL=1.
   28    TT=CR(J)*T
         SS=WAB(J)*TL**MH(J)*CEXP(TT)+SS
  170    CONTINUE
         Y(I)=REAL(SS)
         X(I)=T
  160    T=T+DELT
         WRITE(6,612)
         WRITE(6,613) (X(I),Y(I),I=1,NN)
         GO TO 1000
         END
$DATA
   3    3   11         0..               0.1
   1    1    1
            1.            10.            120.            0.
            0.             0.            100.
            0.            -5.             -5.
            0.         9.746794      -9.746794
```

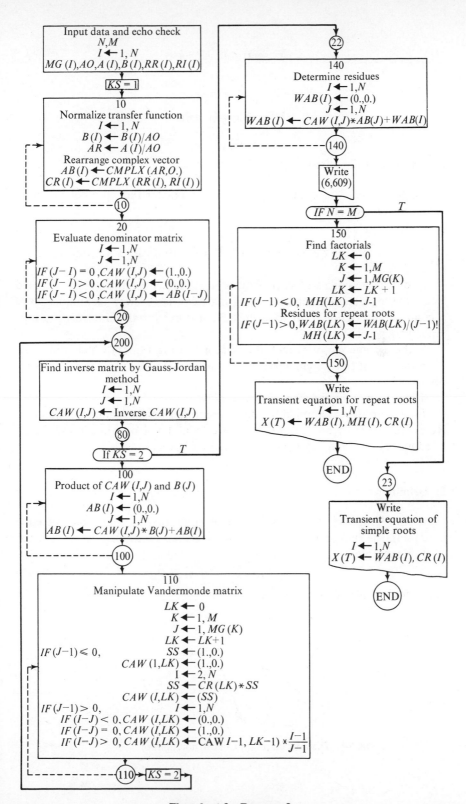

Flow chart for Program 9

IV. Flow Graphs

1. Graphs and Equations

Most problems in control systems are characterized by a set of linear equations of the form

$$a_{ij}x_j = y_i \qquad\qquad i = 1, 2, \ldots, n \qquad (1)$$

The solution for the system amounts to finding the x's in terms of the a's and y's.

In our discussions so far, the solution of the problem has been obtained by evaluating determinants, inverting matrices, etc. These techniques have a common shortcoming in that they do not exhibit the interrelationships among the variables.

Is there some pictorial method of representing Eq. (1) which will show the structure of the system and allow us to evaluate the variables more simply?

There is, by construction of flow graphs.

Flow graphs were applied to analog computer programming by Shannon. Rules for assembly and calculation were formulated by Mason and developed more completely by Coates, Desoer, and others.

2. Definitions

A flow graph represents a system of equations in which one variable is equal to the sum of a number of other variables multiplied by coefficients.

For example,

$$a_{11}x_1 + a_{12}x_2 + a_{13}x_3 = 0 \qquad\qquad (2)$$

has a flow graph as shown in Fig. 7-6.

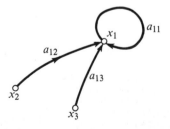

Fig. 7-6. The flow graph (in Coates' notation) of
$$a_{11}x_1 + a_{12}x_2 + a_{13}x_3 = 0$$

The variables x_1, x_2, and x_3 are represented by the small circles called *nodes*. The coefficients a_{11}, a_{12}, and a_{13} are represented by the directed lines between variables called *branches*.

Figure 7-6 represents Eq. (2) in Coates' notation. On the analog computer, Eq. (2) would be programmed as shown in Fig. 7-7.

In the nodal equations, as in (2), the x_1's are represented by voltages and the a_{ij}'s by conductances. A node in a flow graph can be considered as the shaded part in Fig. 7-8.

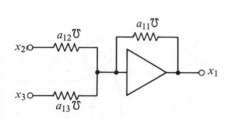

Fig. 7-7. The analog computer program of Eq. (2).

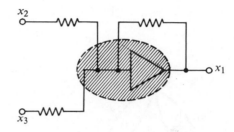

Fig. 7-8. Definition of a node in Coates' sense.

3. Simultaneous Equations

A set of two simultaneous equations can be represented by Fig. 7-9.

$$a_{11}x_1 + a_{12}x_2 = m_1$$
$$a_{21}x_1 + a_{22}x_2 = m_2 \tag{3}$$

To accommodate the nonhomogeneous equations, extra nodes, called *source nodes*, have been added. These nodes have only outgoing branches; if a node has only incoming branches (such as p) it is called a *sink node*.

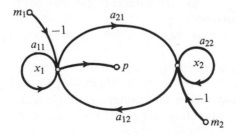

Fig. 7-9. The flow graph of

$$a_{11}x_1 + a_{12}x_2 = m_1$$
$$a_{21}x_1 + a_{22}x_2 = m_2$$

4. A Set of n Simultaneous Equations

For a set of three simultaneous equations, the technique is easily extended (see Fig. 7-10).

$$a_{11}x_1 + a_{12}x_2 + a_{13}x_3 = m_1$$
$$a_{21}x_1 + a_{22}x_2 + a_{23}x_3 = m_2 \qquad (4)$$
$$a_{31}x_1 + a_{32}x_2 + a_{33}x_3 = m_3$$

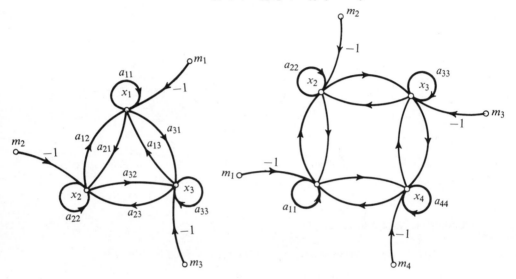

Fig. 7-10. A typical flow graph involving three variables.

Fig. 7-11. A typical flow graph involving four variables.

Similarly, for a set of four simultaneous equations, the corresponding flow graph looks like Fig. 7-11.

For n simultaneous equations, draw n small circles uniformly distributed about a large circle; draw the directed branches between the nodes and finally add the source nodes.

The question is: How can we solve the simultaneous equations from the flow graph?

In the ordinary algebraic solution of a set of equations there are three general approaches: (1) reduction methods—elimination of variables; (2) direct methods—application of Cramer's rule; (3) matrix methods—inversion. There are corresponding techniques for solution through a flow graph.

5. Reduction Method

This requires the use of the following basic rules:

1. The addition rule—see Fig. 7-12.

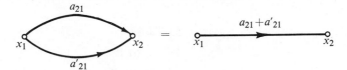

Fig. 7-12. The addition rule.

2. The multiplication rule—see Fig. 7-13.

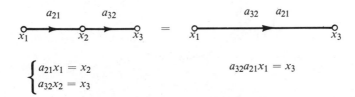

Fig. 7-13. The multiplication rule.

3. Self-loop elimination—see Fig. 7-14.

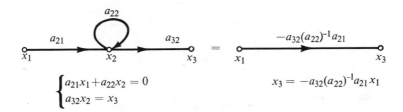

Fig. 7-14. The self-loop elimination rule.

4. Node-duplication rule—see Fig. 7-15.

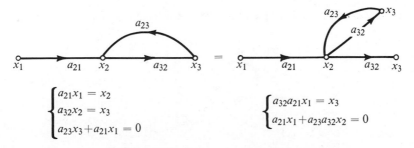

Fig. 7-15. The node-duplication rule.

An example will show the application of these rules in the reduction of a flow graph:

$$3x_1 + 5x_2 = 26$$
$$x_1 + 7x_2 = 30$$
(5)

The corresponding flow graph is shown in Fig. 7-16.

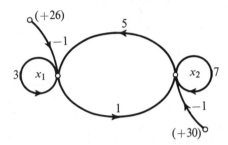

Fig. 7-16. The flow graph of Eq. (5).

Duplicate x_1 by use of rule (4); we have Fig. 7-17.

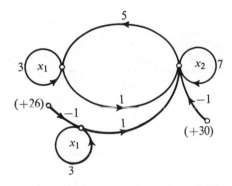

Fig. 7-17. Duplicating x_1 by the use of rule (4).

Eliminate the x_1 self-loop by rule (3); we obtain Fig. 7-18.

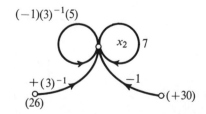

Fig. 7-18. Node x_1 is eliminated by applying rule (3).

Combination of the two loops by rule (2) gives a graph such as Fig. 7-19.

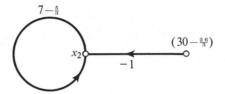

Fig. 7-19. Two loops combined.

By definition,

$$x_2 = \frac{30 - \frac{26}{3}}{7 - \frac{5}{3}} = 4$$

This flow graph reduction is simply a graphic means of eliminating a variable. However, we sometimes solve equations by Cramer's rule. It is advantageous in that it gives a direct evaluation of the unknowns. We now consider the analogous technique on a flow graph.

6. Direct Method

There are two direct formulas: one from Mason and the other from Coates. The latter will be presented here; but first of all, some basic terms must be defined.

1. *The cold graph of* Γ: a graph obtained from Γ by removing source nodes. The symbol is Γ^0.

2. *The complete chain graph of* Γ^0: a chain (opened circle) graph which has all the nodes originally involved in Γ^0.

3. L_u is the number of loops contained in the graph for the uth term.

4. Γ^m is a *modified* graph of Γ which is obtained from Γ^0 by replacing the outgoing branches of the node m by the sources.

5. *The complete chain graph of* Γ^m: a chain graph which necessarily contains all the nodes originally involved in Γ^m.

6. L_v is the number of loops contained in the graph for the vth term.

These symbols appear in Coates' formula,

$$x_m = \frac{\sum_v (-1)^{L_v} \pi_v^m}{\sum_u (-1)^{L_u} \pi_u^0} \tag{6}$$

where π_u^0 is the product of the loop gains of all loops contained in the uth complete chain graph of the cold system, and π_v^m is the product of all loop gains contained in the vth complete chain graph of the Γ^m system.

Consider a set of simultaneous equations:

$$a_{11}x_1 + a_{12}x_2 + a_{13}x_3 = y_1$$
$$a_{21}x_1 + a_{22}x_2 + a_{23}x_3 = y_2 \tag{7}$$
$$a_{31}x_1 + a_{32}x_2 + a_{33}x_3 = y_3$$

Its flow graph is Fig. 7-20.

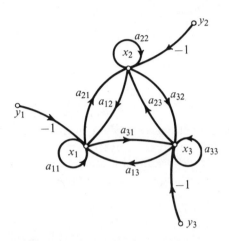

Fig. 7-20. The complete flow graph of Eq. (7).

The cold system is shown in Fig. 7-21.

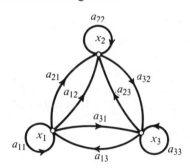

Fig. 7-21. The cold flow graph of Eq. (7).

The complete chain graphs of Γ^0 are given in Fig. 7-22. Suppose it is required to find x_3. Simply replace the gains of outgoing branches by the sources y_i.

$\Gamma^{(3)}$ is shown as Fig. 7-23. By a similar approach we can construct a table, like the one shown in Fig. 7-22, but for $\pi_v^{(3)}$. Substituting the results into (6), we have

$$x_3 = \frac{(-1)^3 a_{11}a_{22}y_3 + (-1)^2 a_{11}y_2 a_{32} + (-1)^2 a_{22}a_{31}y_1 + (-1)^2 y_3 a_{21}a_{12}}{(-1)^3 a_{11}a_{22}a_{33} + (-1)^2 a_{11}a_{23}a_{32} + (-1)^2 a_{22}a_{31}a_{13} + (-1)^2 a_{33}a_{21}a_{12}}$$

$$\cdot \frac{+ (-1)^1 a_{21}y_3 a_{32} + (-1)^1 a_{31}a_{12}y_2}{+ (-1)^1 a_{21}a_{13}a_{32} + (-1)^1 a_{31}a_{12}a_{23}}$$

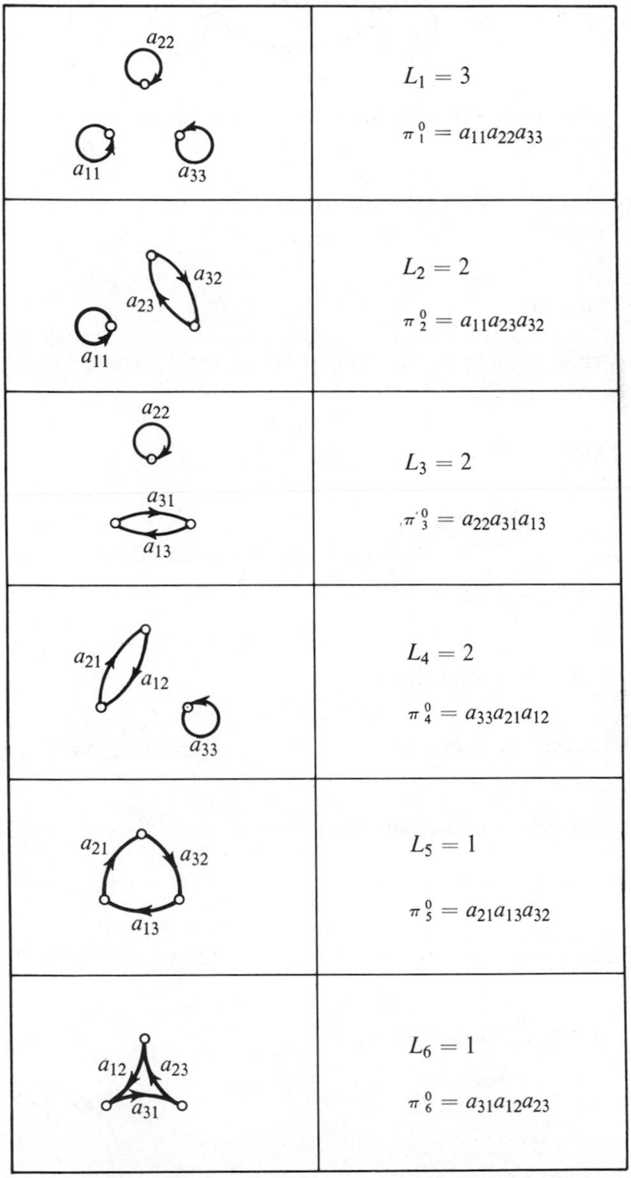

	$L_1 = 3$ $\pi_1^{\,0} = a_{11}a_{22}a_{33}$
	$L_2 = 2$ $\pi_2^{\,0} = a_{11}a_{23}a_{32}$
	$L_3 = 2$ $\pi_3^{\,0} = a_{22}a_{31}a_{13}$
	$L_4 = 2$ $\pi_4^{\,0} = a_{33}a_{21}a_{12}$
	$L_5 = 1$ $\pi_5^{\,0} = a_{21}a_{13}a_{32}$
	$L_6 = 1$ $\pi_6^{\,0} = a_{31}a_{12}a_{23}$

Fig. 7-22. Application of Coates' formula.

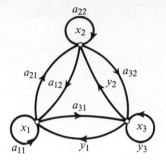

Fig. 7-23. Replacing the gains of the outgoing branches by the sources.

It is interesting to note that this approach is simply a topological version of Cramer's rule.

7. Matrix Method

All the rules shown in Sec. 5 can be directly extended to matrix flow graphs as shown in Fig. 7-24.

(1) The addition rule:

(2) The multiplication rule:

(3) The self-loop elimination rule:

(4) The node duplication rule:

Fig. 7-24. Rules for matrix flow graphs.

A matrix flow graph is shown in Fig. 7-25; it is desired to reduce it to a simple form by using the reduction rules: (1) the addition rule, (2) the multiplication rule, (3) the self-loop elimination rule, (4) the node-duplication rule.

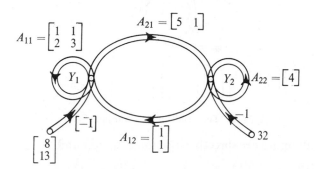

Fig. 7-25. A matrix flow graph example.

We use rule (4), obtaining Fig. 7-26. Applying rule (3), we get Fig. 7-27.

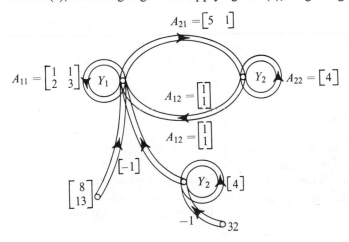

Fig. 7-26. Duplicating the node Y_2.

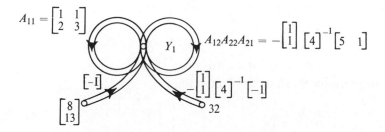

Fig. 7-27. Node Y_2 has been eliminated.

Then using rule (1), we obtain Fig. 7-28.

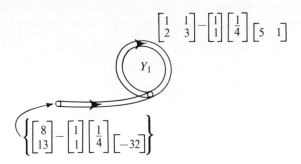

$$\begin{bmatrix} 1 & 1 \\ 2 & 3 \end{bmatrix} - \begin{bmatrix} 1 \\ 1 \end{bmatrix}\begin{bmatrix} \frac{1}{4} \end{bmatrix}\begin{bmatrix} 5 & 1 \end{bmatrix}$$

$$Y_1$$

$$\left\{ \begin{bmatrix} 8 \\ 13 \end{bmatrix} - \begin{bmatrix} 1 \\ 1 \end{bmatrix}\begin{bmatrix} \frac{1}{4} \end{bmatrix}\begin{bmatrix} -32 \end{bmatrix} \right\}$$

Fig. 7-28. The result of two combined loops.

By definition, we can directly write the equation of Y_1.

$$\left\{ \begin{bmatrix} 1 & 1 \\ 2 & 3 \end{bmatrix} - \begin{bmatrix} 1 \\ 1 \end{bmatrix}[\tfrac{1}{4}][5,\ 1] \right\}[Y_1] = \begin{bmatrix} 8 \\ 13 \end{bmatrix} - \begin{bmatrix} 1 \\ 1 \end{bmatrix}[\tfrac{1}{4}][+32]$$

or

$$\begin{bmatrix} -\frac{1}{4} & \frac{3}{4} \\ \frac{3}{4} & \frac{11}{4} \end{bmatrix}[Y_1] = \begin{bmatrix} 0 \\ 5 \end{bmatrix}$$

Therefore,

$$[Y_1] = \begin{bmatrix} 3 \\ 1 \end{bmatrix}$$

8. Partitioning a Matrix Flow Graph

The matrix flow graph, shown in the last example representing the relationships among variables, can also be used with groups of variables. There are no restrictions on the manner of grouping. There is a one-to-one correspondence between partitioning a matrix and forming a flow graph.

As an example, consider the following set of equations.

$$\begin{bmatrix} a_{11} & a_{12} & a_{13} & \vdots & a_{14} & a_{15} \\ a_{21} & a_{22} & a_{23} & \vdots & a_{24} & a_{25} \\ a_{31} & a_{32} & a_{33} & \vdots & a_{34} & a_{35} \\ \cdots\cdots\cdots & \vdots & \cdots\cdots \\ a_{41} & a_{42} & a_{43} & \vdots & a_{44} & a_{45} \\ a_{51} & a_{52} & a_{53} & \vdots & a_{54} & a_{55} \end{bmatrix}\begin{bmatrix} x_1 \\ x_2 \\ x_3 \\ \cdots \\ x_4 \\ x_5 \end{bmatrix} = \begin{bmatrix} m_1 \\ m_2 \\ m_3 \\ \cdots \\ m_4 \\ m_5 \end{bmatrix} \tag{8}$$

We can partition (8) along the dotted lines:

$$\begin{bmatrix} A_{aa} & \vdots & A_{ab} \\ \cdots & \vdots & \cdots \\ A_{ba} & \vdots & A_{bb} \end{bmatrix}\begin{bmatrix} X_a \\ \cdots \\ X_b \end{bmatrix} = \begin{bmatrix} M_a \\ \cdots \\ M_b \end{bmatrix} \tag{8a}$$

where A_{aa} corresponds to

$$\begin{bmatrix} a_{11} & a_{12} & a_{13} \\ a_{21} & a_{22} & a_{23} \\ a_{31} & a_{32} & a_{33} \end{bmatrix}$$

X_a corresponds to

$$\begin{bmatrix} x_1 \\ x_2 \\ x_3 \end{bmatrix}$$

and so forth.

Equation (8) can be represented by the flow graph shown in Fig. 7-29.

If it is desired to group x_1, x_2, and x_3 together, we can find the self matrix loop A_{aa} by deleting all the branches that go out or come into this subgroup. This is shown in the lower left corner of Fig. 7-30. The self matrix loop A_{bb} is similarly formed and illustrated in the upper right corner of Fig. 7-30.

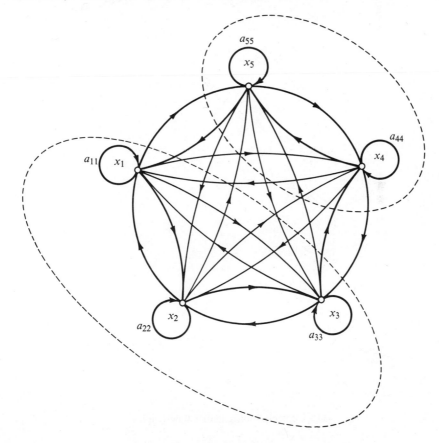

Fig. 7-29. Partitioning a flow graph.

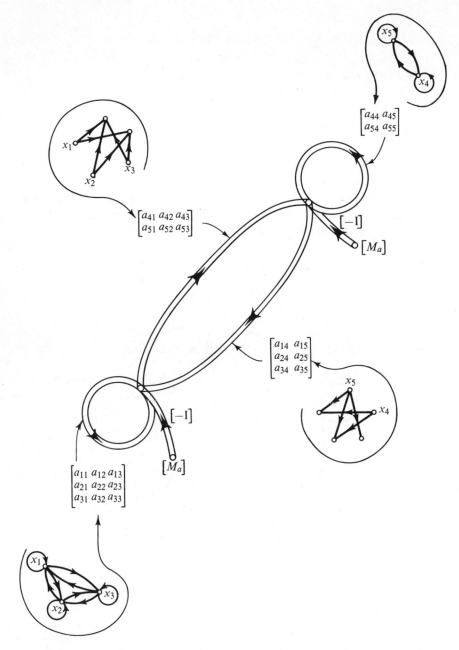

Fig. 7-30. A corresponding matrix flow graph obtained.

The matrix branch A_{ba} can be considered as the branch from the group x_1, x_2, x_3 to the group x_4, x_5. It is formed by the following rule: Retain the outgoing branches from x_1, x_2, and x_3 to x_4 and x_5; discard all others. The subgraph thus obtained can be translated into a rectangular matrix, as shown in the upper left corner of Fig. 7-30. Similarly, A_{ab} can be formed as another rectangular matrix.

After forming Fig. 7-30 we can apply the matrix flow graph rules to evaluate X_a and X_b, which correspond to (x_1, x_2, x_3) and (x_4, x_5), respectively.

V.　Numerical Solution

1. Numerical Integration

State equations may be solved numerically. We will describe the Runge-Kutta method step by step as it would be applied in a program for a computer, first for a single differential equation and then for simultaneous equations. A continuance for n equations should then be apparent and will be commented upon.

A numerical solution of a first order differential equation could be described as a process of finding a series of points on this line. If we substitute a known set of values x_n and t_n into the equation

$$\frac{dx}{dt} = f(t, x)$$

we in effect get the slope of the curve at t_n.

The gist of this numerical integration method is finding an approximate slope of the function at a known point, and using this approximate slope and a small enough increment of time to proceed to the next point. Then,

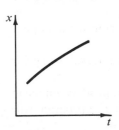

Fig. 7-31. A curve.

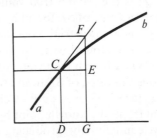

Fig. 7-32. Illustration of numerical integration.

assuming that the new point is now a known point on the line, one proceeds to repeat the operation of getting an approximate slope and incrementing to the next point.

By way of preliminary illustration look upon Figs. 7-31 and 7-32 as a simple problem in plane geometry. Then if we translate this trivial problem into the terminology of our differential equation, line ab is the solution of a differential equation, CE is the increment of t or Δt, CD is one known value of $f(x, t)$, and EF/CE is the approximate slope of the function found by substituting the known value x_n.

$$\frac{dx}{dt} = f(t, x)$$

$$\left.\frac{dx}{dt}\right|_{x=x_n} = f(t_n, x_n)$$

The simple operation above adds to the initial value x_n the product of this approximate slope and the increment. This gives a new approximate value of $f(x, t)$ at $t + \Delta t$.

We are ready to repeat the process, starting with the new point, new slope at this point and another increment of t and proceeding to the next point. The error in this method shows up clearly in Fig. 7-32: The tangent line CF is not the curve and point F is not on the curve ab.

The Runge-Kutta method finds the approximate slope to a higher degree of accuracy. Instead of using the slope found at the beginning of the interval and assuming it constant for the entire interval, we find approximate values of the function at the beginning, the middle, and the end of the interval in an iterative procedure. The final approximate value of $f(x_{n+1}, t_{n+1})$ is a weighted average of these.

We define terms:

SL is the slope at a point.

H is the increment of t.

$Q1$ is the first trial value for x_{n+1} found by using the initial conditions.

$Q2$ is the second trial value of x_{n+1} found at the half interval using for a slope a value of SL found by substituting $Q1$.

$Q3$ is a third trial value found at the half interval, using for a slope a value of SL found with the aid of $Q2$.

$Q4$ is a fourth trial value found at the end of the time interval, using for a slope the value of SL found with the substitution of $Q3$.

Figure 7-33 shows a flow chart for a program that will produce one new point from one known point by the Runge-Kutta method. The program will recycle indefinitely, producing an additional point each time.

Note that each Q produced is used to produce the next one.

TN is the initial value of t.

XN is the initial value of x.

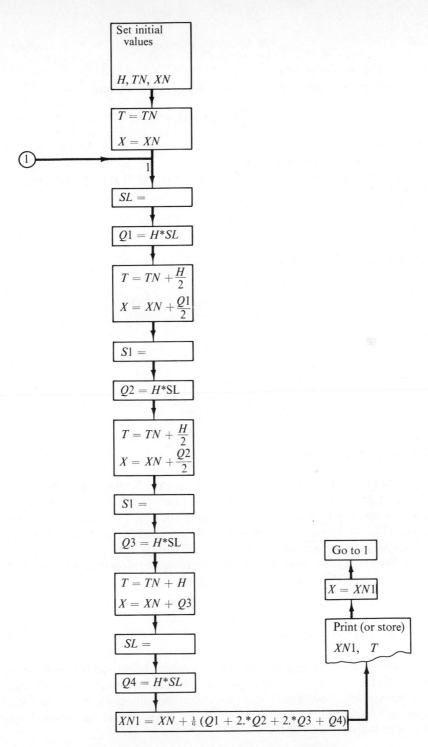

Fig. 7-33. Flow chart of the Runge-Kutta method.

371

H is the increment of t.

SL (or FX) is the number obtained by substituting into

$$\frac{dx}{dt} = f(t, x)$$

the "present values" of x and t. This particular line in the program will change for each equation solved.

The new value x_{n+1} at the end of time interval H is x_n plus a weighted average of the Q's.

Calling

$$\frac{dx}{dt} = SL(t, x)$$

we find that the Runge-Kutta equations are

$$Q1 = HSL(t_n, x_n)$$

$$Q2 = HSL\left(t_n + \frac{H}{2}, x_n + \frac{Q1}{2}\right)$$

$$Q3 = HSL\left(t_n + \frac{H}{2}, x_n + \frac{Q2}{2}\right)$$

$$Q4 = HSL(t_n + H, x_n + Q3)$$

$$x_{n+1} = x_n + \tfrac{1}{6}(Q1 + 2Q2 + 2Q3 + Q4)$$

2. Simultaneous Equations

A program for solving simultaneous equations is much the same, except that there are sets of Q's like the above for each equation in the set. These Q's must be found in the proper order: first, all the $Q1$'s must be found, then all the $Q2$'s, and so on, because all the $Q1$'s are needed to find the first $Q2$, and all the $Q2$'s are needed to find the first $Q3$, and so on.

A flow chart for solving two or more state equations is shown in Fig. 7-34. N is the number of equations.

3. A Computer Program

Program 10 is a digital computer program based on the method of Runge-Kutta. It not only can solve differential equations with constant or time-varying coefficients but also can handle nonlinear differential equations in general. In fact, the solutions of the nonlinear system examples shown in the next section are calculated by the program.

Program 10. SOLVING STATE EQUATIONS BY THE RUNGE-KUTTA METHOD

```
C  C   RUNGE-KUTTA   1 TO 7 EQUATIONS

C      DATA CARDS CONTAIN THE NUMBER OF EQUATIONS
C      THE SIZE OF THE INCREMENT. THE INITIAL VALUE OF X
C      AND THE INITIAL VALUES FOR EACH F(X). THE FIRST
C      NUMBER IN I1. THE REST IN E10.3.
       DIMENSION XN(7),X(7),Q(7,5),FX(7)
       READ 100,N,H,TN,(XN(K),K=1,N)
 100   FORMAT(I1,E10.3,E10.3,7E10.3)
       DO 5 M=1,N
   5   FX(M)=0.0
       PUNCH22
 22    FORMAT(/3X1HT13X2HX111X2HX211X2HX311X2HX4
      111X2HX511X2HX611X2HX7
       PUNCH 200,TN,(XN(K),K=1,N)
 1     L=1
       T=TN
       DO 777 K=1,N
 777   X(K)=XN(K)
       GO TO 101
  10 DO 151 K=1,N
 151   Q(K,L)=H*FX(K)
       T=TN+H/2.
       DO 252 K=1,N
 252   X(K)=XN(K)+Q(K,L)/2.
       L=2
       GO TO 101
  20 DO 251 K=1,N
 251   Q(K,L)=H*FX(K)
       T=TN+H/2.
       DO 352 K=1,N
 352   X(K)=XN(K)+Q(K,L)/2.
       L=3
       GO TO 101
  30 DO 351 K=1,N
 351   Q(K,L)=H*FX(K)
       T=TN+H
       DO 452 K=1,N
 452   X(K)=XN(K)+Q(K,L)
       L=4
       GO TO 101
  40 DO 451 K=1,N
 451   Q(K,L)=H*FX(K)
       GO TO 7
 101 FX(1)=.05*X(1)-.0002*X(1)*X(2)
       FX(2)=-.1*X(2)+.0002*X(1)*X(2)
       GO TO (10,20,30,40),L
 7     TN=TN+H
       DO 8 K=1,N
 8     XN(K)=XN(K)+(1./6.)*(Q(K,1)+2.*Q(K,2)+2.*Q(K,3)
      1+Q(K,4))
       PUNCH 200,TN,(XN(K),K=1,N)
 200   FORMAT(8E11.4)
       GO TO 1
       END
```

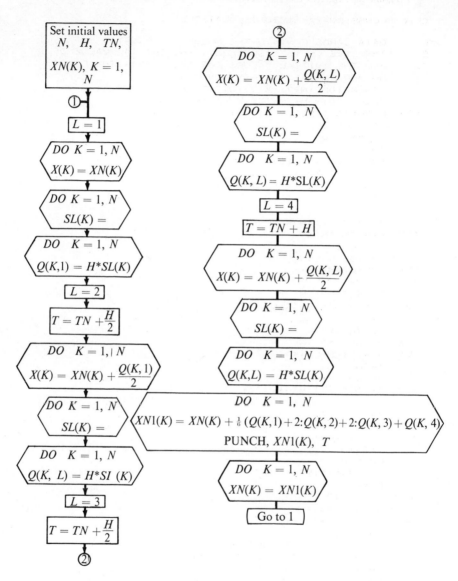

Fig. 7-34. Flow chart for N state equations (Runge-Kutta method).

4. Example 1—Rayleigh's Equation

In classical dynamics, a well-known differential equation is called Rayleigh's equation and is written as follows:

$$\ddot{x} + 2\zeta(1 - \alpha\dot{x}^2)\dot{x} + x = 0 \tag{1}$$

This is an ordinary second order nonlinear differential equation in which ζ and α are two constants of the middle term. In order to solve it by the Runge-Kutta method, it should be reformulated into a set of state equations:

$$\dot{x}_1 = x_2$$
$$\dot{x}_2 = 2\zeta(\alpha x_2^2 - 1)x_2 - x_1 \tag{1a}$$

where $x_1 = x$.

If we give the constants specific numerical values, such as $\zeta = \frac{1}{4}$ and $\alpha = -\frac{1}{3}$:

$$\dot{x}_1 = x_2$$
$$\dot{x}_2 = -0.5x_2 - 0.166x_2^3 - x_1 \tag{1b}$$

it is readily possible to compute the solution by the Runge-Kutta method.

Suppose further that the initial conditions for this particular case are

$$x_1(0) = 2$$
$$x_2(0) = 0$$

Statements in the Runge-Kutta computer program corresponding to these two equations, constants, and initial conditions are as follows:

    101   FX(1) = X(2)

          FX(2) = −0.5*X(2) − 0.166*X(2)*X(2)*X(2) − X(1)

Input data:
    H, time increment, is equal to "1."
    TN, the initial time, is equal to "0."
    XN(1), the initial condition of x_1, is equal to "2."
    XN(2), the initial condition of x_2, is equal to "0."
    The computer solution lists corresponding values of x_1, x_2, and t. From these x_1 vs. t, x_2 vs. t, and x_1 vs. x_2 solution plots can be made as shown in Figs. 7-35, 7-36, and 7-37, respectively.
    The solution can also be plotted on three coordinates, as shown in

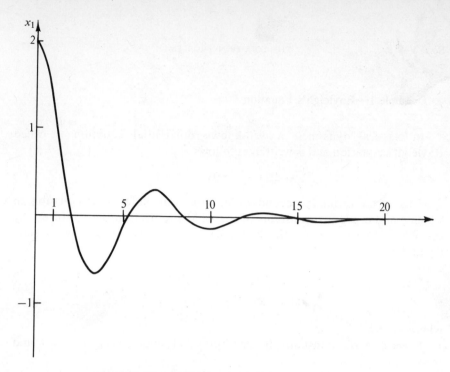

Fig. 7-35. The solution curve of Rayleigh's equation
$\dot{x}_1 = x_2, \dot{x}_2 = -0.5x_2 - 0.166x_2^3 - x_1$ in the x_1-t plane.

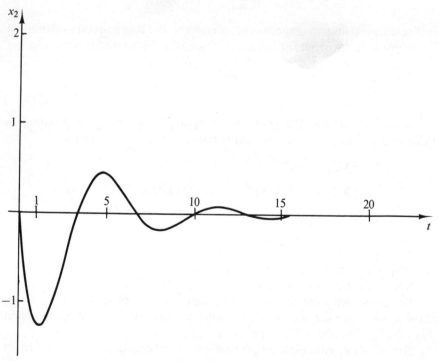

Fig. 7-36. The solution curve in the x_2-t plane.

376

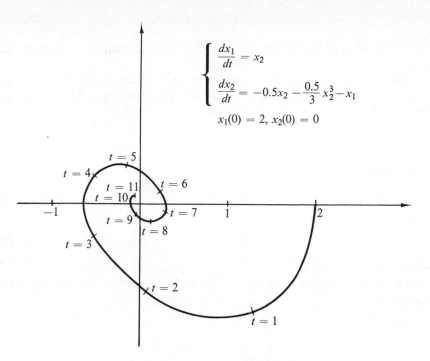

$$\begin{cases} \dfrac{dx_1}{dt} = x_2 \\[2mm] \dfrac{dx_2}{dt} = -0.5x_2 - \dfrac{0.5}{3}x_2^3 - x_1 \end{cases}$$

$$x_1(0) = 2, \ x_2(0) = 0$$

Fig. 7-37. The curve in the phase plane.

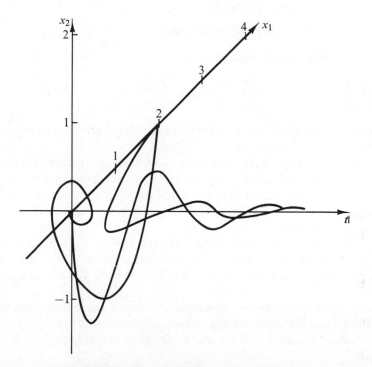

Fig. 7-38. Three projections of the motion space.

377

Fig. 7-38. These three curves are actually the orthographic projections of a single curve in three-dimensional space.

The x_1-t and x_2-t curves are the ordinary expressions of the solution; x_1 vs. x_2 is called a trajectory and the x_1-x_2 plane of Fig. 7-37 is called the phase plane.

5. Example 2—Van der Pol's Equation

The operation of an electronic oscillator can be described by an equation attributed to Van der Pol:

$$\ddot{x} - e(1 - x^2)\dot{x} + x = 0 \tag{2}$$

where e is a constant.

To obtain a numerical solution for $e = 0.8$, change (2) into the state equations:

$$\dot{x}_1 = x_2$$
$$\dot{x}_2 = e(1 - x_1^2)x_2 - x_1 \tag{2a}$$

These will be introduced into the digital computer program as:

$$101 \quad FX(1) = X(2)$$
$$FX(2) = 0.8*(1. -X(1)*X(1))*X(2) - X(1) \tag{2b}$$

Two sets of initial conditions are given:

$$\begin{array}{ccc} x_1(0) = 1 & & x_1(0) = 3 \\ & \text{and} & \\ x_2(0) = 0 & & x_2(0) = 0 \end{array}$$

The computer output resulting from these data has been used to plot curves in the x_1-x_2 plane, Fig. 7-39.

The first trajectory starts from (1, 0) and travels in a path which revolves about the origin and rapidly approaches a certain fixed cycle as a limit. The second trajectory starts from (3, 0) and travels inward to the same cycle as a limit. This cycle indicates the essential behavior of the circuit—it is oscillatory—and is called a limit cycle.

If x_1 vs. t or x_2 vs. t is drawn, it would appear as a periodic wave, but not sinusoidal. In a case where e is equal to zero the phase curve would be a circle and the solution plot would be both periodic and sinusoidal.

Many graphical methods, developed to obtain the phase plane curves for the Van der Pol equation, have been rendered obsolete in the present era of the digital computer. They may be found in earlier texts.

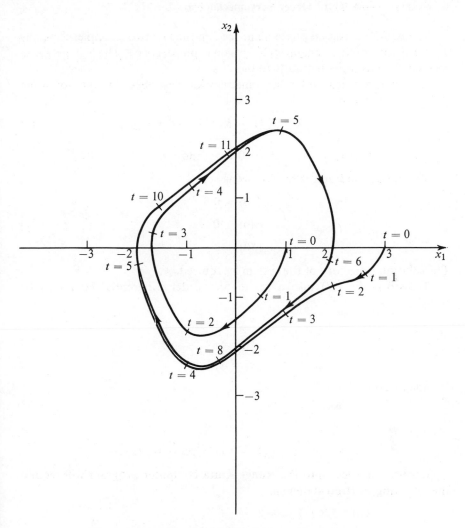

Fig. 7-39. The phase plane curve of the Van der Pol equation

$$\dot{x}_1 = x_2$$
$$\dot{x}_2 = 0.8(1 - x_1^2)x_2 - x_1$$

6. Example 3—A Third Order Servomechanism

The two-dimensional phase plane used in the last two examples is merely a special case of the general n-dimensional phase space. The Runge-Kutta method extends quite naturally to these cases.

Consider a third order servomechanism described by the following differential equation:

$$T\ddot{x} + \ddot{x} + K_2(1 - K_3x^2)\dot{x} + K_1x = 0 \tag{3}$$

Let

$$T = 1, \qquad K_1 = K_3 = 0.915 \quad \text{and} \quad K_2 = 1$$

The initial condition is the following set:

$$x_1(0) = 0.5$$
$$x_2(0) = 0 \tag{4}$$
$$x_3(0) = 0$$

The solution trajectory of the system is required.

Transform Eq. (3) into a set of first order differential equations by using

$$x = x_1$$
$$\dot{x}_1 = x_2 \tag{5}$$
$$\dot{x}_2 = x_3$$

We then have

$$\dot{x}_1 = x_2$$
$$\dot{x}_2 = x_3 \tag{6}$$
$$\dot{x}_3 = -x_3 - (1 - 0.915x_1^2)x_2 - 0.915x_1$$

Introducing these into the Runge-Kutta computer program will require the following Fortran statements.

```
101   FX (1) = X(2)
      FX (2) = X(3)
      FX (3) = −X(3) − (1. −0.915*X(1)*X(1))
            *X(2) − 0.915*X(1)
```

Supplying input data for initial conditions, size of increment, etc. will result in an output listing of time intervals with the corresponding values of $\dot{x}_1$, $\dot{x}_2$, and $\dot{x}_3$.

From these a three-dimensional phase space trajectory can be plotted as in Fig. 7-40. It starts from the given initial condition (0.5, 0, 0) and traces through one cycle in the phase space.

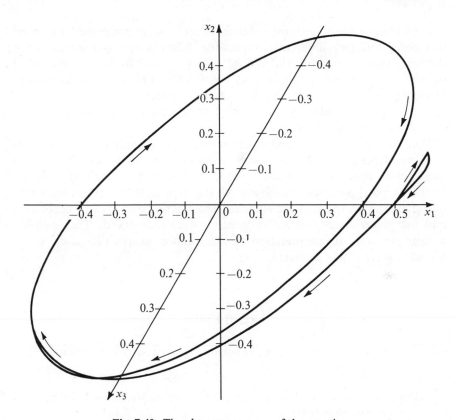

Fig. 7-40. The phase space curve of the equation

$$\dot{x}_1 = x_2$$
$$\dot{x}_2 = x_3$$
$$\dot{x}_3 = -x_3 - (1 - 0.915x_1^2)x_2 - 0.915x_1$$

7. Summary

This chapter deals with the solution of any system governed by a set of first order ordinary differential equations. Three methods for evaluating the transition matrix for the special case governed by linear differential equations with constant coefficients are (1) the Laplace transformation, (2) power series, and (3) linear transformation methods.

Lagrange's parametric variation method was explained, from which the convolution integral was derived.

For the general differential equations, including nonlinear, the solution problem depends on digital computer numerical solutions. One of the most powerful, the Runge-Kutta method, is described in detail, and applied to three typical nonlinear problems. All the numerical results are presented in n-dimensional space as trajectories. The phase plane idea in particular, and the phase space concept in general, were introduced. They provide a clear geometrical interpretation of the solution in question and lead to a qualitative study of a system.

PROBLEMS

7-1. Consider the n-square matrix $[A]$ having the characteristic equation:

$$|sI - A| = 0, \quad \text{or} \quad C(s) = s^n + c_1 s^{n-1} + c_2 s^{n-2} + \cdots + c_n = 0$$

The Cayley-Hamilton theorem states that every square matrix $[A]$ satisfies its characteristic equation, $C(s) = 0$.
(a) Take an arbitrary 3×3 matrix and find its characteristic equation.
(b) Verify the Cayley-Hamilton theorem by using the example in (a).

7-2. The coefficients of the characteristic equation of a square matrix can be determined by the traces of $[A]$, $[A]^2, \ldots, [A]^n$. The relationships between the coefficients and traces, established by Newton and called Newton's formulas, are as follows:

$$\begin{cases} c_1 = -\alpha_1 \\ c_2 = -\tfrac{1}{2}(c_1\alpha_1 + \alpha_2) \\ c_3 = -\tfrac{1}{3}(c_2\alpha_1 + c_1\alpha_2 + \alpha_3) \\ \vdots \\ c_n = -\dfrac{1}{n}(c_{n-1}\alpha_1 + c_{n-2}\alpha_2 + \ldots + c_1\alpha_{n-1} + \alpha_n) \end{cases}$$

where $\alpha_1 = \operatorname{tr}[A]$, $\alpha_2 = \operatorname{tr}[A]^2, \ldots, \alpha_n = \operatorname{tr}[A]^n$.
(a) Use the matrix constructed for Problem 7-1 to verify the Newton formulas.
(b) Show that Newton's formulas can also be written in the following matrix form:

$$
\begin{bmatrix} \alpha_1 \\ \alpha_2 \\ \alpha_3 \\ \cdot \\ \alpha_n \end{bmatrix} = \begin{bmatrix} 1 & 0 & 0 & 0 & \cdot & \cdot & 0 \\ \alpha_1 & 2 & 0 & 0 & \cdot & \cdot & 0 \\ \alpha_2 & \alpha_1 & 3 & 0 & \cdot & \cdot & 0 \\ & \cdot & \cdot & \alpha_1 & \cdot & \cdot & \cdot \\ \alpha_n & \alpha_{n-1} & \cdot & \cdot & \cdot & \alpha_1 & n \end{bmatrix} \begin{bmatrix} -c_1 \\ -c_2 \\ -c_3 \\ \cdot \\ -c_n \end{bmatrix}
$$

7-3. An application of Newton's formulas and Cayley-Hamilton's theorem is to invert a matrix $[A]$.

Consider the characteristic equation:

$$ C(s) = s^n + c_1 s^{n-1} + c_2 s^{n-2} + \cdots + c_{n-1} s + c_n = 0 $$

By using Cayley-Hamilton's theorem, we have:

$$ C([A]) = [A]^n + c_1[A]^{n-1} + \cdots + c_{n-1}[A] + c_n[I] = 0 $$

Multiplying $[A]^{-1}$, we obtain:

$$ [A]^{n-1} + c_1[A]^{n-2} + \cdots + c_n[A]^{-1} = 0 $$

Therefore,

$$ [A]^{-1} = -\frac{1}{c_n}([A]^{n-1} + c_1[A]^{n-2} + \cdots + c_{n-1}[I]) $$

where c_n is easily proved to be $|A|$.

(a) Use the forgoing inverse formula to invert the following matrix:

$$ [A] = \begin{bmatrix} 1 & 2 & 4 \\ 3 & 9 & 7 \\ 5 & 0 & 6 \end{bmatrix} $$

(b) Use the same formula to invert

$$ \{[I]s - [A]\} = \begin{bmatrix} s-1 & -2 & -4 \\ 3 & (s-9) & -7 \\ -5 & 0 & (s-6) \end{bmatrix} $$

7-4. Evaluate the transition matrix for each of the following systems:

$$ \begin{bmatrix} \dot{x}_1 \\ \dot{x}_2 \\ \dot{x}_3 \end{bmatrix} = \begin{bmatrix} 0 & 1 & 0 \\ 0 & 0 & 1 \\ -20 & -24 & -9 \end{bmatrix} \begin{bmatrix} x_1 \\ x_2 \\ x_3 \end{bmatrix}, \qquad \begin{bmatrix} \dot{y}_1 \\ \dot{y}_2 \\ \dot{y}_3 \end{bmatrix} = \begin{bmatrix} 0 & 1 & 0 \\ 0 & 0 & 1 \\ -24 & -26 & -9 \end{bmatrix} \begin{bmatrix} y_1 \\ y_2 \\ y_3 \end{bmatrix} $$

(a) by the basic Laplace transform method;

(b) by the matrizant method;

(c) by the generalized Heaviside expression method.

7-5. Give the Heaviside expansion for each of the following by using the formula involving the Vandermonde matrices and initial vectors:

(a) $ X_1 = \dfrac{s+5}{(s+1)(s+2)(s+3)(s+4)} $

(b) $ Y_1 = \dfrac{(s+2)(s+4)}{(s+1)(s+3)(s+5)} $

(c) $Z_1 = \dfrac{(s+3)}{(s+1)^4(s+2)}$

7-6. Solve the circuits in Fig. P7-6 by using the state variable approach.

(a) In Fig. P7-6(a), consider i_l and v_c as the state variables. Initially, $v_c(0) = 1$ volt; $i_l(0) = 0$.

(b) In Fig. P7-6(b), consider i_1 and i_2 as state variables. Initial conditions are zero.

(c) In Fig. P7-6(c), consider v_c and i_l as state variables. Initial conditions are as shown.

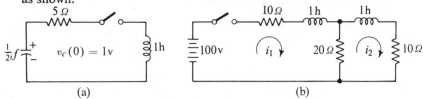

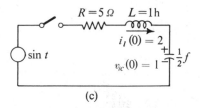

(c)

Fig. P7-6

7-7. Solve the following set of state equations where all initial conditions are equal to zero:

$$\dot{x}_1 = x_2$$
$$\dot{x}_2 = x_3$$
$$\dot{x}_3 = -6x_1 - 11x_2 - 6x_3 + 10$$

(a) by using Lagrange's variation of parameters method;

(b) by the flow graph direct method;

(c) by the matrix flow graph partition method.

7-8. Formulate a set of state equations for the system and then solve for $c(t)$.

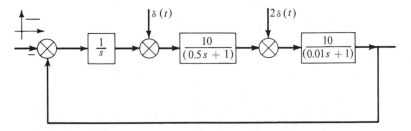

Fig. P7-8

7-9. Write Problems 7-5(a), 7-6(a), and 7-7(a) into Runge-Kutta's formulation. Try to solve them on a digital computer.

CHAPTER 8

The Stability Problem

In Chapter 3 we used a classical definition of stability based on the input-output relationship and built a stability theory for linear systems. Several techniques and criteria were developed, all involving the transfer function as the link.

If state variables are to be used for describing a dynamic system, a more general definition of stability will be required, and is as follows.

Given a dynamic system which is initially in a state of equilibrium: if any disturbance produces an effect which increases with time, the system was in a state of unstable equilibrium; but if the effect decreases, the system is stable.

This definition is consistent with the one given in Chapter 3 but is more general on two counts: (1) it attributes stability to an equilibrium state—the general nonlinear system may have several equilibrium states, some stable and others not; (2) it speaks of a general disturbance, not necessarily a well known driving function.

In this chapter equilibrium points are defined and classified. Stability in the Liapunov sense is defined and illustrated; and Liapunov theorems with related techniques are presented.

I. Equilibrium Points

1. Phase Plane and Phase Space

The solution of a differential equation

$$\dot{x}_1 = x_2$$
$$\dot{x}_2 = x_3$$
$$\cdot$$
$$\cdot \tag{1}$$
$$\cdot$$
$$\dot{x}_n = f_n(x_1, x_2, \ldots, x_n)$$

can be expressed by

$$x_1 = x_1(t)$$
$$x_2 = x_2(t)$$
$$\cdot$$
$$\cdot$$
$$\cdot$$ \hfill (2)
$$x_n = x_n(t)$$

The relationship between x_1 and t is usually drawn as an x_1-t curve; so are the x_2-t, ..., x_n-t curves. If the independent variable t is eliminated from (2), the following equation is obtained.

$$\varphi(x_1, x_2, x_3, \ldots, x_n) = 0 \tag{3}$$

If we interpret $x_1, x_2, \ldots$, and x_n as mutual orthogonal coordinates, Eq. (3) uniquely defines a curve or a trajectory in the $(x_1, x_2, \ldots, x_n)$ coordinates.

A particular case that we usually encounter is that $n = 2$, or

$$\dot{x}_1 = x_2$$
$$\dot{x}_2 = f_2(x_1, x_2) \tag{4}$$

The solution is written as

$$x_1 = x_1(t)$$
$$x_2 = x_2(t) \tag{5}$$

Elimination of t gives

$$\psi(x_1, x_2) = 0 \tag{6}$$

The x_1-x_2 space is two-dimensional and is called the phase plane.

2. State Space

Equation (1) for describing a system is a special form of the following general one:

$$\dot{y}_1 = f_1(y_1, y_2, \ldots, y_n)$$
$$\dot{y}_2 = f_2(y_1, y_2, \ldots, y_n)$$
$$\cdot$$
$$\cdot$$
$$\cdot$$ \hfill (7)
$$\dot{y}_n = f_n(y_1, y_2, \ldots, y_n)$$

Comparing the first $(n - 1)$ equations of (7) and (1), we see that whereas

$\dot{x}_i = x_{i+1}$, $i = 1, 2, \ldots, (n-1)$ in Eq. (1), $\dot{y}_i = f_i(y_1, \ldots, y_n)$, $i = 1, 2, \ldots,$ $(n-1)$, in Eq. (7). The solution of (7) can be also written as

$$
\begin{aligned}
y_1 &= y_1(t) \\
y_2 &= y_2(t) \\
&\;\;\cdot \\
&\;\;\cdot \\
&\;\;\cdot \\
y_n &= y_n(t)
\end{aligned}
\tag{8}
$$

Elimination of t in (8) yields

$$\eta(y_1, y_2, \ldots, y_n) = 0 \tag{9}$$

which also defines a trajectory in a space. To distinguish the x space from the y space, we give y space a general name: "the state space." Obviously, the phase space is a special case of the state space. Because Eq. (7) is so general in form and so comprehensive in representing systems, we may say that *all* solutions of differential equations, and consequently the behavior of any dynamic system may be summarized by a sketch of trajectories in an appropriate state space.

3. Phase Portrait

In the last chapter, we considered two particular problems in which the systems were of second order, namely, Rayleigh's equation and Van der Pol's equation. In those cases the systems are completely defined by the trajectories of the (x_1, x_2) point in the phase plane.

The starting point of a trajectory corresponds to a set of initial conditions, and the trajectory corresponds to a solution. If several sets of initial conditions are given and several trajectories are drawn as in the case of Rayleigh's equation as shown in Fig. 8-1, we have a portrait of the Rayleigh system. Such a family of trajectories in the phase plane, which completely define the dynamic behavior for a variety of specific conditions, is called a phase portrait.

To investigate the relationship between the trajectories, the common destination of the trajectories, etc. is a means of understanding the dynamic behavior of the system and is an important part of the qualitative analysis of a differential equation.

From now on in this section, only the second order system will be treated. First we will define and describe the various types of points of destination or departure on the trajectories of nonlinear systems; next, we will relate these points to similar points found in linear systems in the locating of roots of characteristic equations; finally, we will interpret geometrically the eigenvectors of linear systems.

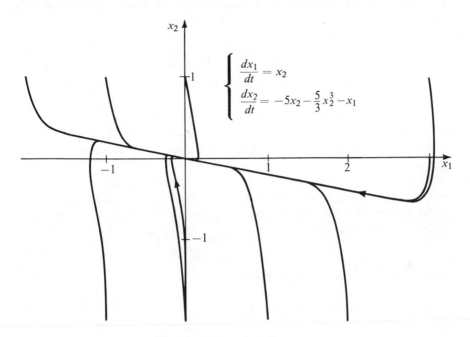

Fig. 8-1. Trajectories of

$$\dot{x}_1 = x_2$$

$$\dot{x}_2 = -5x_2 - \frac{5}{3}x_2^3 - x_1$$

The origin of the plane is a node.

4. Singularities

If there is no proper physical or geometrical interpretation associated with it, a phase portrait is simply a place for storing data. Fortunately Poincaré has analyzed the nature of a family of trajectories in a very detailed manner. He first observed that there is a unique trajectory which passes through each point in the phase plane with the exception of some, which are called singular points. He observed also that either an infinite number, or none, of the trajectories pass through a singular point.

When we investigate the dynamic behavior of a system through the phase plane approach, the first thing that we should do is to locate the singular points.

Consider the following nonlinear differential equation.

$$\dot{x}_1 = x_2$$
$$\dot{x}_2 = -3x_1 - x_1^2 - 2x_2 \tag{10}$$

In order to locate the singular points, let the left-hand members of (10) equal zero:

$$x_2 = 0$$
$$-3x_1 - x_1^2 - 2x_2 = 0 \tag{11}$$

The solution gives two points, $(0, 0)$ and $(-3, 0)$, in the x_1-x_2 plane (the phase plane).

Geometrically speaking, in the neighborhood of the first singular point, $(0, 0)$, the trajectories travel toward the point, whereas in the neighborhood of the second singular point, $(-3, 0)$, the trajectories move away from the point.

Physically speaking, if the differential equation represents a dynamic system, the first point corresponds to one kind of equilibrium and the second point corresponds to another. Singular points, therefore, are sometimes called equilibrium points.

5. Classification of Equilibrium Points

Equilibrium points are of great importance in the analysis of dynamic systems. There are several kinds which are summarized as follows:

(a) CENTER

If a family of curves in the phase plane encircles a point, it is called a center.

Example 1.

$$\frac{dx_1}{dt} = x_2$$
$$\frac{dx_2}{dt} = -\sin x_1 \tag{12}$$

The phase plane trajectories shown in Fig. 8-2 encircle the equilibrium point at the origin, which is therefore called a center.

(b) NODE

A node is a point in the phase plane which a family of trajectories approach or from which they depart.

Example 2.

$$\frac{dx_1}{dt} = x_2$$
$$\frac{dx_2}{dt} = -5x_2 - \frac{5}{3}x_2^3 - x_1 \tag{13}$$

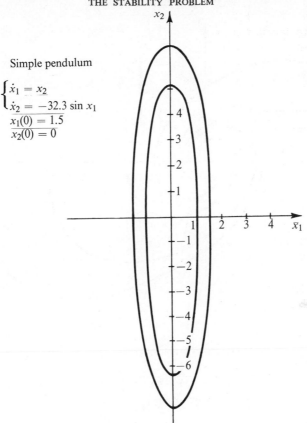

Simple pendulum

$$\begin{cases} \dot{x}_1 = x_2 \\ \dot{x}_2 = -32.3 \sin x_1 \end{cases}$$
$$\frac{\overline{x_1(0) = 1.5}}{x_2(0) = 0}$$

Fig. 8-2. Trajectories of
$\dot{x}_1 = x_2$
$\dot{x}_2 = -32.2 \sin x_1$
The origin of the plane is a center.

Trajectories in the phase plane are shown in Fig. 8-1. Since all the trajectories are approaching the origin, it is a node.

(c) FOCUS

This is a point to which, or from which, the trajectories spiral.

Example 3.

$$\frac{dx_1}{dt} = x_2$$

$$\frac{dx_2}{dt} = -0.5x_2 - 0.166x_2^3 - x_1 \qquad (14)$$

We see that the trajectories shown in Fig. 8-3 approach the origin spirally. The origin is, therefore, a focus.

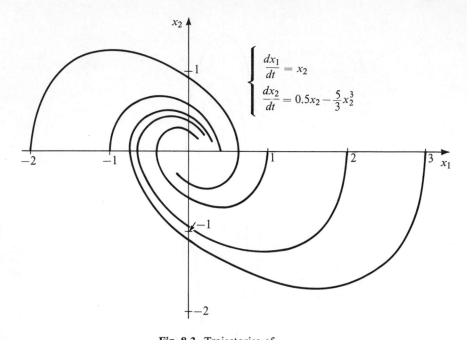

Fig. 8-3. Trajectories of

$$\dot{x}_1 = x_2$$
$$\dot{x}_2 = -0.5x_1 - 0.166x_2^3 - x_1$$

The origin of the plane is a focus.

(d) SADDLE POINT

If the trajectories initially may move toward a point but finally veer off and move away, it is called a saddle point.

Example 4.

$$\frac{dx_1}{dt} = x_2$$

$$\frac{dx_2}{dt} = x_1$$

(15)

The origin of the phase plane shown in Fig. 8-4 is a saddle point.

6. Limit Cycles

If all the trajectories in a region of the phase plane approach a single closed curve or depart from it, this line represents some type of equilibrium and is called a limit cycle. It is an isolated, closed curve in the phase plane. A nonlinear system has a fixed number of limit cycles, each of which separates

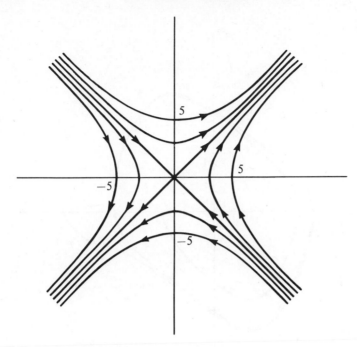

Fig. 8-4. Trajectories of
$$\dot{x}_1 = x_2$$
$$\dot{x}_2 = x_1$$
The origin of the plane is a saddle point.

the phase plane into two distinct regions. All nearby trajectories approach the limit cycle as time approaches infinity, positive or negative.

A classic example of a limit cycle is shown in the Van der Pol equation

$$\dot{x}_1 = x_2$$
$$\dot{x}_2 = 0.8(1 - x_1^2)x_2 - x_1 \tag{16}$$

Figure 8-5 shows the closed curve which is the limit cycle of this system. The trajectory which starts at $x_1(0) = 1$, $x_2(0) = 0$ approaches the cycle from the inside, and the trajectory starting from $x_1(0) = 3$, $x_2(0) = 0$ approaches it from the outside. Thus an equilibrium is established.

7. Linearization

Is there a way to determine the nature of an equilibrium point without drawing the phase portrait?

One of the methods is to approximate the nonlinear system with a linear system. Consider the following nonlinear system:

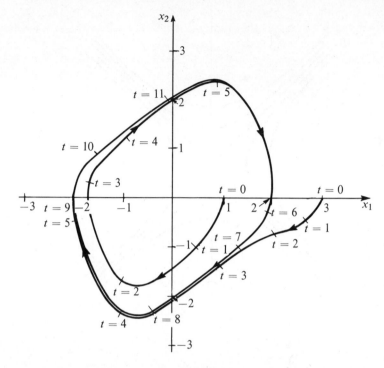

Fig. 8-5. A limit cycle in the phase plane of the Van der Pol equation

$$\dot{x}_1 = x_2$$
$$\dot{x}_2 = 0.8(1 - x_1^2)x_2 - x_1$$

$$[\dot{x}] = [f(x)] \tag{17}$$

The linearized system

$$[\dot{x}] = [A][x] \tag{17a}$$

is obtained by using the Jacobian matrix, where

$$[A] = \left[\frac{\partial f}{\partial x}\right] \tag{17b}$$

The partial differentiation in (17) is defined by the Jacobian matrix, or

$$\left[\frac{\partial f}{\partial x}\right] = \begin{bmatrix} \dfrac{\partial f_1}{\partial x_1} & \dfrac{\partial f_1}{\partial x_2} & \dfrac{\partial f_1}{\partial x_3} & \cdots & \dfrac{\partial f_1}{\partial x_n} \\[2mm] \dfrac{\partial f_2}{\partial x_1} & \dfrac{\partial f_2}{\partial x_2} & \dfrac{\partial f_2}{\partial x_3} & \cdots & \dfrac{\partial f_2}{\partial x_n} \\[2mm] \cdot & \cdot & \cdot & \cdot & \cdot \\[2mm] \dfrac{\partial f_n}{\partial x_1} & \dfrac{\partial f_n}{\partial x_2} & \dfrac{\partial f_n}{\partial x_3} & \cdots & \dfrac{\partial f_n}{\partial x_n} \end{bmatrix} \tag{18}$$

Recall the method of labeling rows and columns in this matrix: The sub-scripts of the functions in the numerator correspond to the rows, whereas the independent variables label the columns.

The Jacobian matrix can be considered as an extension of the technique of using the tangent line to approximate a nonlinearity, as shown in Fig. 8-6.

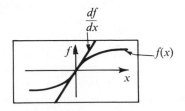

Fig. 8-6. Linearization of a nonlinearity.

To illustrate the use of the Jacobian matrix in linearization, consider Eq. (10) again.

$$\dot{x}_1 = x_2$$
$$\dot{x}_2 = -3x_1 - x_1^2 - 2x_2$$

or

$$\dot{x}_1 = f_1(x)$$
$$\dot{x}_2 = f_2(x)$$

(19)

The linear approximation is given by

$$[A] = \begin{bmatrix} \dfrac{\partial f_1}{\partial x_1} & \dfrac{\partial f_1}{\partial x_2} \\ \dfrac{\partial f_2}{\partial x_1} & \dfrac{\partial f_2}{\partial x_2} \end{bmatrix} = \begin{bmatrix} 0 & 1 \\ -3 - 2x_1 & -2 \end{bmatrix}$$

(20)

Two equilibrium points have been obtained by Eq. (11). They are $(0, 0)$ and $(-3, 0)$.

At the equilibrium point $(0, 0)$, the linear approximation of the system becomes

$$\dot{x}_1 = x_2$$
$$\dot{x}_2 = -3x_1 - 2x_2$$

(21)

For the equilibrium point $(-3, 0)$, the linear approximation becomes

$$\dot{x}_1 = x_2$$
$$\dot{x}_2 = 3x_1 - 2x_2$$

(22)

The two linear approximations, (21) and (22), for the system (19) are easier to analyze. Combining the equations of (21) gives

$$\ddot{x} + 2\dot{x} + 3x = 0$$

The easily obtained solution is

$$x = e^{-t}(\alpha \sin \sqrt{2}\, t + \beta \cos \sqrt{2}\, t)$$

From the above solution we can visualize the phase portrait near the equilibrium point $(0, 0)$ as being a family of spiral curves. This equilibrium point is therefore a focus.

On the other hand, from Eq. (22) we obtain a solution:

$$x = \gamma e^{-3t} + \delta e^t \tag{23}$$

which shows that the equilibrium point $(-3, 0)$ would be a saddle point.

It is a commonly used technique to employ one or several linear models to approximate a nonlinear system. Based on the analysis of the linear systems, the properties of the equilibrium points are determined.

8. Equilibrium Points of Linear Second Order Systems

A further study is necessary. From the last example, we see that a system with a pair of conjugate roots has a focus as an equilibrium point, whereas a system with a positive and a negative root has a saddle as an equilibrium point. It seems to be that there is a one-to-one corresponding relationship between the nature of an equilibrium point and the values of the roots of the characteristic equation.

Let us consider in detail the following cases of second order systems.

(a) CENTER

A set of simple second order differential equations is considered:

$$\frac{dx_1}{dt} = -x_2$$

$$\frac{dx_2}{dt} = x_1 \tag{24}$$

In order to obtain an analytic phase plane equation, the two equations may be divided and then integrated.

$$\frac{dx_2}{dx_1} = -\frac{x_1}{x_2}$$

or

$$x_1\, dx_1 + x_2\, dx_2 = 0$$

with the solution,

$$x_1^2 + x_2^2 = c^2 \tag{25}$$

where c^2 is a constant of integration.

Equation (25) is the phase plane equation; if it is plotted, a family of circles will be obtained which constitute the phase plane portrait for the system described by (24).

The characteristic equation of (24) has two roots:

$$\lambda_2, \lambda_2 = \pm j1 \tag{26}$$

We observe that a center in the phase plane corresponds to a pair of imaginary roots on the imaginary axis of the s plane.

(b) NODE

Consider the following system:

$$\frac{dx_1}{dt} = -x_1$$
$$\frac{dx_2}{dt} = -2x_2 \tag{27}$$

Eliminating t, we have

$$\frac{dx_1}{dx_2} = \frac{x_1}{2x_2}$$
$$\frac{dx_1}{x_1} = \frac{dx_2}{2x_2}$$

and integration yields

$$\log x_1 - \log x_2^2 = \log C$$

or

$$\frac{x_1}{x_2^2} = C \quad \text{or} \quad x_1 = Cx_2^2 \tag{28}$$

which is a family of parabolas.

Examining the characteristic equation of (27), we find that the two roots are

$$\lambda_1 = 1, \qquad \lambda_2 = 2 \tag{29}$$

showing that the characteristic equation of (27) has two real roots with the same algebraic sign.

We make the following observation: A node in the phase plane corresponds to real roots on the real axis of the s plane.

(c) SÁDDLE

Consider the system

$$\frac{dx_1}{dt} = -x_1$$
$$\frac{dx_2}{dt} = x_2 \tag{30}$$

The solution would be

$$x_1 \cdot x_2 = c$$

which corresponds to a family of hyperbolas. The roots of the characteristic equation are

$$\lambda_1 = 1, \qquad \lambda_2 = -1$$

from which we observe that a saddle point in the phase plane corresponds to two roots with opposite algebraic signs.

(d) FOCUS

In the following system,

$$\frac{dx_1}{dt} = -x_1 + x_2$$

$$\frac{dx_2}{dt} = -x_1 - x_2$$

(31)

The two equations are divided.

$$\frac{dx_2}{dx_1} = \frac{-x_1 - x_2}{-x_1 + x_2}$$

$$= \frac{x_1 + x_2}{x_1 - x_2}$$

$$= \frac{1 + \dfrac{x_2}{x_1}}{1 - \dfrac{x_2}{x_1}}$$

(32)

Let

$$\frac{x_2}{x_1} = \tan \varphi \quad \text{and} \quad \frac{dx_2}{dx_1} = \tan \theta$$

Substitution gives

$$\tan \theta = \frac{1 + \tan \varphi}{1 - \tan \varphi}$$

or

$$\theta = 45° + \varphi$$

(32a)

$$\theta - \varphi = 45°$$

Equation (32) is reduced to Eq. (32a), which can be interpreted geometrically as follows: Because the difference between angle θ and angle φ is constant, when φ decreases, θ must also decrease. This means that the slope decreases and that the trajectory moves in spirally. See Fig. 8-7.

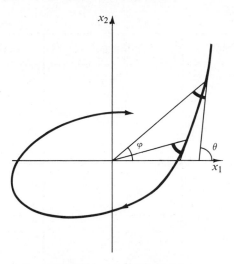

Fig. 8-7. $\theta - \varphi = $ constant.

Let us examine the nature of the characteristic equation of (31). Differentiation gives

$$\frac{d^2x_1}{dt^2} = -\frac{dx_1}{dt} + \frac{dx_2}{dt}$$

$$\frac{d^2x_2}{dt^2} = -\frac{dx_1}{dt} - \frac{dx_2}{dt} \tag{33}$$

Combination of (31) and (33) yields

$$\frac{d^2x}{dt^2} = -2x^2$$

$$\frac{d^2x_1}{dt^2} + 2\left(\frac{dx_1}{dt} + x_1\right) = 0 \tag{34}$$

or

$$\frac{d^2x_1}{dt^2} + 2\frac{dx_1}{dt} + 2x_1 = 0$$

from which we obtain the two roots

$$\lambda_1, \lambda_2 = \frac{-2 \pm \sqrt{4-8}}{2} = -1 \pm j2 \tag{35}$$

Clearly, this is a damped oscillatory system.

The relationships between the roots of characteristic equations and the corresponding phase plane plots can be summarized as shown in Fig. 8-8.

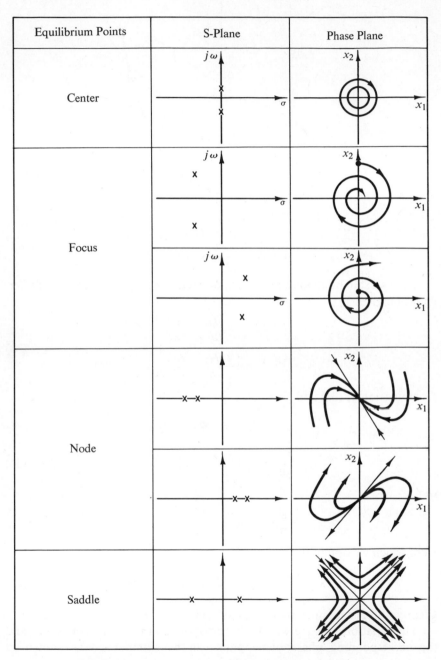

Fig. 8-8. Equilibrium points of a second order differential equation.

9. Geometrical Interpretation of Eigenvectors

It was stated previously that by using the eigenvectors one can diagonalize a state matrix. Here we offer a geometrical interpretation in the phase plane.
Consider the following second order linear system:

$$\frac{d^2x}{dt^2} + 4\frac{dx}{dt} + x = 0 \tag{36}$$

The corresponding state equations are

$$\begin{bmatrix} \dot{x}_1 \\ \dot{x}_2 \end{bmatrix} = \begin{bmatrix} 0 & 1 \\ -1 & -4 \end{bmatrix}\begin{bmatrix} x_1 \\ x_2 \end{bmatrix} \tag{36a}$$

The eigenvalues of this equation are easily found to be

$$\lambda_1, \lambda_2 = -3.73, -0.27$$

from which eigenvectors are evaluated.

$$\left\{ \begin{bmatrix} 0 & 1 \\ -1 & -4 \end{bmatrix} - (-3.73)\begin{bmatrix} 1 & 0 \\ 0 & 1 \end{bmatrix} \right\}\begin{bmatrix} m_{11} \\ m_{21} \end{bmatrix} = \begin{bmatrix} 0 \\ 0 \end{bmatrix} \tag{37}$$

or

$$\begin{bmatrix} 3.73 & 1 \\ -1 & -0.27 \end{bmatrix}\begin{bmatrix} m_{11} \\ m_{21} \end{bmatrix} = \begin{bmatrix} 0 \\ 0 \end{bmatrix} \tag{37a}$$

Two simultaneous equations are then obtained:

$$\begin{aligned} 3.73m_{11} + m_{21} &= 0 \\ -m_{11} - 0.27m_{21} &= 0 \end{aligned} \tag{38}$$

Following the standard procedure, let

$$m_{11} = 1$$

Then

$$m_{21} = -3.73 \tag{39}$$

Thus, an eigenvector is found:

$$\begin{bmatrix} m_{11} \\ m_{21} \end{bmatrix} = \begin{bmatrix} 1 \\ -3.73 \end{bmatrix} \tag{40}$$

Similarly, for $\lambda_2 = -0.27$

$$\left\{ \begin{bmatrix} 0 & 1 \\ -1 & 4 \end{bmatrix} - (-0.27)\begin{bmatrix} 1 & 0 \\ 0 & 1 \end{bmatrix} \right\}\begin{bmatrix} m_{12} \\ m_{22} \end{bmatrix} = \begin{bmatrix} 0 \\ 0 \end{bmatrix} \tag{41}$$

or

$$\begin{bmatrix} 0.27 & 1 \\ -1 & -3.73 \end{bmatrix}\begin{bmatrix} m_{12} \\ m_{22} \end{bmatrix} = \begin{bmatrix} 0 \\ 0 \end{bmatrix}$$

Let $m_{12} = 1$; then $m_{22} = -0.27$, and another eigenvector is obtained as follows:

$$\begin{bmatrix} m_{11} \\ m_{22} \end{bmatrix} = \begin{bmatrix} 1 \\ -0.27 \end{bmatrix} \tag{42}$$

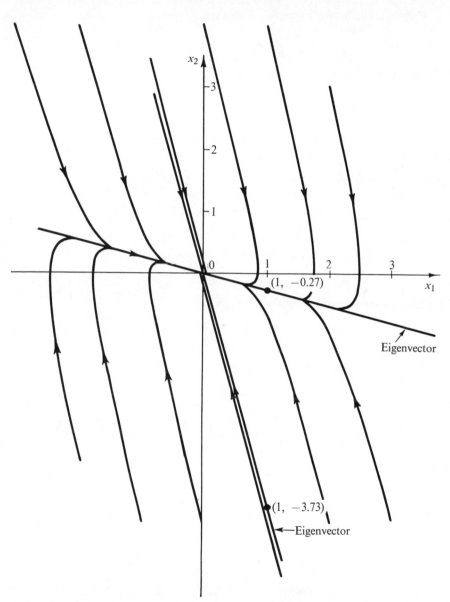

Fig. 8-9. Eigenvectors among other trajectories.

Through the origin of the phase plane, two straight lines can be drawn to represent the eigenvectors $[M]_1$ and $[M]_2$. These are shown in Fig. 8-9. Note that the eigenvectors are a set of trajectories of the differential equation in question. It can also be seen that other trajectories start out as straight lines parallel to an eigenvector and then curve to approach the other eigenvector.

10. Equilibrium Points in a State Space

Consider a general case in which the order of the set of differential equations is more than 2.

$$\frac{dx_1}{dt} = a_{11}x_1 + a_{12}x_2 + a_{13}x_3 + g_1(x_1, x_2, x_3)$$

$$\frac{dx_2}{dt} = a_{21}x_1 + a_{22}x_2 + a_{23}x_3 + g_2(x_1, x_2, x_3) \qquad (43)$$

$$\frac{dx_3}{dt} = a_{31}x_1 + a_{32}x_2 + a_{33}x_3 + g_3(x_1, x_2, x_3)$$

where g_1, g_2, and g_3 are polynomials containing terms of order higher than the first in x_1, x_2, and x_3. Let the left members of (43) equal zero; then the trivial solution of the system is $x_1 = x_2 = x_3 = 0$, which means that the origin is a "singular" or an equilibrium point. In the neighborhood of the singular point, we may consider the following set of linear equations as an approximation of (43).

$$\frac{dx_1}{dt} = a_{11}x_1 + a_{12}x_2 + a_{13}x_3$$

$$\frac{dx_2}{dt} = a_{21}x_1 + a_{22}x_2 + a_{23}x_3 \qquad (44)$$

$$\frac{dx_3}{dt} = a_{31}x_1 + a_{32}x_2 + a_{33}x_3$$

The general trajectories of (44) may be written as

$$\zeta(x_1, x_2, x_3) = 0 \qquad (45)$$

If the three roots of the characteristic equation are λ_1, λ_2, and λ_3, we can determine from them the nature of the equilibrium point in question.

1. If all the roots are real and of the same sign, the equilibrium point is a node.

2. If all the roots are real, but not of the same sign, the equilibrium point is a saddle.

3. If one root is real, two roots are conjugate, and the sum of the two conjugate roots has the same sign as the real root, the equilibrium point is a focal point.

4. If one root is real and two roots are conjugate, and the sum of the

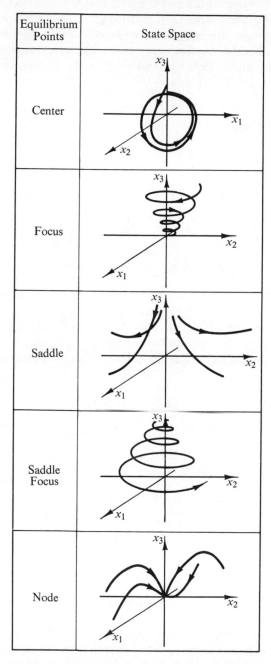

Fig. 8-10. Equilibrium points of third order differential equations.

two conjugate roots has the opposite sign from the real root, the equilibrium point is a saddle-focus.

5. If one root is real and negative, and the other two roots are pure imaginary and conjugate, the equilibrium point is a center.

Sketches of the trajectories of various third order systems in state space are shown in Fig. 8-10.

II.   The Direct Method of Liapunov

1. Stability in the Liapunov Sense

A system is stable if it tends to remain in the neighborhood of its equilibrium point after a small disturbance has been applied. A pendulum is a typical example. On the other hand, a system is considered unstable if it moves far from the equilibrium point after a small disturbance has been applied. An inverted pendulum offers a good illustration (Fig. 8 -11).

Fig. 8-11. Stability and instability in common sense—a pendulum and an inverted pendulum.

To investigate the conditions under which a sufficiently small change in the initial conditions causes only an arbitrarily small change in the solution is to look at the stability problem from the Liapunov viewpoint, which coincides with our intuition.

Liapunov not only extends the intuitive concept but also lays out precise definitions, states theorems in mathematical form, and describes the procedures for constructing the related functions. In other words, he establishes a school of thought as well as a philosophy for attacking the stability problem.

2. Equilibrium Points and Trivial Solutions

It is apparent that there is no motion at an equilibrium point, x_{ei} $(i = 1, 2, \ldots, n)$. This is why we endeavor to determine the equilibrium points of any system, such as

$$\dot{x}_1 = f_1(x_1, x_2, \ldots, x_n)$$
$$\dot{x}_2 = f_2(x_1, x_2, \ldots, x_n)$$
$$\cdot$$
$$\cdot \qquad (1)$$
$$\cdot$$
$$\dot{x}_n = f_n(x_1, x_2, \ldots, x_n)$$

by solving the algebraic equations:

$$f_1(x_1, x_2, \ldots, x_n) = 0$$
$$f_2(x_1, x_2, \ldots, x_n) = 0$$
$$\cdot$$
$$\cdot \qquad (2)$$
$$\cdot$$
$$f_n(x_1, x_2, \ldots, x_n) = 0$$

The number of equilibrium points is independent of n and depends only on the nonlinearities. If the system under consideration is linear, there is only one solution set for (2), which implies that there is only one equilibrium point. If the system under consideration is nonlinear, the number of equilibrium points could be many.

If we wish to analyze a solution in the vicinity of an equilibrium point, $(x_{e1}, x_{e2}, \ldots, x_{en})$, we simply use the following transformation:

$$y_1 = x_1 - x_{e1}$$
$$y_2 = x_2 - x_{e2}$$
$$\cdot$$
$$\cdot \qquad (3)$$
$$\cdot$$
$$y_n = x_n - x_{en}$$

The corresponding set of differential equations becomes

$$\dot{y}_1 = g_1(y_1, y_2, \ldots, y_n)$$
$$\dot{y}_2 = g_2(y_1, y_2, \ldots, y_n)$$
$$\cdot$$
$$\cdot \qquad (4)$$
$$\cdot$$
$$\dot{y}_n = g_n(y_1, y_2, \ldots, y_n)$$

System (4) has an equilibrium point at the origin. The origin $(0, 0, \ldots, 0)$ of y axes corresponds to the equilibrium point $(x_{e1}, x_{e2}, \ldots, x_{en})$ in the x coordinates.

In analytic form, the equilibrium point can be written as follows:

$$y_1 = 0$$
$$y_2 = 0$$
$$.$$
$$.$$
$$.$$
$$y_n = 0$$

(5)

which is also a solution of (4). However, it is so obvious that it is called trivial.

Liapunov's stability study is based on the formulation of (4) to investigate the nature of the zero equilibrium point in the state space, E^n, by constructing a suitable function $v(y_i)$, or devising a sufficient condition for predicting the qualitative behavior of the system in the neighborhood of the origin without finding the solutions of (4).

From now on, $i = 2$ will be used frequently for illustrative examples which are so typical that there is no loss of generality; however, when necessary, higher order systems will be used.

3. Definition of Stability

The point of equilibrium $\mathbf{x} = \mathbf{0}$ is stable in the Liapunov sense if for each $\epsilon > 0$ there can be chosen a $\delta(\epsilon) > 0$ such that from the following relation

$$||\mathbf{x}(t_0)|| < \delta(\epsilon)$$

it follows that

$$||\mathbf{x}(t)|| < \epsilon$$

A graphical interpretation is shown in Fig. 8-12. The x_1-x_2 plane is for $t = t_0$. If $||\mathbf{x}_0||$ is in the interior of the circle of radius δ, the curve $\mathbf{x}(t)$ in

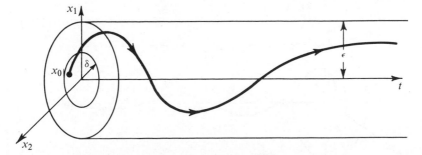

Fig. 8-12. Definition of stability in the Liapunov sense (geometrical interpretation in the motion space).

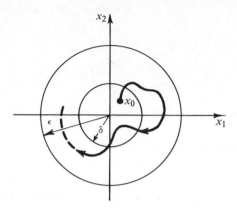

Fig. 8-13. Definition of stability in the Liapunov sense (geometrical interpretation in the state space).

the motion space remains for all future times inside the cylinder of radius ϵ.
The projection of the curve on the x_1-x_2 plane is shown in Fig. 8-13.

4. Definition of Asymptotic Stability

If the equilibrium point is not only stable in the Liapunov sense, but also

$$\lim_{t \to \infty} \| \mathbf{x}(t) \| = 0, \qquad\qquad \text{if } \| \mathbf{x}(t_0) \| < \delta$$

the equilibrium point is said to be asymptotically stable.

An asymptotically stable picture in space is shown in Fig. 8-14. When
$t \to \infty$, it is seen that the curve approaches the t axis. The corresponding
picture in the phase plane is shown in Fig. 8-15, which can be considered as
the projection of the motion space curve on the x_1-x_2 plane.

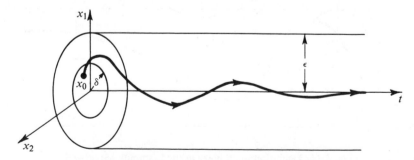

Fig. 8-14. Definition of asymptotic stability in the Liapunov
sense (geometrical interpretation in the motion space).

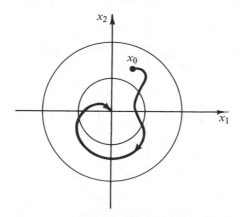

Fig. 8-15. Definition of asymptotic stability in the Liapunov sense (geometrical interpretation in the state space).

5. Reference Grid

From the phase plane viewpoint the stability of a dynamic system in the Liapunov sense as discussed thus far can be summarized as follows:

1. Stable: the trajectories tend toward an orbit as $t \to \infty$ (shown in Fig. 8-13).

2. Asymptotically stable: the trajectories tend toward the origin as $t \to \infty$ (shown in Fig. 8-15).

3. Unstable: the trajectories tend toward infinity as $t \to \infty$ (shown in Fig. 8-16).

Thus a clear geometric picture is obtained as an aid in stability studies.

The following analogy, if not carried too far, is sometimes helpful. Let a stream on a geographical map represent a trajectory in a phase plane. The top of a mountain would be the place where a stream starts and would correspond to an unstable equilibrium point in the phase plane. The bottom of a lake where a stream flow terminates would correspond to a stable equilibrium point in the phase plane.

There is another kind of map which displays contour lines (lines joining points of equal elevation). From these, it is easy to predict the direction of a stream flow (normal to the lines) and to recognize whether a certain spot is the peak of a mountain or the bottom of a valley.

Just as the equipotential contour lines are a better method for showing the direction of stream flow and distinguishing the nature of a spot, so also the techniques of using a potential function to predict the direction of trajectories and to distinguish the nature of the equilibrium point is a very powerful method in the stability study of a dynamic system.

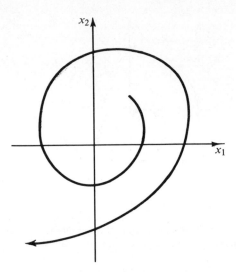

Fig. 8-16. Definition of instability in the Liapunov sense.

Liapunov originated a method of using a function or a reference grid by which the direction of a trajectory can be measured, and the destination of a trajectory can be predicted. The method is called Liapunov's direct method.

6. Liapunov's Asymptotic Stability Theorem

In the neighborhood of the origin of the coordinates x_i, for system (3), if there exists a differentiable, real, variable function v that satisfies the following conditions:

(1)
$$v(x_1, x_2, \ldots, x_n) > 0$$
and $v = 0$ only for $x_i = 0$ $(i = 1, 2, \ldots, n)$

(or v is positive definite).

(2)
$$\frac{dv}{dt} = \sum_{i=1}^{n} \frac{\partial v}{\partial x_i} \cdot \frac{dx_i}{dt} < 0 \qquad \text{for all } x_i \neq 0 \text{ and } t > 0$$

(i.e., dv/dt is negative definite) or

(2a)
$$\frac{dv}{dt} = \sum_{i=1}^{n} \frac{\partial v}{\partial x_i} \cdot \frac{dx_i}{dt} \leq 0$$

(or negative semidefinite) and not identically equal to 0 on the trajectory of the given equation.

(3)
$$v \to \infty \quad \text{for} \quad \|\mathbf{x}\| \to \infty$$

where $\|\mathbf{x}\|$ is the norm of x_i. Then the point $\mathbf{x} = 0$ is asymptotically stable in the sense of Liapunov and v is called a Liapunov function for the system.

Liapunov's asymptotic stability theorem can be proved geometrically. There is no loss of generality if $n = 2$; i.e., a set of two first order equations is considered:

$$\frac{dx_1}{dt} = f_1(x_1, x_2)$$
$$\frac{dx_2}{dt} = f_2(x_1, x_2)$$
(6)

First of all, a positive definite function of the following form is taken:

$$v = a_1 x_1^2 + a_2 x_2^2$$

which is assumed as a tentative Liapunov function. If some values are assigned to v, for example, 1, 2, 5, we obtain

$$a_1 x_1^2 + a_2 x_2^2 = 1$$
$$a_1 x_1^2 + a_2 x_2^2 = 2$$
$$a_1 x_1^2 + a_2 x_2^2 = 5$$
(7)

The first equation of (7) corresponds to a closed curve around the origin; the second and the third equations correspond to two-layer closed curves around the origin.

Taking the derivative of v with respect to t gives

$$\frac{dv}{dt} = 2a_1 x_1 f_1(x_1, x_2) + 2a_2 x_2 f_2(x_1, x_2)$$

If $dv/dt < 0$, at all points of the given phase space or state space in general, except at the origin of coordinates, a point P in the state space will move toward decreasing values of v. In other words, P will tend toward the origin of the state space penetrating the closed curves one after another, as shown in Fig. 8-17. This signifies attenuation with time of all deviations

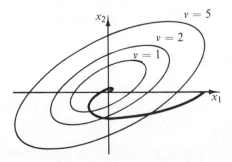

Fig. 8-17. A trajectory of an asymptotically stable system and equal v contours.

x_1 and x_2 in the transient. Or (2) is asymptotically stable in the sense of Liapunov.

7. An Asymptotically Stable Nonlinear System Example

A system is given as shown in Fig. 8-18. By Liapunov's procedure, the stability of the system is to be investigated as follows: First, write the governing equation in terms of a set of first order differential equations

$$\frac{dx_1}{dt} = -x_2 - x_1^3$$

$$\frac{dx_2}{dt} = x_1 - x_2$$

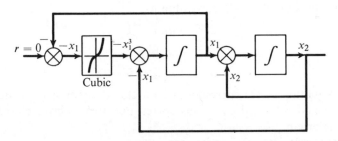

Fig. 8-18. A nonlinear system.

Second, for this system assume a tentative Liapunov function:

$$v = x_1^2 + x_2^2 \tag{8}$$

Third, take the derivative function:

$$\begin{aligned}
\frac{dv}{dt} &= 2x_1\dot{x}_1 + 2x_2\dot{x}_2 \\
&= 2x_1(-x_2 - x_1^3) + 2x_2(x_1 - x_2) \\
&= -2x_1x_2 - 2x_1^4 + 2x_1x_2 - 2x_2^2 \\
&= -(2x_1^4 + 2x_2^2)
\end{aligned} \tag{9}$$

which is negative definite. Fourth,

$$v \to \infty \quad \text{for} \quad ||\mathbf{x}|| \to \infty \tag{10}$$

Equations (8), (9), and (10) satisfy the conditions (1), (2), and (3) of Liapunov's asymptotical stability, respectively. Therefore, system (7) is asymptotically stable.

Figure 8-19 shows one of the trajectories of the solution curves as well as the constant v contours. It is noted that the trajectory tends toward the origin, which is asymptotically stable.

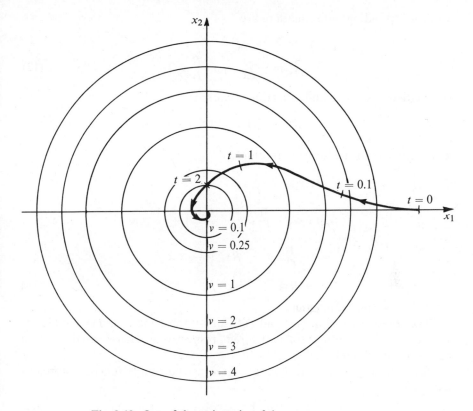

Fig. 8-19. One of the trajectories of the system
$$\dot{x}_1 = -x_2 - x_1^3$$
$$\dot{x}_2 = -x_1 - x_2$$

8. An Asymptotically Stable Linear System Example

An *RLC* circuit is shown in Fig. 8-20; its stability in the sense of Liapunov is to be studied.

The governing equation in the second order form is

$$L\frac{di}{dt} + Ri + \frac{1}{C}\int i\,dt = 0 \tag{11}$$

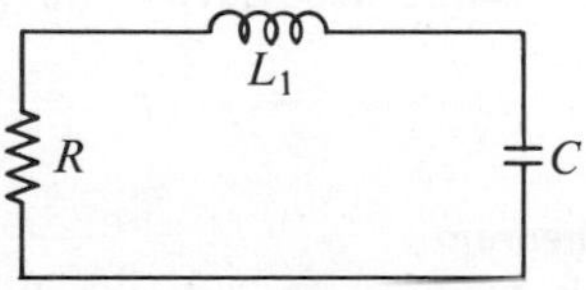

Fig. 8-20. An *RLC* circuit.

The corresponding state equations are

$$\dot{x}_1 = x_2$$

$$\dot{x}_2 = -\frac{1}{LC}x_1 - \frac{R}{L}x_2 \tag{12}$$

A tentative positive definite function is chosen:

$$v = \frac{1}{2}Lx_2^2 + \frac{1}{2C}x_1^2 \tag{13}$$

The derivative of the function is found:

$$\frac{dv}{dt} = Lx_2\dot{x}_2 + \frac{1}{C}x_1\dot{x}_1$$

$$= Lx_2\left(-\frac{1}{LC}x_1 - \frac{R}{L}x_2\right) + \frac{1}{C}x_1x_2$$

$$= -\frac{1}{C}x_1x_2 - Rx_2^2 + \frac{1}{C}x_1x_2$$

$$= -Rx_2^2 \tag{14}$$

which is negative semidefinite

The results shown in (13) and (14) satisfy the conditions (1) and (2a) of Liapunov's asymptotic stability theorem, respectively. Therefore, the system (11) is asymptotically stable.

It is noted that (13) can be interpreted as the energy function stored in the system, or

$$v = \frac{1}{2}Lx_1^2 + \frac{1}{2C}x_1^2 = \frac{1}{2}Li^2 + \frac{1}{2C}q^2 \tag{15}$$

Energy stored in a physical system is, of course, a scalar quantity, as is a Liapunov function. However, the idea of Liapunov functions is more general and more widely applicable than that of an energy function. It should be noted also that the energy function of a system is not necessarily a Liapunov function, and a Liapunov function is not necessarily the energy function.

The derivative of v with respect to t for the system being studied is a power function. Equation (14) shows that the rate of change of energy, or power, is proportional to the resistance R. Of course, this is only offering an interpretation of the derivative of this Liapunov function. A power function or a dissipation function is not necessarily the derivative of a Liapunov function either.

9. Liapunov's Stability Theorem

For system (3), if there exists a differentiable function v in the neighborhood of the origin of the coordinate x_i, it satisfies the following conditions:

(1) $$v(x_1, x_2, \ldots, x_n) > 0$$

$$\text{and } v = 0 \quad \text{only for } x_i = 0 \quad (i = 1, 2, \ldots, n)$$

(or v is positive definite).

(2) $$\frac{dv}{dt} = \sum_{i=1}^{n} \frac{\partial v}{\partial x_i} \frac{dx_i}{dt} \leq 0 \qquad \text{for all } x_i \neq 0 \text{ and } t > 0$$

(or dv/dt is negative semidefinite). Then the point $x_i = 0$ is stable in the sense of Liapunov.

10. A Stable Nonlinear System Example

In order to illustrate how to apply the stability theorem of Liapunov, the following example is investigated.

$$\frac{dx_1}{dt} = -x_1 x_2^4$$

$$\frac{dx_2}{dt} = x_2 x_1^4 \tag{16}$$

Try the following positive definite function as a tentative Liapunov function.

$$v = x_1^4 + x_2^4 \tag{17}$$

Then, examining its derivative function, we have

$$\frac{dv}{dt} = -4x_1^4 x_2^4 + 4x_1^4 x_2^4 = 0 \tag{18}$$

which is negative semidefinite. Therefore, the origin $x_i = 0$ is stable in the sense of Liapunov.

11. A Stable Linear System Example

Consider the system

$$\frac{d^3 x}{dt^3} + \frac{d^2 x}{dt^2} + \frac{dx}{dt} + x = 0 \tag{19}$$

The corresponding state equations are

$$\dot{x}_1 = x_2$$
$$\dot{x}_2 = x_3 \tag{20}$$
$$\dot{x}_3 = -x_1 - x_2 - x_3$$

A positive definite function is assumed:

$$v = (x_1 + x_2)^2 + (x_2 + x_3)^2 + (x_1 + x_3)^2 \tag{21}$$

Now we examine the derivative function

$$\frac{dv}{dt} = 2(x_1 + x_2)(\dot{x}_1 + \dot{x}_2) + 2(x_2 + x_3)(\dot{x}_2 + \dot{x}_3)$$
$$+ 2(x_1 + x_3)(\dot{x}_1 + \dot{x}_3) = -2(x_1 + x_3)^2 \tag{22}$$

which is negative semidefinite. Therefore, by the stability theorem system (20) is stable in the sense of Liapunov.

Figure 8-21 shows the Liapunov function contours of (21) and one of the trajectories. It is interesting to note that the trajectory tends toward an orbit as $t \to \infty$.

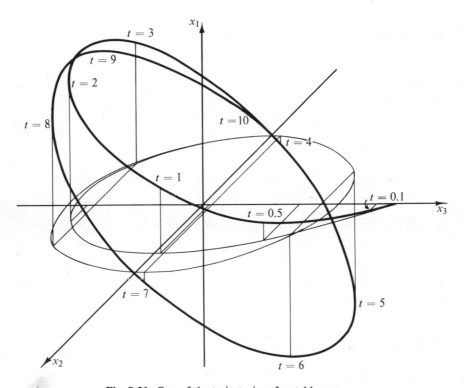

Fig. 8-21. One of the trajectories of a stable system.

12. Chetayev's Instability Theorem

For system (3), if there exists a real, differentiable function v in the neighborhood of the origin of the coordinates x_i, which satisfies the following:
1. $v > 0$ in a certain domain near the origin.
2. The derivative

$$\frac{dv}{dt} = \sum_{i=1}^{n} \frac{\partial v}{\partial x_i} f_i(x_1, x_2, \ldots, x_n) > 0$$

for $t > 0$ in that domain.

3. $v = 0$ at the boundary point of that domain.

4. The origin is a boundary point of that domain.

then the origin is unstable.

To prove the theorem, start at condition (2) or

$$\text{for } t > 0, \qquad \frac{dv}{dt} > 0$$

This is to say that dv/dt is equal to or is larger than a real positive number η, or

$$\frac{dv}{dt} > \eta$$

Integration yields

$$v(t) - v(0) > \eta t$$

When $t \to \infty$, the function v increases without bound along the trajectory, which contradicts the original assumption (3), so the trajectory cannot cross the boundary of the domain from assumption (4), since the origin is on the boundary of the domain, the point can be found arbitrarily near the origin from which trajectories go outward. This means that the origin is unstable (Fig. 8-22).

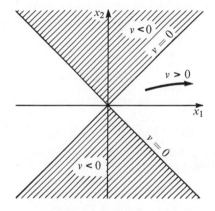

Fig. 8-22. One of the trajectories of an unstable system.

13. An Unstable System Example

Consider the following system:

$$\frac{dx_1}{dt} = x_1^5 + x_2^3$$

$$\frac{dx_2}{dt} = x_1^3 + x_2^5 \tag{23}$$

A function $v = x_1^4 - x_2^4$ satisfies all the conditions shown above, because

1. $v > 0$, in the domain $|x_1| > |x_2|$.
2. $dv/dt = 4(x^8 - y^8) > 0$ in the domain $|x_1| > |x_2|$ for $t > 0$.
3. At the boundary $|x_1| = |x_2|$, $v = 0$.
4. The origin is a boundary point.

Therefore, the origin is unstable.

III. Construction of Liapunov Functions
for Linear Systems

1. Classification of Construction Methods

In Sec. I we have shown the use of Liapunov functions in studying the stabilities of systems. The procedures we applied fall into a single category that can be summarized as follows:

1. Choose a tentative function $v(x)$.
2. Calculate $\dot{v}(x)$.
3. Examine the definiteness of $\dot{v}(x)$.

This can be called the forward approach. However, for many systems it is difficult to construct a tentative function as an initial step. We then resort to a backward approach as follows:

1. Choose a tentative function $\dot{v}(x)$.
2. Calculate $v(x)$.
3. Examine the definiteness of $v(x)$.

This second approach is very powerful in the construction of Liapunov functions for systems.

Many techniques have been developed to implement the backward procedure. We will explain them one by one, starting in this section with those applicable to linear systems.

2. A Procedure Based on Undetermined Elements

A linear system is described by

$$\frac{d[x]}{dt} = [A][x] \tag{1}$$

where $[A]$ is a constant matrix.

The undetermined element procedure has the following steps:

Step 1. Consider a quadratic form

$$v(x) = [x]^T[P][x] = \sum_{i,j=1}^{n} P_{ij}x_i x_j \tag{2}$$

Name the undetermined elements of $[P]$.

As an example, for a second order system, assume

$$[P] = \begin{bmatrix} p_{11} & p_{12} \\ p_{12} & p_{22} \end{bmatrix} \tag{3}$$

Step 2. Taking the derivative of $v(x)$ yields

$$\frac{dv}{dt} = \frac{d[x]^T}{dt} \cdot [P][x] + [x]^T[P]\frac{d(x)}{dt} \tag{4}$$

$$= [x]^T\{[A]^T[P] + [P][A]\}[x]$$

Step 3. Constrain the $\dot{V}$ function to be negative definite; i.e., let

$$[x]^T\{[A]^T[P] + [P][A]\}[x] = \text{negative definite} \tag{5}$$

Step 4. Determine the value of each element of the $[P]$ matrix and then check the sign definiteness of $[P]$. If $[P]$ is positive definite $[x]^T[P][x]$ is a Liapunov function for the system.

The following problem is considered as an example for which a Liapunov function is to be found.

$$\begin{bmatrix} \dot{x}_1 \\ \dot{x}_2 \end{bmatrix} = \begin{bmatrix} 0 & 1 \\ -3 & -4 \end{bmatrix} \begin{bmatrix} x_1 \\ x_2 \end{bmatrix} \tag{6}$$

Step 1. Assume a $v(x)$ function and name its elements, or

$$v(x) = [x_1, x_2] \begin{bmatrix} p_{11} & p_{12} \\ p_{12} & p_{22} \end{bmatrix} \begin{bmatrix} x_1 \\ x_2 \end{bmatrix} \tag{7}$$

Step 2. Take the derivative with respect to t:

$$\frac{dv}{dt} = [x]^T\{[A]^T[P] + [P][A]\}[x]$$

$$= [x_1, x_2] \left\{ \begin{bmatrix} 0 & -3 \\ 1 & -4 \end{bmatrix} \begin{bmatrix} p_{11} & p_{12} \\ p_{12} & p_{22} \end{bmatrix} + \begin{bmatrix} p_{11} & p_{12} \\ p_{12} & p_{22} \end{bmatrix} \begin{bmatrix} 0 & 1 \\ -3 & -4 \end{bmatrix} \right\} \begin{bmatrix} x_1 \\ x_2 \end{bmatrix} \tag{8}$$

Step 3. Let

$$\begin{bmatrix} 0 & -3 \\ 1 & -4 \end{bmatrix} \begin{bmatrix} p_{11} & p_{12} \\ p_{12} & p_{22} \end{bmatrix} + \begin{bmatrix} p_{11} & p_{12} \\ p_{12} & p_{22} \end{bmatrix} \begin{bmatrix} 0 & 1 \\ -3 & -4 \end{bmatrix} = \begin{bmatrix} -1 & 0 \\ 0 & -1 \end{bmatrix} \tag{9}$$

The right-hand matrix was arbitrarily chosen; it was the simplest matrix by which a negative definite function can be formed.

After equating both sides, one obtains

$$-6p_{12} = -1$$

$$p_{11} - 4p_{12} - 3p_{22} = 0 \tag{10}$$

$$-8p_{22} + 2p_{12} = -1$$

The solution of (10) is

$$p_{12} = \tfrac{1}{6}$$

$$p_{22} = \tfrac{1}{6}$$

$$p_{11} = \tfrac{7}{6}$$

The v function is constructed as follows:

$$v(x) = [x_1, x_2]\begin{bmatrix} \frac{7}{6} & \frac{1}{6} \\ \frac{1}{6} & \frac{1}{6} \end{bmatrix}\begin{bmatrix} x_1 \\ x_2 \end{bmatrix} \tag{11}$$

Step 4. The sign definiteness of (11) could be examined in several ways.

Program 11. EXPANDING $[P][A] + [A]^T[P] = -[D]$.

```
C         SOLVING   P A + TRANSPOSED P A = -I

          DIMENSION P(55,56),B(10,10),XX(10)
100       READ(5,500)N,((B(L,K),K=1,N),L=1,N),(XX(L),L=1,N)
          WRITE(6,600)
          WRITE(6,501) N,((B(L,K),K=1,N),L=1,N),(XX(L),L=1,N)
          M=0
          NI=N*(N+1)/2
          NJ=NI+1
          DO 1 II=1,N
          DO 2 JJ=II,N
          M=M+1
          P(M,NJ)=0.
          IF(II.EQ.JJ) P(M,NJ)=-XX(II)/2.
          LM=0
          DO 4 K=1,N
          DO 7 L=K,N
          LM=LM+1
          IF(K.NE.II.AND.K.NE.JJ.AND.L.NE.II.AND.L.NE.JJ) GO TO 10
          IF(K.EQ.II.AND.L.NE.JJ) GO TO 11
          IF(K.NE.II.AND.L.EQ.JJ) GO TO 12
          IF(K.NE.II.AND.L.NE.JJ.AND.K.EQ.JJ.AND.L.NE.II) GO TO 13
          IF(K.NE.II.AND.L.NE.JJ.AND.K.NE.JJ.AND.L.EQ.II) GO TO 14
          IF(K.EQ.II.AND.L.EQ.JJ.AND.K.EQ.L) GO TO 15
          IF(K.EQ.II.AND.L.EQ.JJ.AND.K.NE.L) P(M,LM)=B(K,K)+B(L,L)
          GO TO 7
15        P(M,LM)=B(K,K)
          GO TO 7
14        P(M,LM)=B(K,JJ)
          GO TO 7
13        P(M,LM)=B(L,II)
          GO TO 7
12        P(M,LM)=B(K,II)
          GO TO 7
11        P(M,LM)=B(L,JJ)
          GO TO 7
10        P(M,LM)=0.
7         CONTINUE
4         CONTINUE
2         CONTINUE
1         CONTINUE
          WRITE(6,602)
          WRITE(6,601) ((P(II,JJ),JJ=1,NJ),II=1,NI)
500       FORMAT(I2/(8F10.4))
501       FORMAT(5X,I2,(10F10.3)/)
600       FORMAT(6X,15HGIVEN MATRIX IS)
601       FORMAT(5X,(11F10.3)/)
602       FORMAT(6X,18HREQUIRED EQUATIONS)
          GO TO 100
          END
```

The simplest is by using Sylvester's theorem:

$$\tfrac{7}{6} > 0$$

$$\begin{vmatrix} \tfrac{7}{6} & \tfrac{1}{6} \\ \tfrac{1}{6} & \tfrac{1}{6} \end{vmatrix} > 0$$

This means that (11) is positive definite and is therefore a suitable Liapunov function for the system.

For linear cases Liapunov's asymptotic stability theorem can be restated as follows: A sufficient condition for (1) to be asymptotically stable is that $[x]^T[P][x]$ be a positive definite quadratic form, where $[P]$ satisfies the equation

$$[P][A] + [A]^T[P] = -[D] \tag{12}$$

and $[x]^T[D][x]$ is any positive definite form.

3. A Technique for Expanding $[P][A] + [A]^T[P] = -[D]$

To expand the matrix equation (12) into simultaneous algebraic equations is very tedious. If the dimension of A is n, we would have to write $n(n + 1)/2$ equations.

A simple technique is developed for the expansion. A computer program based on that technique can be easily written (see Program 11).

Consider that $[P]$ is a symmetrical matrix. We use the elements of $[P]$ to form a vector, $\mathbf{p}_v$, whose components are $p(1, 1), p(1, 2) \ldots p(1, n), p(2, 2),$ $p(2, 3) \ldots p(2, n) \ldots p(n - 1, n - 1), p(n - 1, n), p(n, n)$. Also, using the elements of $[D]$, we construct a vector, $\mathbf{d}_v$, whose components are $d(1, 1),$ $d(1, 2) \ldots d(1, n), d(2, 2) \ldots d(n - 1, n - 1), d(n - 1, n), d(n, n)$.

It is desirable to convert (12) into the following set of equations.

$$[\alpha]\mathbf{p}_v = -\tfrac{1}{2}\mathbf{d}_v \tag{13}$$

The problem is then reduced to how to construct an $[\alpha]$ matrix. $[\alpha]$ is an $n(n + 1)/2$ by $n(n + 1)/2$ matrix. We use (K, L) and (I, J) as the row and column subscripted index, respectively. The following restrictions should be satisfied when forming the $[\alpha]$ matrix directly from the $[A]$ matrix.

(1)   if $K = I, L \neq J \longrightarrow A(L, J)$
         $K \neq I, L = J \longrightarrow A(K, I)$

(2) $K \neq I, L \neq J$
         $K = J, L \neq I \longrightarrow A(L, I)$
         $K \neq J, L = I \longrightarrow A(K, J)$ (14)
         $K \neq J, L \neq I \longrightarrow 0$

(3) $K = I, L = J$
         $K = J, L = I \longrightarrow A(K, I)$
         $K \neq J, L \neq I \longrightarrow A(K, I) + A(L, J).$

Once $[\alpha]$ is formed, finding $\mathbf{p}_v$ from (13) is a straightforward procedure.

Consider a third-order system $(n = 3)$. We form $\mathbf{p}_\nu$ vector and the $\mathbf{d}_\nu$ vector respectively

$$\mathbf{p}_\nu = \begin{bmatrix} p_{11} \\ p_{12} \\ p_{13} \\ p_{22} \\ p_{23} \\ p_{33} \end{bmatrix} \quad \text{and} \quad \mathbf{d}_\nu = \begin{bmatrix} d_1 \\ 0 \\ 0 \\ d_2 \\ 0 \\ d_3 \end{bmatrix}$$

Construct the following table by using (14):

		1-1	1-2	1-3	2-2	2-3	3-3
	1-1	A_{11}	A_{21}	A_{31}	0	0	0
	1-2	A_{12}	$A_{11} + A_{22}$	A_{32}	A_{21}	A_{31}	0
(I,J)	1-3	A_{13}	A_{23}	$A_{11} + A_{33}$	0	A_{21}	A_{31}
	2-2	0	A_{12}	0	A_{22}	A_{32}	0
	2-3	0	A_{13}	A_{12}	A_{23}	$A_{22} + A_{33}$	A_{32}
	3-3	0	0	A_{13}	0	A_{23}	A_{33}

(The column group header is $(K-L)$.)   (15)

The application of the restriction rules is not difficult. For example, we want to find the element X.

	$(K-L)$
$(I-J)$	$Z = ?$

We consider (1) if $K = I = L = J$, we have $Z = A(K, I)$

             (2) if $K \neq I$, $L \neq J$ and $K \neq J$, $L \neq I$, then $Z = 0$,

etc.

The result obtained in (10) is the required $[\alpha]$.

4. A Procedure Based on Integration by Parts

Choosing a dv/dt function as negative definite or negative semidefinite, and using integration by parts, we can evaluate the corresponding v function. Then the definiteness of $v(x)$ can be examined: if it is positive definite, we have a Liapunov function for the system in question. The procedure is best explained by the following example.

A third order linear system is given.

$$\begin{aligned} \dot{x}_1 &= x_2 \\ \dot{x}_2 &= x_3 \\ \dot{x}_3 &= -a_1 x_1 - a_2 x_2 - a_3 x_3 \end{aligned} \tag{16}$$

For this method, the integral form of (16) is needed:

$$\int x_1 \, dt = -\frac{a_2}{a_1} x_1 - \frac{a_3}{a_1} x_2 - \frac{1}{a_1} x_3$$

$$\int x_2 \, dt = x_1 \tag{17}$$

$$\int x_3 \, dt = x_2$$

Assuming $dv/dt = -x_3^2$, find v by using the technique of integration by parts.

$$v = \int -x_3^2 \, dt$$

$$-v = x_3 \int x_3 \, dt - \iint x_3 \, dt \, dx_3$$

$$= x_3 x_2 - \int x_2 [-a_1 x_1 - a_2 x_2 - a_3 x_3] \, dt$$

$$= x_3 x_2 + \int a_1 x_1 \, dx_1 + \int a_2 x_2^2 \, dt + \int a_3 x_2 x_3 \, dt$$

$$= x_2 x_3 + a_1 \frac{x_1^2}{2} + \int a_2 x_2^2 \, dt + a_3 \frac{x_2^2}{2} \tag{18}$$

Integrating the middle term by parts gives

$$a_2 \int x_2^2 \, dt = a_2 x_2 \int x_2 \, dt - a_2 \iint x_2 \, dt \, dx_2$$

$$= a_2 x_2 x_1 - a_2 \int x_1 \, dx_2$$

$$= a_2 x_1 x_2 - a_2 \int x_1 x_3 \, dt$$

$$= a_2 x_1 x_2 - a_2 \int x_3 \left(-\frac{a_2}{a_1} \dot{x}_1 - \frac{a_3}{a_1} \dot{x}_2 - \frac{1}{a_1} \dot{x}_3 \right) dt$$

$$= a_2 x_1 x_2 + \frac{a_2}{a_1} \frac{x_3^2}{2} + a_2 \int \frac{a_2}{a_1} \dot{x}_2 x_2 \, dt + \frac{a_2 a_3}{a_1} \int x_3 \dot{x}_2 \, dt$$

$$= a_2 x_1 x_2 + \frac{a_2}{2a_1} x_3^2 + \frac{a_2^2}{a_1} \frac{x_2^2}{2} + \frac{a_2 a_3}{a_1} \int x_3^2 \, dt \tag{19}$$

Substitution of (19) into (18) yields

$$\int x_3^2 \, dt = x_2 x_3 + a_1 \frac{x_1^2}{2} + a_3 \frac{x_2^2}{2} + a_2 x_1 x_2 + \frac{a_2}{2a_1} x_3^2 + \frac{a_2^2}{a_1} \frac{x_2^2}{2} + \frac{a_2 a_3}{a_1} \int x_3^2 \, dt$$

$$\left(1 - \frac{a_2 a_3}{a_1} \right) \int x_3^2 \, dt = x_2 x_3 + \frac{a_1 x_1^2}{2} + \frac{a_3 x_2^2}{2} + a_2 x_1 x_2 + \frac{a_2}{2a_1} x_3^2 + \frac{a_2^2}{a_1} \frac{x_2^2}{2}$$

Therefore,

$$\int -x_3^2 \, dt = v = \frac{1}{\frac{a_2 a_3}{a_1} - 1} \left[\frac{a_1 x_1^2}{2} + a_2 x_1 x_2 + \left(\frac{a_2^2}{a_1} + a_3 \right) \frac{x_2^2}{2} + x_2 x_3 + \frac{a_2}{2a_1} x_3^2 \right]$$

or

$$v(x) = \frac{1}{2} \frac{a_1}{(a_2 a_3 - a_1)} \cdot [x_1, x_2, x_3] \begin{bmatrix} a_1 & a_2 & 0 \\ a_2 & \dfrac{a_2^2}{a_1} + a_3 & 1 \\ 0 & 1 & \dfrac{a_2}{a_1} \end{bmatrix} \begin{bmatrix} x_1 \\ x_2 \\ x_3 \end{bmatrix} \tag{20}$$

which is positive definite, and dv/dt is negative semidefinite; therefore, the function v is a Liapunov function.

It is noted that the necessary conditions for v to be positive definite coincide with those of Routh-Hurwitz. From this viewpoint, the Routh-Hurwitz criterion can also be derived from a Liapunov function.

5. A Procedure Based on Dynamic Variables

A $v(x)$ function can be assumed in the form of

$$v(x) = [\dot{x}]^T [P][\dot{x}] \tag{21}$$

or

$$v(x) = [f(x)]^T [P][f(x)]$$

for the system

$$[\dot{x}] = [f(x)] \tag{22}$$

where $[P]$ is a constant matrix whose elements are to be determined.

Take the time derivative of both sides:

$$\frac{dv}{dt} = [f(x)]^T \frac{d}{dt}\{[P][f(x)]\} + \frac{d}{dt}\{[f(x)]^T [P]\}[f(x)]$$

$$= [f(x)]^T [P] \frac{d}{dt}[f(x)] + \frac{d}{dt}\{[f(x)]^T\}[P][f(x)]$$

$$= [f(x)]^T [P] \left[\frac{\partial f(x)}{\partial x}\right]\left[\frac{dx}{dt}\right] + \left[\frac{dx}{dt}\right]^T \left[\frac{\partial f(x)}{\partial x}\right]^T [P][f(x)]$$

$$= [f(x)]^T \left\{[P]\left[\frac{\partial f(x)}{\partial x}\right] + \left[\frac{\partial f(x)}{\partial x}\right]^T [P]\right\}[f(x)] \tag{23}$$

where $[\partial f(x)/\partial x]$ is the Jacobian matrix of the system. It is defined as follows:

$$\left[\frac{\partial f(x)}{\partial x}\right] = \begin{bmatrix} \dfrac{\partial f_1}{\partial x_1} & \dfrac{\partial f_1}{\partial x_2} & \cdots & \dfrac{\partial f_1}{\partial x_n} \\ \dfrac{\partial f_2}{\partial x_1} & \dfrac{\partial f_2}{\partial x_2} & \cdots & \cdots \\ \cdot & & & \cdot \\ \cdot & & & \cdot \\ \cdot & & & \cdot \\ \dfrac{\partial f_n}{\partial x_1} & \dfrac{\partial f_n}{\partial x_2} & \cdots & \dfrac{\partial f_n}{\partial x_n} \end{bmatrix} \tag{24}$$

In order to make $v(x)$ a Liapunov function, its derivative is constrained to be negative definite; i. e.,

$$[P]\left[\frac{\partial f(x)}{\partial x}\right] + \left[\frac{\partial f(x)}{\partial x}\right][P] = [-I] \tag{25}$$

where $[I]$ is the identity matrix.

Solving (25) gives a set of simultaneous equations with the elements of $[P]$ determined. An example will show the detailed procedure.

Consider the system shown in Fig. 8-23. Its governing equations are as follows:

$$\begin{aligned} \dot{x}_1 &= x_2 \\ \dot{x}_2 &= -16x_1 - 10x_2 \end{aligned} \tag{26}$$

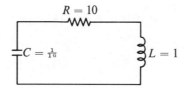

$$R = 10$$
$$C = \tfrac{1}{16} \qquad L = 1$$

Fig. 8-23. An *RLC* circuit.

In matrix form:

$$\begin{bmatrix} \dot{x}_1 \\ \dot{x}_2 \end{bmatrix} = \begin{bmatrix} 0 & 1 \\ -16 & -10 \end{bmatrix} \begin{bmatrix} x_1 \\ x_2 \end{bmatrix} \tag{26a}$$

A $v(x)$ function is then assumed:

$$\begin{aligned} v(x) &= [x_1, x_2] \begin{bmatrix} f_1(x) \\ f_2(x) \end{bmatrix}^T \begin{bmatrix} p_{11} & p_{12} \\ p_{12} & p_{22} \end{bmatrix} \begin{bmatrix} f_1(x) \\ f_2(x) \end{bmatrix} \begin{bmatrix} x_1 \\ x_2 \end{bmatrix} \\ &= [x_1, x_2] \begin{bmatrix} 0 & -16 \\ 1 & -10 \end{bmatrix} \begin{bmatrix} p_{11} & p_{12} \\ p_{12} & p_{22} \end{bmatrix} \begin{bmatrix} 0 & 1 \\ -16 & -10 \end{bmatrix} \begin{bmatrix} x_1 \\ x_2 \end{bmatrix} \end{aligned} \tag{27}$$

Substituting (26a) into (3) gives

$$\frac{dv}{dt} = [f(x)]^T \left\{ [P]\left[\frac{\partial f(x)}{\partial x}\right] + \left[\frac{\partial f(x)}{\partial x}\right]^T [P] \right\} [f(x)]$$

Forcing the function to a negative definite form yields

$$\begin{bmatrix} p_{11} & p_{12} \\ p_{12} & p_{22} \end{bmatrix} \begin{bmatrix} \dfrac{\partial f_1}{\partial x_1} & \dfrac{\partial f_1}{\partial x_2} \\ \dfrac{\partial f_2}{\partial x_1} & \dfrac{\partial f_2}{\partial x_2} \end{bmatrix} + \begin{bmatrix} \dfrac{\partial f_1}{\partial x_1} & \dfrac{\partial f_2}{\partial x_1} \\ \dfrac{\partial f_1}{\partial x_2} & \dfrac{\partial f_2}{\partial x_2} \end{bmatrix} \begin{bmatrix} p_{11} & p_{12} \\ p_{12} & p_{22} \end{bmatrix} = \begin{bmatrix} -1 & 0 \\ 0 & -1 \end{bmatrix}$$

or $\tag{28}$

$$\begin{bmatrix} p_{11} & p_{12} \\ p_{12} & p_{22} \end{bmatrix} \begin{bmatrix} 0 & 1 \\ -16 & -10 \end{bmatrix} + \begin{bmatrix} 0 & -16 \\ 1 & -10 \end{bmatrix} \begin{bmatrix} p_{11} & p_{12} \\ p_{12} & p_{22} \end{bmatrix} = \begin{bmatrix} -1 & 0 \\ 0 & -1 \end{bmatrix}$$

Expanding the matrix equation into a set of simultaneous equations, we have

$$p_{11} - 10p_{12} - 16p_{22} = 0$$
$$-16(p_{12} + p_{12}) = -1$$
$$2p_{12} - 20p_{22} = -1$$

The solutions are as follows:

$$p_{11} = \tfrac{372}{320}$$
$$p_{12} = \tfrac{1}{32} \tag{29}$$
$$p_{22} = \tfrac{17}{320}$$

The $v(x)$ function is then

$$v(x) = [x_1, x_2]\begin{bmatrix} 0 & -16 \\ 1 & -10 \end{bmatrix}\begin{bmatrix} p_{11} & p_{12} \\ p_{12} & p_{22} \end{bmatrix}\begin{bmatrix} 0 & 1 \\ -16 & -10 \end{bmatrix}\begin{bmatrix} x_1 \\ x_2 \end{bmatrix}$$

$$= [x_1, x_2]\begin{bmatrix} 0 & -16 \\ 1 & -10 \end{bmatrix}\begin{bmatrix} \tfrac{372}{320} & \tfrac{1}{32} \\ \tfrac{1}{32} & \tfrac{17}{320} \end{bmatrix}\begin{bmatrix} 0 & 1 \\ -16 & -10 \end{bmatrix}\begin{bmatrix} x_1 \\ x_2 \end{bmatrix} \tag{30}$$

$$= [x_1, x_2]\begin{bmatrix} \tfrac{68}{5} & 8 \\ 8 & \tfrac{117}{20} \end{bmatrix}\begin{bmatrix} x_1 \\ x_2 \end{bmatrix}$$

The function is examined in the light of Sylvester's theorem:

$$\left| \tfrac{68}{5} \right| > 0$$

$$\begin{vmatrix} \tfrac{68}{5} & 8 \\ 8 & \tfrac{117}{20} \end{vmatrix} > 0$$

Therefore, $v(x)$ is positive definite.

$$\frac{dv}{dt} = [x_1, x_2]\begin{bmatrix} -256 & -160 \\ -160 & -101 \end{bmatrix}\begin{bmatrix} x_1 \\ x_2 \end{bmatrix} \tag{31}$$

which is negative definite. The system, then, is asymptotically stable.

6. Summary of Procedures

The technique for constructing a Liapunov function for a linear system has two alternate approaches:

1. Choose $v(x)$; then calculate $\dot{v}(x)$ and examine the properties of $\dot{v}(x)$.
2. Choose $\dot{v}(x)$; then calculate $v(x)$ and examine the properties of $v(x)$.

The method presented in Sec. I, which was to construct a $v(x)$ function by intuition or inspection, belongs to the first category. The initial part requires experience; the steps following are comparatively easy.

The method shown in this section, which consists of three techniques, belongs to the second category. The only difficulty here is performing the cumbersome calculations. With a digital computer available it should prove to be the easier.

IV. Routh-Hurwitz' Criterion and Liapunov Functions

1. Three Direct Methods

Routh's algorithm, Hurwitz' determinants, and Liapunov's functions belong to one catagory in their approach to the stability problem, namely, the direct method, which uses the coefficients of a system as a basis for predicting the stability of the system without actually finding its solutions.

Historically, the three theories were developed independently. In the investigation of a linear system, each attacks the same problem, but each one obtains different forms and has different means of derivation. But obviously they should all arrive at the same conclusion.

The similarity of Routh's algorithm and Hurwitz' determinants has been pointed out in Chapter 3. In this section, the following topics will be covered: (1) the procedures for constructing a Liapunov function by using Routh's algorithm and (2) a derivation of Routh's criterion from the direct method of Liapunov. Also, a very important form in state variable technique— Schwarz' form—will be explained and a by-product—the symmetrical form of Routh-Hurwitz criterion—will be given.

2. Schwarz' Form

Schwarz developed a canonical form to represent a linear system. His form looks like Fig. 8-24.

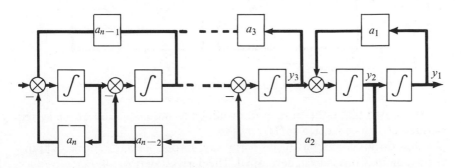

Fig. 8-24. A general Schwarz' form.

The corresponding matrix form is

$$
\begin{bmatrix} \dot{y}_1 \\ \dot{y}_2 \\ \dot{y}_3 \\ \dot{y}_4 \\ \cdot \\ \cdot \\ \cdot \\ \dot{y}_n \end{bmatrix} = \begin{bmatrix} 0 & 1 & 0 & 0 & \cdot & \cdot & \cdot & 0 & 0 \\ -a_1 & 0 & 1 & 0 & \cdot & \cdot & \cdot & 0 & 0 \\ 0 & -a_2 & 0 & 1 & \cdot & \cdot & \cdot & 0 & 0 \\ \cdot & & \cdot & \cdot & \cdot & \cdot & \cdot & & \cdot \\ & & \cdot & \cdot & \cdot & \cdot & & & \\ 0 & 0 & 0 & 0 & 0 & 0 & 0 & 0 & 1 \\ 0 & 0 & 0 & 0 & \cdot & \cdot & \cdot & -a_{n-1} & -a_n \end{bmatrix} \begin{bmatrix} y_1 \\ y_2 \\ y_3 \\ \cdot \\ \\ \\ \cdot \\ y_n \end{bmatrix} \quad (1)
$$

If n is odd (say $n = 5$) the Schwarz' form is as shown in Fig. 8-25.

If n is even (say $n = 6$), the corresponding configuration is as shown in Fig. 8-26.

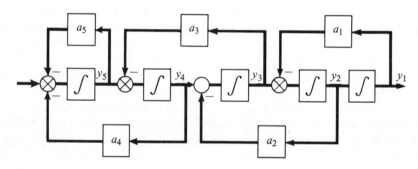

Fig. 8-25. Schwarz' form ($n = 5$).

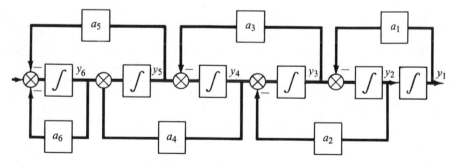

Fig. 8-26. Schwarz' form ($n = 6$).

It is noted that either $n =$ even or odd; the state equation of the system in question can be written as (1).

How do we convert a Bush's form, say, into a Schwarz' form? A transformation matrix has been developed. The derivation is as follows:

Consider a second order system

$$\ddot{x} + b_2\dot{x} + b_1 x = 0$$

The corresponding phase variable form or Bush's form is

$$\begin{bmatrix} \dot{x}_1 \\ \dot{x}_2 \end{bmatrix} = \begin{bmatrix} 0 & 1 \\ -b_1 & -b_2 \end{bmatrix} \begin{bmatrix} x_1 \\ x_2 \end{bmatrix} \quad \text{or} \quad \dot{x} = bx \tag{2}$$

The Schwarz' form is

$$\begin{bmatrix} \dot{y}_1 \\ \dot{y}_2 \end{bmatrix} = \begin{bmatrix} 0 & 1 \\ -a_1 & -a_2 \end{bmatrix} \begin{bmatrix} y_1 \\ y_2 \end{bmatrix} \quad \text{or} \quad \dot{y} = ay \tag{3}$$

There is no difference between (2) and (3). We can interpret the identity
by

$$[P_2][B][P_2]^{-1} = [A]$$

where

$$[P_2] = \begin{bmatrix} 1 & 0 \\ 0 & 1 \end{bmatrix} \tag{4}$$

and is called the transformation matrix for converting a Bush's form to a
Schwarz' form.

For a third order system

$$\dddot{x} + b_3\ddot{x} + b_2\dot{x} + b_1 x = 0 \tag{5}$$

The Bush form is

$$\begin{bmatrix} \dot{x}_1 \\ \dot{x}_2 \\ \dot{x}_3 \end{bmatrix} = \begin{bmatrix} 0 & 1 & 0 \\ 0 & 0 & 1 \\ -b_1 & -b_2 & -b_3 \end{bmatrix} \begin{bmatrix} x_1 \\ x_2 \\ x_3 \end{bmatrix}$$

Before formulating the Schwarz' form, we establish a Routh's array first.

$$\begin{array}{cc} 1 & b_2 \\ b_3 & b_1 \\ b_2 - \dfrac{b_1}{b_3} & \\ b_1 & \end{array} \tag{6}$$

The positions of the elements in the Routh array can be indicated as
matrix double subscript notations or

$$\begin{array}{cc} C_{11} & C_{12} \\ C_{21} & C_{22} \\ C_{31} & \\ C_{41} & \end{array} \tag{6a}$$

A $[P_3]$ matrix is formed as follows:

$$[P_3] = \begin{bmatrix} 1 & 0 & 0 \\ 0 & 1 & 0 \\ \dfrac{C_{22}}{C_{21}} & 0 & 1 \end{bmatrix} \tag{7}$$

Then $[P_3]^{-1}$ is calculated as

$$[P_3]^{-1} = \begin{bmatrix} 1 & 0 & 0 \\ 0 & 1 & 0 \\ \dfrac{-C_{22}}{C_{21}} & 0 & 1 \end{bmatrix} \tag{8}$$

When we operate the similar transformation on $[B]$ matrix,

$$[P_3][B][P_3]^{-1} = [A] \tag{9}$$

or

$$\begin{bmatrix} 1 & 0 & 0 \\ 0 & 1 & 0 \\ \dfrac{C_{22}}{C_{21}} & 0 & 1 \end{bmatrix} \begin{bmatrix} 0 & 1 & 0 \\ 0 & 0 & 1 \\ -b_1 & -b_2 & -b_3 \end{bmatrix} \begin{bmatrix} 1 & 0 & 0 \\ 0 & 1 & 0 \\ \dfrac{-C_{22}}{C_{21}} & 0 & 1 \end{bmatrix} = \begin{bmatrix} 0 & 1 & 0 \\ \dfrac{-b_1}{b_3} & 0 & 1 \\ 0 & \left(\dfrac{b_1}{b_3} - b_2\right) & -b_3 \end{bmatrix} \tag{9a}$$

Observing (9a) shows that the $[A]$ matrix can be expressed in terms of the Routh array, or

$$\begin{bmatrix} 0 & 1 & 0 \\ \dfrac{-b_1}{b_3} & 0 & 1 \\ 0 & \left(+\dfrac{b_1}{b_3} - b_2\right) & -b_3 \end{bmatrix} = \begin{bmatrix} 0 & 1 & 0 \\ \dfrac{-C_{41}}{C_{21}} & 0 & 1 \\ 0 & \dfrac{-C_{31}}{C_{11}} & \dfrac{-C_{21}}{C_{11}} \end{bmatrix} \tag{10}$$

Equation (10) is Schwarz' form in terms of Routh's elements.

For a fourth order system:

$$\ddddot{x} + b_4\dddot{x} + b_3\ddot{x} + b_2\dot{x} + b_1 x = 0 \tag{11}$$

The necessary $[P_4]$ matrix can be derived as

$$[P_4] = \begin{bmatrix} 1 & 0 & 0 & 0 \\ 0 & 1 & 0 & 0 \\ \dfrac{C_{32}}{C_{31}} & 0 & 1 & 0 \\ 0 & \dfrac{C_{22}}{C_{21}} & 0 & 1 \end{bmatrix} \tag{12}$$

For a fifth order system, following a similar reasoning yields

$$[P_5] = \begin{bmatrix} 1 & 0 & 0 & 0 & 0 \\ 0 & 1 & 0 & 0 & 0 \\ \dfrac{C_{42}}{C_{41}} & 0 & 1 & 0 & 0 \\ 0 & \dfrac{C_{32}}{C_{31}} & 0 & 1 & 0 \\ \dfrac{C_{23}}{C_{24}} & 0 & \dfrac{C_{22}}{C_{21}} & 0 & 1 \end{bmatrix} \tag{13}$$

Comparing (4), (7), (12), and (13), we see the pattern of $[P]$ matrix. It has the following properties:

1. $[P_2]$, $[P_3]$, $[P_4]$, $\ldots$, and $[P_{n-1}]$ are principal minors of $[P_n]$.
2. The $[P]$ matrix is always a triangular matrix.
3. Each element of the main diagonal is unity.
4. Each element of the first lower diagonal is zero.
5. The elements of the second lower diagonal are determined by respective ratios of the second column and first column of the Routh array.
6. Each element of the third lower diagonal is zero.
7. The elements of the fourth lower diagonal are determined by the respective ratios of the third column and the first column of the Routh array, etc. In other words, the general $[P_n]$ matrix is shown as follows:

$$\begin{bmatrix} 1 & 0 & 0 & 0 & 0 & 0 & 0 & 0 \\ \cdot & \cdot & \cdot & \cdot & \cdot & \cdot & \cdot & \cdot \\ 0 & 1 & 0 & 0 & 0 & 0 & 0 & 0 \\ \cdot & 0 & 1 & 0 & 0 & 0 & 0 & 0 \\ 0 & \dfrac{C_{62}}{C_{61}} & 0 & 1 & 0 & 0 & 0 & 0 \\ \cdot & 0 & \dfrac{C_{52}}{C_{51}} & 0 & 1 & 0 & 0 & 0 \\ 0 & \dfrac{C_{43}}{C_{41}} & 0 & \dfrac{C_{42}}{C_{41}} & 0 & 1 & 0 & 0 \\ \cdot & 0 & \dfrac{C_{33}}{C_{31}} & 0 & \dfrac{C_{32}}{C_{31}} & 0 & 1 & 0 \\ 0 & \dfrac{C_{24}}{C_{21}} & 0 & \dfrac{C_{23}}{C_{21}} & 0 & \dfrac{C_{22}}{C_{21}} & 0 & 1 \end{bmatrix} \tag{14}$$

It is noted that for the second order system, $[P_2]$ is taken from the right-hand side lower square, etc. The matrix $[P_n]$ can be called the Chen-Chu transformation matrix.

The method is best illustrated by the following numerical example:

For a given system in Bush's form:

$$
\begin{bmatrix} \dot{x}_1 \\ \dot{x}_2 \\ \dot{x}_3 \\ \dot{x}_4 \end{bmatrix} = \begin{bmatrix} 0 & 1 & 0 & 0 \\ 0 & 0 & 1 & 0 \\ 0 & 0 & 0 & 1 \\ -24 & -50 & -35 & -10 \end{bmatrix} \begin{bmatrix} x_1 \\ x_2 \\ x_3 \\ x_4 \end{bmatrix} \tag{15}
$$

The Routh array for the system is

$$
\begin{array}{ccc|ccc}
1 & 35 & 24 & C_{11} & C_{12} & C_{13} \\
10 & 50 & & C_{21} & C_{22} & \\
30 & 24 & & C_{31} & C_{32} & \\
42 & & & C_{41} & & \\
24 & & & C_{51} & &
\end{array} \tag{16}
$$

Substituting the values of the elements in Routh's array into (14) one obtains

$$
P = \begin{bmatrix} 1 & 0 & 0 & 0 \\ 0 & 1 & 0 & 0 \\ 0.8 & 0 & 1 & 0 \\ 0 & 5 & 0 & 1 \end{bmatrix} \tag{17}
$$

The Schwarz' form is

$$
[\ddot{y}] = [A][y] \tag{18}
$$

where

$$
[A] = [P][B][P]^{-1}
$$

$$
= \begin{bmatrix} 0 & 1 & 0 & 0 \\ -8 & 0 & 1 & 0 \\ 0 & -4.2 & 0 & 1 \\ 0 & 0 & -30 & -10 \end{bmatrix}
$$

3. Kalman-Bertram's Liapunov Functions

In using Schwarz' variable, Kalman-Bertram developed a Liapunov function as follows:

$$
v = [y]^T \begin{bmatrix} a_n a_{n-1} & \cdots & a_2 a_1 & & \\ & \cdot & & & \\ & & \cdot & & \\ & & & \cdot & \\ & & & a_n a_{n-1} a_{n-2} & \\ & & & & a_n a_{n-1} \\ & & & & & a_n \end{bmatrix} [y] \tag{19}
$$

and its derivative function reads

$$\frac{dv}{dt} = [y] \begin{bmatrix} 0 & \cdot & \cdot & \cdot & \cdot & \cdot & \cdot & \cdot & \cdot & 0 \\ \cdot & & & & & & & & \\ \cdot & \cdot & 0 & & & & & & \\ \cdot & & \cdot & & & & & & \\ \cdot & & & & \cdot & & 0 & & \\ \cdot & & & & & \cdot & & & \\ \cdot & & & & & & \cdot & & \\ \cdot & & & & & & & \cdot & \\ 0 & & & & & & & & -2a_n^2 \end{bmatrix} [y] \qquad (20)$$

Form (19) can also be expressed by the elements of Routh's array:

$$v = [y]^T \begin{bmatrix} C_{n,1}C_{n+1,1} & & & & \\ & \cdot & & & \\ & & \cdot & & \\ & & & C_{31}C_{41} & \\ & & & & C_{21}C_{31} \\ & & & & & C_{11}C_{21} \end{bmatrix} [y] \qquad (21)$$

If the original phase variables are used, the Liapunov function is expressible as follows:

$$v = [x]^T[P]^T \begin{bmatrix} C_{n,1}C_{n+1,1} & & & \\ & \cdot & & \\ & & \cdot & \\ & & & C_{21}C_{31} \\ & & & & C_{11}C_{21} \end{bmatrix} [P][x] \qquad (22)$$

$$= [x]^T[R][x]$$

where $[R]$ is called the symmetrical form of the Routh-Hurwitz criterion, or Ralston's form.

4. Example

For our example problem (18), the Kalman-Bertram Liapunov function is

$$v = [y]^T \begin{bmatrix} 1008 & & & \\ & 1260 & & \\ & & 300 & \\ & & & 10 \end{bmatrix} [y] \qquad (23)$$

and the derivative function is

$$\frac{dv}{dt} = -200 y_4^2 \tag{24}$$

The symmetrical form of Routh-Hurwitz criterion is taken from the core of the Liapunov function or

$$v = [x]^T \begin{bmatrix} 1200 & 0 & 240 & 0 \\ 0 & 1510 & 0 & 50 \\ 240 & 0 & 300 & 0 \\ 0 & 50 & 0 & 10 \end{bmatrix} [x] \tag{25}$$

Equation (22) shows that we can derive a Liapunov function from Routh's algorithm. For our example, the Liapunov function is (25). As mentioned before, obtaining a Liapunov function for a linear system, even one of fairly low order, is not very easy. The form of (22) offers a direct answer.

5. A Derivation of the Routh-Hurwitz Criterion from the Liapunov Functions

Consider Eq. (21) again. The definiteness of the function $v(y)$ is based on the *signs* of the elements of the *diagonal*. But if any element of the first column of Routh's array is negative, the positive definite property of (21) is necessarily destroyed, because each C_{i1} appears twice in the diagonal. Therefore, the values in the first column of the Routh array must have the same signs. In other words, the fact that there is no change of sign in the C_{i1}'s of a Routh array is Routh's criterion of stability.

V. Controllability and Observability

1. Definition of Controllability

The investigation of the properties of a system does not end with that very important stability analysis. Other properties need to be studied. Two of these, controllability and observability, seem to be destined to play an increasingly important role.

Controllability has to do with the transfer of a system from one state to another by appropriate inputs in finite time. Such a transfer is not always possible for all systems without the imposition of some additional conditions.

Before we discuss these conditions, a few remarks on the meaning of the rank of a matrix are in order.

2. Rank of a Matrix

In a matrix $[A]$ if at least one of its r square minors is not equal to zero, while every $(r + 1)$ square minor is zero, $[A]$ is said to have rank r.

Example 1. The rank of

$$[A] = \begin{bmatrix} 0 & 0 & 1 \\ 0 & 1 & -6 \\ 1 & -6 & 31 \end{bmatrix}$$

is $r = 3$ since

$$\begin{vmatrix} 0 & 0 & 1 \\ 0 & 1 & -6 \\ 1 & -6 & 31 \end{vmatrix} \neq 0$$

Example 2. The rank of

$$[B] = \begin{bmatrix} 3 & 0 & 0 \\ 4 & -2 & 10 \\ 1 & -2 & 10 \end{bmatrix}$$

is $r = 2$ since

$$\begin{vmatrix} 3 & 0 & 0 \\ 4 & -2 & 10 \\ 1 & -2 & 10 \end{vmatrix} = 0$$

while

$$\begin{vmatrix} 3 & 0 \\ 4 & -2 \end{vmatrix} \neq 0$$

Example 3. The rank of

$$[C] = \begin{bmatrix} 1 & 3 & 5 \\ 2 & 4 & 7 \end{bmatrix}$$

is $r = 2$ since

$$\begin{vmatrix} 1 & 3 \\ 2 & 4 \end{vmatrix} \neq 0$$

and there are no minors of order three.

Example 4. The rank of

$$[D] = \begin{bmatrix} 0 & 1 & 2 \\ 0 & 3 & 6 \\ 0 & 2 & 4 \end{bmatrix}$$

is $r = 1$ since

$$\begin{vmatrix} 0 & 1 & 2 \\ 0 & 3 & 6 \\ 0 & 2 & 4 \end{vmatrix} = 0, \qquad \begin{vmatrix} 0 & 1 \\ 0 & 3 \end{vmatrix} = 0, \qquad \begin{vmatrix} 0 & 3 \\ 0 & 2 \end{vmatrix} = 0$$

$$\begin{vmatrix} 1 & 2 \\ 3 & 6 \end{vmatrix} = 0, \qquad \begin{vmatrix} 3 & 6 \\ 2 & 4 \end{vmatrix} = 0, \qquad \begin{vmatrix} 1 & 2 \\ 2 & 4 \end{vmatrix} = 0, \qquad \begin{vmatrix} 0 & 2 \\ 0 & 4 \end{vmatrix} = 0$$

$$\begin{vmatrix} 0 & 1 \\ 0 & 2 \end{vmatrix} = 0, \qquad \begin{vmatrix} 0 & 2 \\ 0 & 6 \end{vmatrix} = 0, \quad \text{and} \quad \begin{vmatrix} 0 & 6 \\ 0 & 4 \end{vmatrix} = 0$$

but not every element is zero.

3. Controllability

As we mentioned before, the concept of controllability involves the dependence of the state variables of the system on the inputs. Consider the following system:

$$[\dot{x}] = [A][x] + [b]u \tag{1}$$

where $[\dot{x}]$, $[x]$ are the velocity vector and the state vector, respectively, $[b]$ is the input vector, and u is a scalar; $[A]$ is a square matrix.

For clarity, (1) is written as follows:

$$\dot{x} = Ax + bu \tag{1a}$$

The initial state can be expressed as $x(0)$ and a final desirable state is expressed as $x(n)$, where n is a finite number. Controllability is concerned with the conditions under which a control can drive the system from an initial state, $x(0)$, to a desirable final state $x(n)$ (see Fig. 8-27). Mathematically, this

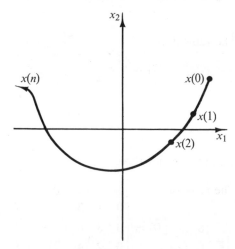

Fig. 8-27. States and the trajectory.

implies that the following relations hold:

$$x(1) = Ax(0) + bu(0)$$
$$x(2) = Ax(1) + bu(1) = A[Ax(0) + bu(0)] + bu(1)$$
$$\vdots$$
$$x(n) = A^n x(0) + A^{n-1} bu(0) + A^{n-2} bu(1) + \cdots + bu(n-1)$$

Therefore,

$$x(n) - A^n x(0) = [A^{n-1}b, A^{n-2}b, A^{n-3}b, \ldots, b] \begin{bmatrix} u(0) \\ u(1) \\ u(2) \\ \vdots \\ u(n-1) \end{bmatrix} \tag{2}$$

The left-hand side of (2) involves the initial and the final state. It is apparent that for a unique solution to exist:

$$[A^{n-1}b, A^{n-2}b, A^{n-3}b, \ldots, b] \text{ must have rank } n \tag{3}$$

The matrix shown in (3) is called the test matrix for controllability.

4. Example

The following example is used to clarify the concept of controllability. A system is shown in Fig. 8-28.

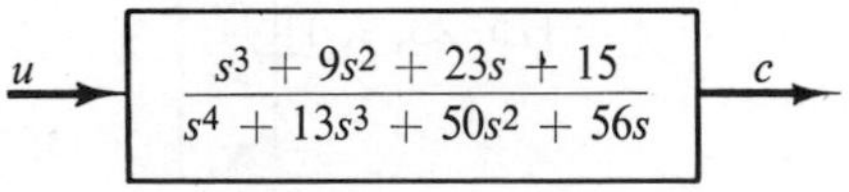

$$u \longrightarrow \boxed{\dfrac{s^3 + 9s^2 + 23s + 15}{s^4 + 13s^3 + 50s^2 + 56s}} \longrightarrow c \tag{4}$$

Fig. 8-28. A given system.

We can use Bush's form to expand it as shown in Fig. 8-29. The describing equations are

$$\dot{x}_1 = x_2$$
$$\dot{x}_2 = x_3$$
$$\dot{x}_3 = x_4 \tag{4a}$$
$$\dot{x}_4 = -56x_2 - 50x_3 - 13x_4 + u$$
$$c = 15x_1 + 23x_2 + 9x_3 + x_4 \tag{5}$$

The corresponding matrix equations are

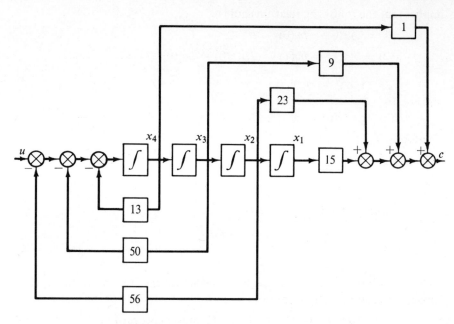

Fig. 8-29. Bush's form or the phase variable form.

$$\begin{bmatrix} \dot{x}_1 \\ \dot{x}_2 \\ \dot{x}_3 \\ \dot{x}_4 \end{bmatrix} = \begin{bmatrix} 0 & 1 & 0 & 0 \\ 0 & 0 & 1 & 0 \\ 0 & 0 & 0 & 1 \\ 0 & -56 & -50 & -13 \end{bmatrix} \begin{bmatrix} x_1 \\ x_2 \\ x_3 \\ x_4 \end{bmatrix} + \begin{bmatrix} 0 \\ 0 \\ 0 \\ 1 \end{bmatrix} u \qquad (4b)$$

$$c = [15, 23, 9, 1] \begin{bmatrix} x_1 \\ x_2 \\ x_3 \\ x_4 \end{bmatrix} \qquad (5a)$$

For this example, we have

$$[b] = \begin{bmatrix} 0 \\ 0 \\ 0 \\ 1 \end{bmatrix}$$

$$[A][b] = \begin{bmatrix} 0 & 1 & 0 & 0 \\ 0 & 0 & 1 & 0 \\ 0 & 0 & 0 & 1 \\ 0 & -56 & -50 & -13 \end{bmatrix} \begin{bmatrix} 0 \\ 0 \\ 0 \\ 1 \end{bmatrix} = \begin{bmatrix} 0 \\ 0 \\ 1 \\ -13 \end{bmatrix}$$

$$[A]^2[b] = \begin{bmatrix} 0 & 1 & 0 & 0 \\ 0 & 0 & 1 & 0 \\ 0 & 0 & 0 & 1 \\ 0 & -56 & -50 & -13 \end{bmatrix} \begin{bmatrix} 0 & 1 & 0 & 0 \\ 0 & 0 & 1 & 0 \\ 0 & 0 & 0 & 1 \\ 0 & -56 & -50 & -13 \end{bmatrix} \begin{bmatrix} 0 \\ 0 \\ 0 \\ 1 \end{bmatrix} = \begin{bmatrix} 0 \\ 1 \\ -13 \\ 119 \end{bmatrix}$$

and

$$[A]^3[b] = \begin{bmatrix} 1 \\ -13 \\ 119 \\ -953 \end{bmatrix}$$

The test matrix is then formed as follows:

$$\begin{bmatrix} 0 & 0 & 0 & 1 \\ 0 & 0 & 1 & -13 \\ 0 & 1 & -13 & 119 \\ 1 & -13 & 119 & -953 \end{bmatrix} \tag{6}$$

The rank of the test matrix (6) is equal to its order: $r = 4$ while $n = 4$. Therefore, the system is completely controllable.

5. Physical Interpretation

System (4a) can also be decomposed into the parallel programming, as shown in Fig. 8-30. The matrix equation is as follows:

$$\begin{bmatrix} \dot{y}_1 \\ \dot{y}_2 \\ \dot{y}_3 \\ \dot{y}_4 \end{bmatrix} = \begin{bmatrix} 0 & 0 & 0 & 0 \\ 0 & -2 & 0 & 0 \\ 0 & 0 & -4 & 0 \\ 0 & 0 & 0 & -7 \end{bmatrix} \begin{bmatrix} y_1 \\ y_2 \\ y_3 \\ y_4 \end{bmatrix} + \begin{bmatrix} 1 \\ 1 \\ 1 \\ 1 \end{bmatrix} u \tag{7}$$

or

$$[\dot{y}] = [\Lambda][y] + [\beta]u \tag{7a}$$

The test matrix for (7) is then:

$$\begin{bmatrix} 1 & 0 & 0 & 0 \\ 1 & -2 & 4 & -8 \\ 1 & -4 & 16 & -64 \\ 1 & -7 & 49 & -343 \end{bmatrix} \tag{8}$$

the rank of which is also $r = 4$. This means that the system is completely controllable.

Examining (7), we see that there are four uncoupled equations representing the four basic modes of the system. If all modes are excited by u, the

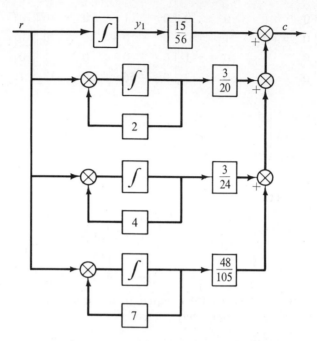

Fig. 8-30. Foster's form or parallel programming.

control vector $[\beta]$ must not have any zero components. From this viewpoint we can say that a *controllable system* is one whose input u directly affects each of the canonical state variables of the system.

If the condition [3] is satisfied, then $[A]$, $[b]$ is called a controllable pair. Clearly, controllability is completely independent of the way in which the outputs of the system are formed. It is a property of the couple $[A]$, $[b]$.

6. Geometrical Interpretation

The test matrix

$$[b, Ab, A^2b, \ldots, A^{n-1}b]$$

can be considered as a collection of n vectors. Each column is a vector. In order to evaluate the controllability, check the linear independence of these vectors. The theorem concerning controllability is converted into the following:

System (1) is completely controllable if and only if the vectors

$$b, Ab, A^2b, \ldots, A^{n-1}b$$

are linearly independent.

In higher dimensional space, the determination of linear dependence can be stated as follows: By changing the notations of the vectors, we have

$$[b] = \begin{bmatrix} a_{11} \\ a_{21} \\ a_{31} \\ \cdot \\ \cdot \\ \cdot \\ a_{n1} \end{bmatrix}, \quad [A]^2[b] = \begin{bmatrix} a_{12} \\ a_{22} \\ a_{32} \\ \cdot \\ \cdot \\ \cdot \\ a_{n2} \end{bmatrix}, \quad [A]^3[b] = \begin{bmatrix} a_{13} \\ a_{23} \\ a_{33} \\ \cdot \\ \cdot \\ \cdot \\ a_{n3} \end{bmatrix} \cdots \qquad (9)$$

If the vectors are linearly dependent, then there exist λ_i, not all zero, such that

$$\sum_{i=1}^{n} \lambda_i a_i = 0 \qquad (10)$$

or

$$\begin{aligned} a_{11}\lambda_1 + \cdots + a_{1n}\lambda_n &= 0 \\ a_{21}\lambda_1 + \cdots + a_{2n}\lambda_n &= 0 \\ \cdot \qquad \cdot \qquad \cdot \qquad \cdot \\ a_{n1}\lambda_1 + \cdots + a_{nn}\lambda_n &= 0 \end{aligned} \qquad (10a)$$

If (10) can be satisfied only with the trivial solution, $\lambda_i = 0$, the vectors are linearly independent.

From the algebraic viewpoint, the system (10a) is a set of n simultaneous linear equations with n unknowns, λ_i which are to be solved. If the determinant of these homogeneous linear equations is not equal to zero, or

$$\begin{vmatrix} a_{11} & a_{12} & \cdots & a_{1n} \\ a_{21} & a_{22} & \cdots & a_{2n} \\ \cdot & & & \\ \cdot & & & \\ a_{n1} & a_{n2} & \cdots & a_{nn} \end{vmatrix} \neq 0 \qquad (11)$$

The only solution of (10a) is that $\lambda_i = 0$; other solutions cannot exist. It follows that the vectors a_i are linearly independent. However, (11) is exactly the determinant of our test matrix for controllability. Therefore, a corresponding controllability theorem is established as follows: The system (1) is completely controllable if

$$\det (b, Ab, A^2b, A^3b, \ldots, A^{n-1}b) \neq 0 \qquad (11a)$$

7. Observability

Observability is concerned with the problem of determining the state by observing the outputs. We will investigate the conditions under which this is possible.

Consider the following system:

$$[\dot{x}] = [A][x] + [b]u \tag{12}$$
$$[z] = [c]^T[x] \tag{13}$$

where $[z]$ represents the outputs.

By definition,

$$z(0) = c^T x(0)$$

at the next instant, we write

$$z(1) = c^T x(1) = c^T A x(0)$$

Similarly,

$$z(2) = c^T A^2 x(0)$$
$$\cdot$$
$$\cdot$$
$$\cdot$$
$$z(n) = c^T A^{n-1} x(0)$$

or

$$\begin{bmatrix} z(0) \\ z(1) \\ \cdot \\ \cdot \\ z(n) \end{bmatrix} = \begin{bmatrix} c^T \\ c^T A \\ \cdot \\ \cdot \\ c^T A^{n-1} \end{bmatrix} \tag{14}$$

or

$$\begin{bmatrix} c^T \\ c^T A \\ \cdot \\ \cdot \\ c^T A^{n-1} \end{bmatrix}^{-1} \begin{bmatrix} z(0) \\ z(1) \\ \cdot \\ \cdot \\ z(n) \end{bmatrix} = x(0) \tag{14a}$$

If $x(0)$ is uniquely determined, the matrix

$$\begin{bmatrix} c^T \\ c^T A \\ \cdot \\ \cdot \\ c^T A^{n-1} \end{bmatrix} \tag{15}$$

must have its inverse. This implies that (15) should have rank n, or be non-singular.

If this condition is satisfied, then A, c is called an observable pair.

Condition (15) can be rewritten into a row form:

$$[c, A^T c, \ldots, (A^T)^{n-1}c] \text{ should have rank } n \tag{16}$$

For system (46) we have

$$[A] = \begin{bmatrix} 0 & 1 & 0 & 0 \\ 0 & 0 & 1 & 0 \\ 0 & 0 & 0 & 1 \\ 0 & -56 & -50 & -13 \end{bmatrix}$$

$$c^T = [15, 23, 9, 1]$$

The test matrix for observability would be as follows:

$$\begin{bmatrix} 15 & 0 & 0 & 0 \\ 23 & -41 & 224 & -1400 \\ 9 & -27 & 159 & -1026 \\ 1 & -4 & 25 & -166 \end{bmatrix} \tag{17}$$

From this we can get the determinant of the matrix and find that $r = 4$. Therefore, the system is completely observable.

8. A Not Completely Observable Example

There is a system whose transfer function is

$$\frac{s^2 + 4s + 3}{s^3 + 6s^2 + 5s} \tag{18}$$

If it is decomposed into Bush's form, we have the configuration shown in Fig. 8-31.

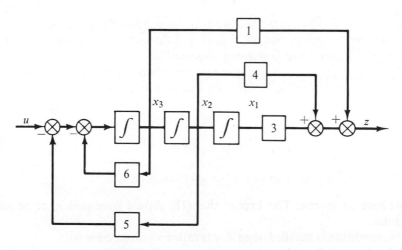

Fig. 8-31. A partially observable example.

$$\dot{x}_1 = x_2$$
$$\dot{x}_2 = x_3 \tag{18a}$$
$$\dot{x}_3 = r - 5x_2 - 6x_3$$

or

$$\begin{bmatrix} \dot{x}_1 \\ \dot{x}_2 \\ \dot{x}_3 \end{bmatrix} = \begin{bmatrix} 0 & 1 & 0 \\ 0 & 0 & 1 \\ 0 & -5 & -6 \end{bmatrix} + \begin{bmatrix} 0 \\ 0 \\ 1 \end{bmatrix} u \tag{18b}$$

$$z = [3, 4, 1] \begin{bmatrix} x_1 \\ x_2 \\ x_3 \end{bmatrix} \tag{18c}$$

from which we have

$$[A] = \begin{bmatrix} 0 & 1 & 0 \\ 0 & 0 & 1 \\ 0 & -5 & -6 \end{bmatrix}, \quad [b] = \begin{bmatrix} 0 \\ 0 \\ 1 \end{bmatrix} \quad \text{and} \quad [c] = \begin{bmatrix} 3 \\ 4 \\ 1 \end{bmatrix} \tag{19}$$

The matrix for testing controllability is

$$\begin{bmatrix} 0 & 0 & 1 \\ 0 & 1 & -6 \\ 1 & -6 & 31 \end{bmatrix} \tag{20}$$

which is of rank 3. The system is completely controllable. The matrix for testing observability is

$$\begin{bmatrix} 3 & 0 & 0 \\ 4 & -2 & 10 \\ 2 & -2 & 10 \end{bmatrix} \tag{21}$$

which has $r = 2$. Therefore, the system is not completely observable.

From the example, we see that if a pole and a zero have a common factor, the system is not completely observable.

If a set of equations is completely controllable and completely observable, there is a unique transfer function which describes the system completely.

PROBLEMS

8-1. What are the singular points of the following system?

$$\frac{dx_1}{dt} = x_2$$

$$\frac{dx_2}{dt} = \frac{1}{3 - x_1} - x_1$$

8-2. For the system

$$\begin{bmatrix} \dot{x}_1 \\ \dot{x}_2 \\ \dot{x}_3 \end{bmatrix} = \begin{bmatrix} 0 & 1 & 0 \\ 0 & 0 & 1 \\ -4 & -8 & -5 \end{bmatrix} \begin{bmatrix} x_1 \\ x_2 \\ x_3 \end{bmatrix}$$

is the following function a Liapunov function?

$$v = [x_1 \quad x_2 \quad x_3] \begin{bmatrix} 1.84722 & 1.19444 & 0.12500 \\ 1.19444 & 2.07292 & 0.21181 \\ 0.12500 & 0.21181 & 0.14236 \end{bmatrix} \begin{bmatrix} x_1 \\ x_2 \\ x_3 \end{bmatrix}$$

If so, what is the function $\dot{v}$?

8-3. If the $\dot{v}$ function for

$$\begin{bmatrix} \dot{x}_1 \\ \dot{x}_2 \\ \dot{x}_3 \\ \dot{x}_4 \end{bmatrix} = \begin{bmatrix} 0 & 1 & 0 & 0 \\ 0 & 0 & 1 & 0 \\ 0 & 0 & 0 & 1 \\ -24 & -50 & -35 & -10 \end{bmatrix} \begin{bmatrix} x_1 \\ x_2 \\ x_3 \\ x_4 \end{bmatrix}$$

$= -x_1^2 - x_2^2 - x_3^2 - x_4^2$, find v.

8-4. Verify that a Liapunov function of the following system

$$\begin{bmatrix} \dot{x}_1 \\ \dot{x}_2 \\ \dot{x}_3 \\ \dot{x}_4 \end{bmatrix} = \begin{bmatrix} 0 & 1 & 0 & 0 \\ 0 & 0 & 1 & 0 \\ 0 & 0 & 0 & 1 \\ -10000 & -10500 & -1035 & -70.5 \end{bmatrix} \begin{bmatrix} x_1 \\ x_2 \\ x_3 \\ x_4 \end{bmatrix}$$

is

$$v = 872.94968x_1^2 + 967.65361x_2^2 + 15.05144x_3^2 + 0.00839x_4^2$$
$$+ 1831.09188x_1x_2 + 167.82248x_1x_3 + 0.00010x_1x_4$$
$$+ 188.50730x_2x_3 + 0.17448x_2x_4 + 0.18310x_3x_4.$$

Use Program 11 and assume $\dot{v} = -(x_1^2 + x_2^2 + x_3^2 + x_4^2)$ to find v.

8-5. Verify that either

(a) $v_1 = \frac{1}{2}x_1^4 + x_2^2$,

(b) $v_2 = \frac{1}{2}x_1^4 + (x_1 + x_2)^2$, or

(c) $v_3 = x_1^4 + 2x_2^2 + 2x_1x_2 + x_1^2$

is a Liapunov function of the following system:

$$\dot{x}_1 = x_2$$
$$\dot{x}_2 = -x_2 - x_1^3$$

8-6. For the following system

$$\dot{x}_1 = x_2$$
$$\dot{x}_2 = x_3$$
$$\dot{x}_3 = -(x_1 + 2x_2)^3 - x_3$$

if we assume $\dot{v} = -x_3^2$, find v by using the integration by parts technique. Is what you found a Liapunov function of the system?

8-7. For the following system

$$\dot{x}_1 = x_2$$
$$\dot{x}_2 = x_3$$
$$\dot{x}_3 = -24x_1 - 26x_2 - 9x_3$$

give:

(a) the Kalman-Bertram Liapunov function in Schwarz variables;
(b) the Kalman-Bertram Liapunov function in Bush variables;
(c) a Liapunov function whose derivative is negative definite.

8-8. Investigate for controllability and observability of the following system:

$$\dot{x}_1 = -5x_1 + u$$
$$\dot{x}_2 = x_3$$
$$\dot{x}_3 = -6x_2 - 7x_3 + u$$
$$c = -3x_1 + 4x_2 + 4x_3$$

8-9. A set of state equations in Bush's form is given as follows:

$$[\dot{x}] = [B][x]$$

where

$$[B] = \begin{bmatrix} 0 & 1 & 0 & 0 & 0 \\ 0 & 0 & 1 & 0 & 0 \\ 0 & 0 & 0 & 1 & 0 \\ 0 & 0 & 0 & 0 & 1 \\ -10 & -24 & -25 & -15 & -5 \end{bmatrix}$$

(a) Derive that the corresponding Schwarz' form state equations are

$$[\dot{y}] = [A][y]$$

where

$$[A] = \begin{bmatrix} 0 & 1 & 0 & 0 & 0 \\ -5/7 & 0 & 1 & 0 & 0 \\ 0 & -52/35 & 0 & 1 & 0 \\ 0 & 0 & -14/5 & 0 & 1 \\ 0 & 0 & 0 & -10 & -5 \end{bmatrix}$$

(b) What is the $[P]$ matrix you used and what is its inverse?
(c) Write the Kalman-Bertram Liapunov function for the system.

FORTRAN Programming Language

For readers who are completely unfamiliar with the FORTRAN language, the following notes are included so that they may satisfy their curiosity as to the programs found in the text. This is not a training manual, intended to give a *complete* set of rules or develop proficiency in the use of FORTRAN. (There are many such available.) Rather, it is a few remarks written by a programmer to give some useful impressions to one who would consider himself an interested spectator. These notes, although accurate, are deliberately brief and nonrigorous.

1. Looping

The capacity of a computer to perform mountains of computations is due to its ability to *loop*. This concept is one which is most striking to the uninitiated. It simply means that the machine can be instructed to take in some data, perform a series of computations with these data, turn out some answers, and then go back to the beginning, where it will take in new data to replace the old and will execute the same operations on the new data as it did before.

2. The "Equals" (=) Sign

A statement such as $A = B + C$ is best described by the phrase, "A will be replaced by the sum of B and C." Whatever number was in location A before this statement was executed will be erased, and the new number resulting from the operations prescribed on the right side will be stored in A.

3. Arithmetic Operations

The FORTRAN statements which perform addition, subtraction, multiplication, division, and involution (raising to a power) have been made so simple as to require no explanation:

$$A = B + C$$
$$A = B - C$$
$$A = B * C$$
$$A = B / C$$
$$A = B ** C$$

4. Hierarchy of Operations

When the computer is confronted with an arithmetic expression requiring several operations, it does these in a certain order. Among those listed above, exponentiation(**) has the highest priority and will be executed first. On the next level of priority are multiplication and division. If there are two or more such operations in the same expression, they will be done in order from left to right across the expression. On the lowest level are addition and subtraction, also taken from left to right.

5. Parentheses

Top priority goes to the removal of parentheses. Because of this any conceivable order of operations can be built into an expression, despite the prearranged hierarchy, by the liberal insertion of parentheses. The machine will remove the innermost of a nest of parentheses first (by executing all the operations enclosed by them) and replace them in the expression by the single number obtained in the execution. If there is more than one set, they will be dealt with in turn from left to right.

6. Mode

Probably the most peculiar concept in FORTRAN is the division of numbers into "fixed point" and "floating point." The machine handles these two "modes" in different ways, and therefore they are the cause of some restrictions in programming. Any number, of four digits or less, *without a decimal point* is called "fixed point"; any number *containing a decimal point*, regardless of its magnitude or sign, is called "floating point." Large quantities can be written in a notation that is much like that used by engineers but modified somewhat to conform with the restricted type forms available on the key punch. For example, 1234964.352 is customarily written as 1.2349×10^6. It can be punched on a card as 1.2349643E + 06. The machine will *ordinarily* retain eight significant digits.

The principal restriction caused by "mode" is that a FORTRAN expression (all to the right of an "equal" sign) must be in only one mode.

An advantage can sometimes be gained by exploiting the restriction on fixed point numbers. Since they cannot be larger than four digits and cannot contain any decimal parts, arithmetic operations, especially division, can be used to truncate a number and utilize the fragments.

7. Variables

The computer deals exclusively with numbers, but the programmer may write instructions containing improvised alpha-numeric combinations called variables, which represent numbers to be read in later as data or to be produced by computation. Variables can consist of any combination which the programmer considers appropriate, as long as it contains no more than five letters and numbers, and as long as the first is a letter. Variables which are to represent "fixed point" numbers must start with the letters I, J, K, L, M, or N. All other starting letters produce "floating point" variables. For example, "ITEM" is a fixed point variable and "AL203" is a floating point.

To the left of an equal sign there are no operands; only a single variable, sometimes preceded by a statement number.

8. Transfer

A computer is an "ordered" machine in the sense that it executes the instructions in the sequence in which they are read into it. However, a transfer command is available which will tell it to jump from one place in a program to any other place.

9. Statement Numbers

If it is desirable to make a jump in the regular sequence, the destination must be especially identified by placing a statement number (any fixed point number) at the beginning of the FORTRAN statement. For example, "GO TO 107" will cause the computer to do just that.

10. Conditional Transfer

Computers are often credited by the uninitiated with the ability to "think" and make decisions. As a matter of fact, all decisions are made by the program writer, and conditions under which they are to be reached must be carefully and literally stipulated by him. He does this by means of conditional transfer statements such as "IF (exp.) 1, 2, 7." When the computer reaches this statement it will examine the value of what is in the parentheses (it can be any arithmetic expression). If it turns out to be negative, the computer will transfer to the statement numbered 1 and continue from there. If the expression in the parentheses is equal to zero, the machine will jump to the statement numbered 2. If "exp." is positive, the transfer will be to 7. Statements numbered 1, 2, and 7 can each be the first of long lists of commands. The "IF" statement is the most commonly used conditional transfer in FORTRAN.

Another is available which is sometimes likened to a multibranched fork. It is called the "Computed 'GO TO'" and is exemplified by "GO TO (1, 2, 9, 4, 7), L." The numbers in the parentheses are statement numbers. In the above example, L can be 1, 2, 3, 4, or 5, since there are five numbers in the parentheses. There could be more and then L could be higher.

If L were 3 at the moment when the computer reached this command in the program, the computer would transfer to the statement which has the third number in the parentheses (which happens to be statement number 9). If L were 5, the computer would transfer to the statement that has the number 7.

Although a computer apparently makes decisions, or thinks, actually all the decisions have been made ahead of time by the programmer and are dependent upon projected conditions. The IF statement is the heart of the decision-making process. It plays the role most often by using a simple comparison.

Suppose that A is a certain definite and important quantity and B is some other quantity that grows and diminishes during computation. Suppose it has been decided that if B ever gets to be the same as A, then the program must take an entirely new course.

$$IF (A - B) \ 1, 2, 1$$

The above statement will produce the desired results, namely that if they are not equal the program will branch to 1 and its normal course, but if they are ever equal it will branch to 2, which can be the first of an entirely new set of events.

Any decision that can be reduced to a simple comparison or a series of simple comparisons can be incorporated into a program in this manner.

Actually, the decision has been reduced to the simple detection of sign after subtraction.

11. Subroutines

When a program requires that a routine series of operations be executed several times and at various places in the program, they may be incorporated into a "subroutine." It is given a name much like any variable and the entire package may be called for by name.

For example, in Chapter 3 there is a program for determining stability by Routh's criterion. This could be made into a subroutine by placing after the "READ 20, N" the following:

SUBROUTINE ROUTH(N)

and replacing the last ("END") card with "RETURN."

At places in the program where one wished to execute the Routh criterion subroutine, one would merely write

CALL ROUTH(N)

and continue on with the main program.

The parentheses contain the variable or variables representing the data needed in the subroutine. Routh is the name invented for this particular subroutine.

Whenever the computer arrives at a "CALL ROUTH" statement, it branches to the subroutine, executes it, and then returns to the statement following "CALL ROUTH."

12. Subscripts

When repetitious computations are to be done with a list of numbers, or when a matrix is to be employed, it is convenient to use notation similar to the subscripts used in mathematics. However, the limitations of the key punch symbols require that the subscript in FORTRAN be slightly modified. We can write A(1, 3) for $a_{1,3}$. The numbers must be capitals and they must appear on the same level with the A; therefore, it is necessary to specify them as subscripts by enclosing them in parentheses. There can be no A_0; the subscript must be a fixed point number. Three-dimensional arrays are possible, and the subscripts may be variables as well as numbers.

If you plan to use subscripted variables, you must tell the machine how many places you want reserved for the elements of the array. You may or may not use all the spaces reserved during the program, but you must reserve *enough* spaces at compiling time. This is done by a statement such as:

$$\text{DIMENSION A}(20,20)$$

This statement, where A stands for any variable, will allow the use of A(I, J), where I and J can be any number from 1 to 20 and represent elements in a 20 by 20 array.

13. The DO Loop

Since looping and repetition are so common in computery, a special command structure has been developed to expedite the coding as much as possible. A DO loop comprises a special opening statement, a closing statement and all in between. For example, in

$$\text{DO 5 J} = \text{I,K,L}$$

J is called the index of the DO loop. The closing statement (anything executable) is numbered 5. All the statements in between will be executed in turn, first with J equal to I; then J will be changed to I plus L and all will be repeated from the DO statement down to and including the one numbered 5. J will then be again increased by L and so on. This will continue until J becomes greater than K, at which time the program will proceed to the next statement after the DO loop—after the one numbered 5.

J, I, K, and L can be any fixed point variables or numbers. Often J turns up as a subscript in the statements of the DO loop. However, it may serve any purpose so long as its own value is not changed inside the loop.

14. Functions

If a set of arithmetic operations is often used, each time with new variables but with essentially the selfsame set of operations, then these had best be condensed into a function and given a name. Then whenever one wishes to obtain the number that will result from the series of computations, he merely calls for it with one simple statement.

Suppose, for example, that it is necessary to take the "square root of the sum of the squares" many times and one wishes to reduce the operation to a function. The statement

$$\text{FUNCTION SSSQ(A, B)} = \text{SQRT(A**2} + \text{B**2)}$$

will make it possible, from now on, to obtain the square root of the sum of the squares of any two numbers (say x and y) by merely writing

$$Z = \text{SSSQ(X, Y)}$$

Every computer facility has certain functions (and subroutines) that it needs especially, and these comprise a library within the compiler. The most common, included in most compilers, are:

Type of Function	FORTRAN Name
Natural logarithm	LOG( )
Trigonometric sine	SIN( )
Trigonometric cosine	COS( )
Exponential	EXP( )
Square root	SQRT( )
Arctangent	ATAN( )
Absolute value	ABS( )

To illustrate how these are used: find the log of a number called A and store it as F. This is done simply by writing

$$F = LOG(A)$$

In executing this or any other function, the computer may perform a long series of steps, but the programmer need be concerned only with supplying the desired argument in the parentheses.

The FORTRAN statements so far described give the essential operations of the language. There are others which embellish it; the most commonly used will be described briefly.

15. IF (SENSE SWITCH)

Some machines have external switches that may be turned on or off by the operator. These can be "interrogated" in a program with a statement much like the IF statement. Suppose the programmer writes "IF (SENSE SWITCH 3) 50, 80." When the machine comes to this command in the program, it will branch to statement 50 if sense switch 3 is on, or it will branch to statement 80 if switch 3 is off.

"PAUSE" is a command which causes the machine to halt, giving the operator a chance to make decisions, change input devices, etc. When he pushes "START," the program will go to the next statement. "STOP" is a command which is put at the end of a program. It causes the machine to stop permanently so far as the present program is concerned. "CONTINUE" is a dummy statement. It causes the machine to perform no operation except to pass on to the next command in line. It is useful in many ways; for example, it may be numbered and thereby become a target for one of the jumps taken in a transfer statement. "END" must appear as the last statement in any FORTRAN language program. It is a message to the compiler rather than a command in the program that is to be executed later. During compile time, when the FORTRAN statements are being converted by the compiler (master translator) into chains of machine language commands, the compiler is informed by "END" that the final statement requiring translation has been furnished. "READ, ACCEPT, PRINT, PUNCH, TYPE" are input or output commands which can be understood literally. "FORMAT" statements, like the "END" statement, are messages to the compiler. They specify what the input data will look like when fed into the machine at time of execution of the program, or they specify the form that is desired for the output data. The details of FORMAT are quite complicated and often form the principal obstacle for a beginner in programming. For one who is merely interested in reading FORTRAN programs without necessarily acquiring the technique of programming, the FORMAT statements can ordinarily be ignored without clouding the sense of the program.

Bibliography

The literature on control systems analysis is extensive and repetitious. The list of books and papers given in the following sections is a collection of material which may well be helpful to the reader. Numbers in parentheses correspond to those in the references at the end of this bibliography.

General and historical aspects

For historical aspects and classical documents, see Bellman and Kalaba (8); Wiener (138); Newton, Gould, and Kaiser (101). For conventional feedback theory, see Tsien (136); James, Nichols, and Phillips (73); Truxal (135); Aizerman (1); and Lauer, Lesnick, and Matson (88). For modern aspects of control systems, see Kalman and Bertram (76); Bellman and Kalaba (9); Tou (133); Gupta (66); Kuo (83); DeRusso, Roy, and Close (44); Dorf (47); Ogata (102); and Elgerd (49).

Chapter 1

For comprehensive treatments of the cause-effect approach, see Garder and Barnes (57); Goldman (61); Trimmer (134); Guillemin (63, 65); and Chen (27). Suggested references for the energy approach are Elsgolts (51); Ogar and Dazzo (103); Landau and Lifshitz (85); Gelfand and Fomin (58); and Langhaar (86). For the generalized circuit approach, see Paskusz and Russel (108); LePage and Seely (93). References for block diagrams are Cheng (38); Chestnut and Mayer (39). More solved problems can be found in the outline books of Distefano, Stubberud, and Williams (46) and Ayres (4).

Chapter 2

The classical method for solving differential equations is discussed in many texts including Elsgolts (50); Pontryagin (111); Brand (15); and Coddington and Levinson (43). References for the Laplace transformed method are Van Valkenburg (137); Cheng (38); Aseltine (2); Brown and Nilsson (17); Gardner and Barnes (57); and Churchill (41). Additional readings on analog computation are Bush (19); Seely (125); and Thaler and Brown (132). The outline book by Ayres and Spiegel (128) gives more solved problems; Langill (87) shows many detailed analog computer programs.

Chapter 3

A great many books exist in the field of the stability problems. The well-known works with the classical approach are Brown and Campbell (16); Chestnut and Mayer (39); Popov (113); Thaler and Brown (132); Truxal (135); Seifert and Steeg (126); and Lehnigk (92). The derivation of the Routh-Hurwitz criterion is based on Trimmer (134). The main references for the root locus of Evans are Savant (121); Evans (52); Mikhailov (99); Takahashi (130); and Takai (131). The source papers of the root locus of Routh are Bendrikov and Teodorchik (11); Chen and Hsu (27); and Shen and Chen (127). For the Nyquist criterion, see Oldenburger (104); Chen and Tsang (35); and Chen and Shen (32). For the performance studies, see Gibson (59); Chen and Haas (26); Oppelt (105); Takai (131); and Oldenburger (104).

Chapter 4

Theorems of Fourier series and Fourier integrals are discussed by Papoulis (109); Ku (82); Goldman (61); Karman and Biot (78); and Chen (21, 30). For Bode's decomposition, see Bode (13); Dudnikov (48); and Chen (23). For Wiener-Lee's decomposition, see Guillemin (64); Bohn (14); Chen (22); and Bellman, Kalaba, and Lockett (10). For Bush's decomposition see Bush and Caldwell (19); Brown and Campbell (16); Chen and Shen (32); Chen and Lipinski (28); and Chen and Philip (31). For Levy's complex curve-fitting technique, see Levy (95); Sanathanan and Koerner (120); and Kidd, Edgerton, and Chen (79).

Chapter 5

Three excellent references for general matrix theory are Gantmacher (56); Bellman (7); and Pipes (110). For elementary matrices, see Frazer, Duncan, and Collar (55); Ficken (53); Forsythe and Moler (54); and Chen and Haas (25). For quadratic forms, see Ayres (4); Bellman (7); and Yefimov (139). For vectors, see Hadley (67). For the discussions on the general state concept, see Kalman (74, 75); Bertram (12); Zadeh and Desoer (140); Schwarz and Friedland (123); DeRusso, Roy, and Close (44); and Dorf (47).

Chapter 6

For additional studies of the energy method, read Gelfand and Fomin (58); Lanczos (84); Pontryagin, Boltyanskii, Gamkrelidze, and Mishchengo (112); and Landau and Lifshitz (85). For the topological approach see Kron (81); Bashkow (5). For the state diagrams, see Tou (133); Schwarz and Friedland (123); Dorf (47); Bush (19); and Cauer (20).

Chapter 7

The solution problem is covered in many books and papers including Tou (133); Schwarz and Friedland (123); Pipes (110); Ogata (102); and Chen and Parker (29). For the general matrix for the inverse Laplace transformation, see Chen and Yates (36) and Chen and Shieh (33). For the signal flow graphs, see Mason (98); Coates (42); Desoer (45); Robichaud, Boisvert, and Robert (119); Sedlar and Bekey (124). For the numerical method, see Ralston and Wilf (116). For the transformation matrix from a general form to a phase-variable form, see Raue and Johnson (117).

Chapter 8

For the stability theory of differential equations in general, see Bellman (6). For equilibrium points studies, see Lefschetz (91); Minorsky (100); Hayashi (69). For the general concept of stability in the Liapunov sense, see LaSalle (89). For Liapunov theorems, see Liapunov (96); Lure (97); Hahn (68); LaSalle and Lefschetz (90); Letov (94); and Chetayev (40). For the methods of constructing Liapunov functions, see Krasovskii (80); Ingwerson (72); Schultz and Gibson (122); Puri and Weygandt (114); Lehnigk (92); Reiss and Geiss (118); and Chen and Shieh (34). For proving Hurwitz-Routh's criterion by the second method of Liapunov, see Parks (107); Puri and Weygandt (114); Ralston (115); and Chen and Chu (24). For controllability and observability, see Kalman (74); Kalman, Ho, and Narendra (77); Ho (70); Gilbert (60); and Ogata (102).

Appendix

For the FORTRAN language, see Organick (106).

REFERENCES

1. Aizerman, M. A., *Theory of Automatic Control,* Addison-Wesley Publishing Company, Inc., Reading, Mass., 1963.
2. Aseltine, John A., *Transform Method in Linear System Analysis,* McGraw-Hill Book Company, New York, 1958.
3. Ayres, F., Jr., *Theory and Problems of Differential Equations,* Schaum Publishing Company, New York, 1952.
4. Ayres, F., Jr., *Theory and Problems of Matrices,* Schaum Publishing Company, New York, 1962.
5. Bashkow, T. R., "The A Matrix, a New Network Description," *IRE Trans. on Circuit Theory,* vol. CT-4, pp. 117–120, Sept. 1957.
6. Bellman, R., *Stability Theory of Differential Equations,* McGraw-Hill Book Company, New York, 1953.
7. Bellman, R., *Introduction to Matrix Analysis,* McGraw-Hill Book Company, New York, 1960.
8. Bellman, R., and R. Kalaba, *Mathematical Trends in Control Theory,* Dover Publications, Inc., New York, 1964.
9. Bellman, R., and R. Kalaba, *Dynamic Programming and Modern Control Theory* (paperback), Academic Press, Inc., New York, 1965.
10. Bellman, R., R. Kalaba, and J. A. Lockett, *Numerical Inversion of the Laplace Transform,* American Elsevier Publishing Company, Inc., New York, 1966.
11. Bendrikov, G. A., and K. F. Teodorchik, "A Simplified Method for the Analytical Construction of Root Trajectories," Automatika i Telemekhanika, vol. 24, no. 2, pp. 268–270, Feb. 1963.
12. Bertram, J. E., "The Concept of State in the Analysis of Discrete Time Control Systems," 1962 Joint Automatic Control Conference, New York University, June 27–29, 1962.

13. Bode, H. W., *Network Analysis and Feedback Amplifier Design,* D. Van Nostrand Company, Inc., Princeton, N. J., 1945.
14. Bohn, Erik V., *The Transform Analysis of Linear Systems,* Addison-Wesley Publishing Company, Inc., Reading, Massachusetts, 1963.
15. Brand, L., *Differential and Difference Equations,* John Wiley & Sons, Inc., New York, 1966.
16. Brown, G. S., and D. P. Campbell, *Principles of Servomechanisms,* John Wiley & Sons, Inc., New York, 1958.
17. Brown, R. G., and J. W. Nilsson, *Introduction to Linear Systems Analysis,* John Wiley & Sons, Inc., New York, 1962.
18. Bultot, F., *Elements of Theoretical Mechanics for Electronic Engineers,* Pergamon Press, New York, 1965, pp. 29–145.
19. Bush, V., and S. H. Caldwell, "A New Type Differential Analyzer," *Journal of Franklin Institute,* vol. 240, pp. 255–326, 1945.
20. Cauer, Wilhelm, *Synthesis of Linear Communication Networks,* McGraw-Hill Book Company, New York, 1958.
21. Chen, C. F., "A New Approach to Fourier Coefficient Evaluation," *IRE Trans. on Education,* vol. E-5, no. 1, pp. 27–29, Mar. 1962.
22. Chen, C. F., "A New Formula for Obtaining the Inverse Laplace Transformations in Terms of Laguerre Functions," *1966 IEEE International Convention Record,* New York, pp. 281–287, Mar. 1966.
23. Chen, C. F., "A Remark on Polonnikov's Approach to Generalized Bode Diagrams," *IEEE Trans. on Aerospace and Electronic Systems,* vol. AES-3, no. 1, Jan. 1967.
24. Chen, C. F., and Hsin Chu, "A Matrix for Evaluating Schwarz's Form," *IEEE Trans. on Automatic Controls,* vol. AC-11, no. 2, pp. 303–305, Apr. 1966.
25. Chen, C. F., and I. J. Haas, "An Electrical Interpretation of the Algorithm of Gauss," *IRE Trans. on Circuit Theory,* vol. CT-9, no. 3, pp. 298–299, Sept. 1962.
26. Chen, C. F., and I. J. Haas, "An Extension of Oppelt's Stability Criterion Based on the Method of Two Hodographs," *IEEE Trans. on Automatic Control,* vol. AC-10, no. 1., Jan. 1965.
27. Chen, C. F. and C. Hsu, "The Determination of Root Loci Using Routh's Algorithm," *Journal of the Franklin Inst.,* vol. 281, no. 2., Feb. 1966.
28. Chen, C. F., and W. Lipinski, "A Method for Determining EBWR Nuclear Reactor Transfer Functions from Frequency Response Data," presented at the Region III Energy Conversion Meeting, Clearwater, Florida, May 1964.
29. Chen, C. F., and R. R. Parker, "Generalization of Heaviside's Expansion Technique to Transition Matrix Evaluation," *IEEE Trans. on Education,* vol. E-9, no. 4, pp. 209–212, Dec. 1966.
30. Chen, C. F., and B. L. Philip, "Graphical Determination of Transfer Function Coefficients of a System from Its Frequency Response," *IEEE Trans. on Applications in Industry,* vol. 83, pp. 42–45, Mar. 1963.
31. Chen, C. F., and B. L. Philip, "Accurate Determination of Complex Root Transfer Functions from Frequency Response Data," *IEEE Trans. on Automatic Control,* vol. AC-10, No. 3, pp. 356–358, July 1965.
32. Chen, C. F., and D. W. C. Shen, "A New Chart Relating Open-loop and Closed-loop Frequency Responses of Linear Control Systems," *Trans. AIEE on Applications in Industry,* vol. 78, pt. II, pp. 252–255, Sept. 1959.
33. Chen, C. F., and L. S. Shieh, "Generalization and Computerization of the Heaviside Expansion for Performing the Inverse Laplace Transform of High Order Systems," *IEEE Region III Convention Record,* pp. 285–294, Mar. 1967.

34. Chen, C. F., and L. S. Shieh, "A Note on Expanding $PA + A^t P = -Q$," *IEEE Trans. on Automatic Control*, 1967.

35. Chen, C. F., and N. F. Tsang, "A Stability Criterion Based on the Return Difference," *IEEE Trans. on Education*, vol. E-10, no. 3, pp. 180–182, 1967.

36. Chen, C. F., and R. E. Yates, "A New Matrix Formula for the Inverse Laplace Transformation," *ASME Trans., Journal of Basic Engineering*, pp. 269–272, June 1967.

37. Chen, Wayne H., *The Analysis of Linear Systems*, McGraw-Hill Book Company, New York, 1963.

38. Cheng, David K., *Analysis of Linear Systems*, Addison-Wesley Publishing Company, Inc., Reading, Mass., 1959.

39. Chestnut, H. and R. W. Mayer, *Servomechanisms and Regulating System Design*, 2nd. ed., John Wiley & Sons, Inc., New York, 1963, pp. 270–351.

40. Chetayev, N. G., *The Stability of Motion*, Pergamon Press, New York, 1961, pp. 25–31.

41. Churchill, R. W., *Operational Mathematics*, 2nd ed., McGraw-Hill Book Company, New York, 1958.

42. Coates, C. L., "Flow Graph Solutions of Linear Algebraic Equations," *IRE Trans. on Circuit Theory*, vol. CT-6, pp. 170–187, June 1959.

43. Coddington, E. A., and N. Levinson, *Theory of Ordinary Differential Equations*, McGraw-Hill Book Company, New York, 1955.

44. DeRusso, P. M., R. J. Roy, and C. M. Close, *State Variables for Engineers*, John Wiley & Sons, Inc., New York, 1966.

45. Desoer, C. A., "The Optimum Formula for the Gain of a Flow Graph or a Simple Derivation of Coates' Formula," *Proceedings IRE*, vol. 48, pp. 883–889, May 1960.

46. Distefano, J. J., III, A. R. Stubberud, and I. J. Williams, *Theory and Problems of Feedback and Control Systems*, Schaum Publishing Co., New York, 1967.

47. Dorf, R., *Time Domain Analysis and Design of Control Systems*, Addison-Wesley Publishing Company, Inc., Reading, Mass., 1965.

48. Dudnikov, E. E., "Determination of Transfer Function Coefficients of a Linear System from the Initial Portion of an Experimentally Obtained Amplitude-phase Characteristic," *Automatic and Remote Control*, vol. 20, pp. 552–576, May, 1959.

49. Elgerd, O. I., *Control Systems Theory*, McGraw-Hill Book Company, New York, 1967.

50. Elsgolts, L. E., *Differential Equations*, Gordon and Breach, Science Publishers, Inc., New York, 1961.

51. Elsgolts, L. E., *Calculus of Variations*, Addison-Wesley Publishing Company, Inc., Reading, Mass., 1962.

52. Evans, W. R., "Control System Synthesis by the Root Locus Method," *AIEE Trans.*, vol. 69, pp. 66–69, 1950.

53. Ficken, F. A., *Linear Transformations and Matrices*, Prentice-Hall, Inc., Englewood Cliffs, N.J., 1967, pp. 141–163.

54. Forsythe, G., and C. B. Moler, *Computer Solution of Linear Algebraic Systems*, Prentice-Hall, Inc., Englewood Cliffs, N.J., 1967.

55. Frazer, R. A., W. J. Duncan, and A. R. Collar, *Elementary Matrices*, Cambridge University Press,, New York, 1964, pp. 64–197.

56. Gantmacher, F. R., *Matrix Theory*, vols. 1 and 2, Chelsea Publishing Company, New York, 1959.

57. Gardner, M. F., and J. L. Barnes, *Transients in Linear Systems*, John Wiley & Sons, Inc., New York, 1942.

58. Gelfand, I. M., and S. V. Fomin, *Calculus of Variations,* Prentice-Hall, Inc., Englewood Cliffs, N.J., 1963.

59. Gibson, J. E., *Nonlinear Automatic Control,* McGraw-Hill Book Company, New York, 1963, pp. 324–326.

60. Gilbert, E. G., "Controllability and Observability in Multivariable Control Systems," *J. Soc. Ind. Appl. Math.,* Ser. A, Control, vol. 1, no. 2, pp. 128–151, 1963.

61. Goldman, S., *Transformation Calculus and Electrical Transients,* Prentice-Hall, Inc., Englewood Cliffs, N.J., 1949.

62. Graham, D., and D. McRuer, *Analysis of Nonlinear Control Systems,* John Wiley & Sons, Inc., New York, 1961, pp. 201–213.

63. Guillemin, E. A., *Introductory Circuit Theory,* John Wiley & Sons, Inc., New York, 1953.

64. Guillemin, E. A., *Synthesis of Passive Networks,* John Wiley & Sons, Inc., New York, 1957, pp. 308–313.

65. Guillemin, E. A., *Theory of Linear Physical Systems,* John Wiley & Sons, Inc., New York, 1963.

66. Grupta, S. C., *Transform and State Variable Methods in Linear Systems,* John Wiley & Sons, Inc., New York, 1966.

67. Hadley, G., *Linear Algebra,* Addison-Wesley Publishing Company, Inc., Reading, Mass., 1961, pp. 17–52.

68. Hahn, W., *Theory and Application of Liapunov's Direct Method,* Prentice-Hall, Inc., Englewood Cliffs, N.J., 1963, pp. 14–15.

69. Hayashi, C., *Nonlinear Oscillations in Physical Systems,* McGraw-Hill Book Company, New York, 1964.

70. Ho, Y. C., "What Constitutes a Controllable System," *IRE Trans. on Automatic Control,* vol. AC-7, no. 3, p. 76, Apr. 1962.

71. Householder, A. S., *Principles of Numerical Analysis,* McGraw-Hill Book Company, New York, 1953.

72. Ingwerson, D. R., "A Modified Liapunov Method for Nonlinear Stability Analysis," *Trans. IRE,* vol. AC-6, May 1961.

73. James, H. M., N. B. Nichols, and R. S. Phillips, *Theory of Servomechanisms,* Dover Publications, Inc., New York, 1965.

74. Kalman, R. E., "On the General Theory of Control Systems," *Proceedings IFAC,* Butterworth Scientific Publications, London, 1960.

75. Kalman, R. E., "Mathematical Description of Linear Dynamical Systems," *J. Soc. Ind. Appl. Math.,* Ser. A, Control, vol. 1, no. 2, pp. 152–192, 1963.

76. Kalman, R. E., and J. E. Bertram, "Control System Analysis and Design via the Second Method of Lyapunov: I, Continuous-Time Systems; II, Discrete Time Systems," *Trans. ASME,* Ser. D., Journal Basic Engineering, vol. 82, 1960.

77. Kalman, R. E., Y. C. Ho, and K. S. Narendra, "Controllability of Linear Dynamical Systems," *Contributions to Differential Equations,* John Wiley & Sons, Inc., New York, 1963.

78. Karman, V. T., and M. A. Biot, *Mathematical Methods in Engineering,* McGraw-Hill Book Company, New York, 1940.

79. Kidd, J. B., T. E. Edgerton and C. F. Chen, "Transfer Function Synthesis in the Time Domain—An Extension of Levy's Method," *IEEE Trans. on Education,* vol. E-8, nos. 2, 3, pp. 62–67, June–Sept. 1965.

80. Krasovskii, N. N., *Stability of Motion,* Stanford University Press, Stanford, Calif., 1963, pp. 44–125.

81. Kron, G., *Tensor Analysis of Networks,* John Wiley & Sons, Inc., New York, 1939.

82. Ku, Y. H., *Transient Circuit Analysis,* D. Van Nostrand Company, Inc., Princeton, N.J., 1961.

83. Kuo, B. C., *Linear Networks and Systems,* McGraw-Hill Book Company, New York, 1967.

84. Lanczos, C., *The Variational Principles of Mechanics,* University of Toronto Press, Toronto, 1962.

85. Landau, L. D., and E. M. Lifshitz, *Mechanics,* J. B. Sykes and J. S. Bell, trans., Addison-Wesley Publishing Company, Inc., Reading, Mass., 1960.

86. Langhaar, H. L., *Energy Methods in Applied Mechanics,* John Wiley & Sons, Inc., New York, 1962.

87. Langill, A. W., *Automatic Control Systems Engineering,* vols. I and II, Prentice-Hall, Inc., Englewood Cliffs, N.J., 1965.

88. Lauer, H., R. Lesnick, and L. E. Matson, *Servomechanism Fundamentals,* Mc-Graw-Hill Book Company, New York, 1947.

89. LaSalle, J., "Stability and Control," *J. Soc. Ind. Appl. Math., Ser. A, Control,* vol. 1, no. 1, pp. 3–15, 1962.

90. LaSalle, J., and S. Lefschetz, *Stability by Lyapunov's Direct Method with Applications,* Academic Press, Inc., New York, 1961, pp. 37–38.

90a. Lee, Y. W., *Statistical Theory of Communication,* John Wiley & Sons, Inc. New York, 1960, pp. 459–501.

91. Lefschetz, S., *Differential Equations: Geometric Theory,* Interscience Publishers, Inc., New York, 1957, p. 78.

92. Lehnigk, S. H., *Stability Theorems for Linear Motions,* Prentice-Hall, Inc., Englewood Cliffs, N.J., 1966.

93. LePage, W. R., and S. Seely, *General Network Analysis,* McGraw-Hill Book Company, N.Y., 1952.

94. Letov, A. M., *Stability in Nonlinear Control Systems,* Princeton University Press, Princeton, N.J. 1961, pp. 1–19.

95. Levy, E. C., "Complex Curve Fitting," *IRE Trans. on Automatic Control,* vol. AC-4, pp. 37–43, May 1959.

96. Liapunov, A. M., *Stability of Motion,* Academic Press, Inc., New York, 1966.

97. Lure, A. I., *Some Nonlinear Problems in the Theory of Automatic Control,* Gostekhizdat, 1951, English trans., Her Majesty's Stationery Office, London, 1957.

98. Mason, S. J., "Feedback Theory — Some Properties of Signal Flow Graphs," *Proceedings IRE,* vol. 41, pp. 1144–1156, Sept. 1953.

99. Mikhailov, N. N., "Plotting of Root Loci of Automatic Control Systems," *Automatika i Telemekhanika,* vol. 24, no. 2, pp. 661–674, July 1958.

100. Minorsky, N., *Nonlinear Oscillations,* D. Van Nostrand Company, Inc., Princeton, N.J., 1962, pp. 77–80.

101. Newton, G., L. Gould, and J. Kaiser, *Analytical Design of Linear Feedback Controls,* John Wiley & Sons, Inc., New York, 1957.

102. Ogata, K., *State Space Analysis of Control Systems,* Prentice-Hall, Inc., Englewood Cliffs, N.J., 1967.

103. Ogar, G. W., and J. J. Dazzo., "A Unified Procedure for Deriving the Differential Equations of Electrical and Mechanical Systems," *IRE Trans. on Education,* vol. E-5, no. 1, pp. 18–26, Mar. 1962.

104. Oldenburger, R. (ed.), *Frequency Response,* The Macmillan Company, New York, 1956.

105. Oppelt, W., "A Stability Criterion Based on the Method of Two Hodographs," *Automatika i Telemekhanika,* vol. 22, pp. 1175–1178, Sept. 1961.

106. Organick, E. I., *A FORTRAN IV Primer,* Addison-Wesley Publishing Company, Inc., Reading, Mass., 1966.

107. Parks, P. C., "A New Proof of the Routh-Hurwitz Stability Criterion Using Second Method of Lyapunov," *Proc. Cambridge Phil. Soc.,* vol. 58, pt. IV, pp. 694–702, 1962.

108. Paskusz, G. G., and B. Russel, *Linear Circuit Analysis,* Prentice-Hall, Inc., Englewood Cliffs, N.J., 1964.

109. Papoulis, A., *The Fourier Integral and Its Applications,* McGraw-Hill Book Company, New York, 1962.

110. Pipes, L. A., *Matrix Methods for Engineering,* Prentice-Hall, Inc., Englewood Cliffs, N.J., 1963.

111. Pontryagin, L. S., *Ordinary Differential Equations,* Addison-Wesley Publishing Company, Inc., Reading, Mass., 1962, pp. 201–213.

112. Pontryagin, L. S., V. G. Bolyanskii, R. V. Gamkrelidze, and E. F. Mishchengo, *The Mathematical Theory of Optimal Processes,* John Wiley & Sons, Inc., New York, 1962.

113. Popov, E. P., *The Dynamics of Automatic Control Systems,* (English trans.) Addison-Wesley Publishing Company, Inc., Reading, Mass., 1962, pp. 403–444.

114. Puri, N. N. and C. N. Weygandt, "Second Method of Lyapunov and Routh's Canonical Form," *J. Franklin Institute,* vol. 276, no. 5, pp. 365–384, Nov. 1963.

115. Ralston, A., "A Symmetric Matrix Formulation of the Hurwitz-Routh Stability Criterion," *IRE Trans. on Automatic Control,* vol. AC-7, pp. 50–51, July 1962.

116. Ralston, A., and H. S. Wilf, *Mathematical Methods for Digital Computers,* John Wiley & Sons, Inc., New York, 1966, pp. 95–120.

117. Raue, D. S., "A Simplified Transformation to Phase Variable Form," and Johnson, C. D., "Another Note on the Transformation to Canonical Form," *IEEE Trans. on Automatic Control,* vol. AC-11, no. 3, pp. 608–610, 1966.

118. Reiss, R., and G. Geiss, "The Construction of Liapunov Functions," *IEEE Trans. on Automatic Control,* vol. AC-8, no. 4, pp. 382–383, Oct. 1963.

119. Robichaud, L. P. A., M. Boisvert, and J. Robert, *Signal Flow Graphs and Applications,* Prentice-Hall, Inc., Englewood Cliffs, N.J., 1962.

120. Sanathanan, D. K., and J. Koerner, "Transfer Function Synthesis as a Ratio of Two Complex Polynomials," *IEEE Trans. on Automatic Control,* vol. AC-8, pp. 56–68, Jan. 1963.

121. Savant, C. J. Jr., *Control System Design,* 2nd ed., McGraw-Hill Book Company, New York, 1964.

122. Schultz, D. G., and J. E. Gibson, "The Variable Gradient Method for Generating Liapunov Functions," *Trans. AIEE,* vol. 81, pt. II, pp. 203–210, 1962.

123. Schwarz and Friedland, *Linear Systems,* McGraw-Hill Book Company, New York, 1965.

124. Sedlar, M., and G. A. Bekey, "Signal Flow Graphs of Sampled-data Systems: a New Formulation," *IEEE Trans. on Automatic Control,* vol. AC-12, no. 2., Apr. 1967.

125. Seely, S., *Dynamic Systems Analysis,* Reinhold Publishing Corporation, New York, 1964.

126. Seifert, W. W., and Steeg, C. W., *Control System Engineering,* McGraw-Hill Book Company, New York, 1960.

127. Shen, D. W. C., and C. F. Chen, "Routh's Method for Approximating Oscil-

latory Modes in Linear Systems," *IEEE Inter. Conv. Rec.*, pt. 2, pp. 10–15, Mar. 1963.

128. Spiegel, M. R., *Theory and Problems of Laplace Transforms*, Schaum Publishing Co., New York, 1965.

129. Stanton, R. G., *Numerical Methods for Science and Engineering*, Prentice-Hall, Inc., Englewood Cliffs, N.J., 1961.

130. Takahashi, T., *Mathematics of Automatic Control*, Holt, Rinehart-Winston, Inc., New York, 1966.

131. Takai, H., *Theory of Automatic Control*, Illiffe Books Ltd., London, England, 1966.

132. Thaler, G. J., and R. G. Brown, *Analysis and Design of Feedback Control Systems*, 2nd ed., McGraw-Hill Book Company, New York, 1960.

133. Tou, J. T., *Modern Control Theory*, McGraw-Hill Book Company, New York, 1964.

134. Trimmer, J. D., *Response of Physical Systems*, John Wiley & Sons, Inc., New York; Chapman & Hall, Ltd., London, 1950.

135. Truxal, John G., *Automatic Feedback Control Synthesis*, McGraw-Hill Book Company, New York, 1955.

136. Tsien, H. S., *Engineering Cybernetics*, McGraw-Hill Book Company, New York, 1954.

137. Van Valkenburg, M. E., *Introduction to Modern Network Synthesis*, John Wiley & Sons, Inc., New York, 1960.

138. Wiener, N., *Cybernetics, or Control and Communication in the Animal and the Machine*, John Wiley & Sons, Inc., New York, 1948.

139. Yefimov, N. V., *Quadratic Forms and Matrices* (paperback), Academic Press, Inc., New York, 1964.

140. Zadeh, L. A., and C. A. Desoer, *Linear System Theory: The State Space Approach*, McGraw-Hill Book Company, New York, 1963.

Index

Note: Numbers in *italic* in this index indicate numbers of references in the bibliography.

A

acceleration error constant, 161
addition:
  in analog computation, 87
  of matrices, 242
  of vectors, 272
addition rule, 358
additivity, 7
adjoint matrix, 322
Aizerman, M. A., *1*
analog computation, 86
analog computer program, 357
  of a first order differential equation, 89
  of a high order differential equation, 91
  of simultaneous differential equations, 93
analogy, electric circuit, 12
angle, departure, 125
angles of asymptotic lines, 124
angular velocity, 37

B

bandwidth, 157
Barnes, J. L., *57*
Bashkow, T. R., *5*
basic equations of Evans, 119
Bekey, G. A., *124*
Bellman, R., *6, 7, 8, 9, 10*
Bendrikov, G. A., *11, 137*
Bertram, J. E., *12, 76*
Bessel equations, 280
bilinear transformation, 195
Biot, M. A., *78*

block diagram:
  of a direct current motor, 35
  of a voltage regulator, 36
block diagram identities, 29
block diagrams, 3, 25
Bode, H. W., *13*
  basic equations, 185
  decomposition, 185, 203, 213
  diagrams, 186
Bohn, E. V., *14*
Boltyanskii, V. G., *112*
branches:
  incoming, 357
  outgoing, 357
Brand, L., *15*
breakaway point, 124
Brown, G. S., *16*
Brown, R. G., *17*
Bultot, F., *18*
Bush, V., 298, *19*
  canonical form, 313
  form, 298, 314, 338, 437

C

Caldwell, S. H., *19*
Campbell, D. P., *16*
canonical variables, 346
capacitance of a condensor, 21
capacitor microphone, 21
Cauchy theorem, 103, 107
Cauer, W., *20*
  form, 303, 314
cause-effect approach, 3
Cayley-Hamilton:
  method, 323
  theorem, 383
center, 396, 405

center of gravity, 390
chain rule, 282
characteristic equations, 339
characteristic roots, 337
characteristics, 6
Chen, C. F., *21, 22, 23, 24, 25, 26, 27, 28, 29, 30, 31, 32, 33, 34, 35, 36*
  Chen-Chu transformation, 431
  Chen-Shen chart, 207
Chen, W. H., *37*
Cheng, D., *38*
Chestnut, H., *39*
Chetayev, N. G., *40*
  instability theorem, 416
Chu, Hsin, *24*
Churchill, R. V., *41*
circuit diagram, mechanical, 11
classical method, 44, 330
closed loop transfer function, 203
closed set, 275
Coates, C. L., 356, 361, *42*
  formula, 363
  notation, 356
Coddington, E. A., *43*
coefficients of harmonics, 173
cold graph, 361
Collar, A. R., *55*
combination form of Fourier series, 174
combined loop and node method, 295
commutative property, 328
companion matrix, 301
complementary function, 48
complete chain graph, 361
complex translation theorem, 58
complex-root branch, 123
components of a vector, 271
conditional transfer, 450
conservative system, 291
constraint equations, 297
constraints, 15
construction of Liapunov function, 418
continued fraction programming, 214, 303
contour C, 103
contour C_0, 105
contour Γ, 103
contour Γ_0, 105
controllability, 434, 436
controlled element, 34
controller, 34
convolution integral, 68, 335
coordinates, 14
  cylindrical, 14
  generalized, 14
  spherical, 14
corner frequencies, 190
Cramer's rule, 244, 358, 361

critically damped case, 64

D

damping factor, 155, 162
damping ratio, 155, 162, 163
Dazzo, J. J., *103*
degrees of freedom, definition of, 15
departure angle, 125
DeRusso, P. M., *44*
Desoer, C. A., *45, 140,* 356
determinant of square matrix, 243
diagonalization, 337
diagonalized matrix, 242, 338
diagram, schematic, 26
differential equations, 7
  homogeneous, 46
  nonhomogeneous, 48
differentiation in analog computation, 88
digital computer program:
  Fourier coefficient evaluation, 178
  Gauss' reduction, 251
  general inverse Laplace transformation, 360
  inverse Laplace transform based on poles and zeros, 79
  Levy's curve fitting, 229
  Lin's method, 71
  Newton's method, 75
  Routh's criterion, 115
  Runge-Kutta's method, 374
direct-current motor, 34
direct method, 427
  in flow graphs, 358
  of Liapunov, 405
  of Routh-Hurwitz, 106
direct programming, 298
dissipation function:
  of a loud speaker, 290
  Rayleigh, 18
Distefano, J. J., III, *46*
distinct roots, 46, 323
DO loop, 452
dominant poles, 163
Dorf, R., *47*
double pendulum, 14
double subscript notation for Routh's criterion, 14
Dudnikov, E. E., *48*
Duncan, W. J., *55*

E

Edgerton, T. E., *79*
eigenvalues, 339, 348
eigenvectors, 339
  geometrical interpretation of, 401

elastic pendulum, 20
  spherical, 24
electric circuit analogy, 12
electric circuits, analysis of, 10
electrical interpretation of Gauss'
  reduction, 254
electrokinetic momentum, 5
electromechanical system, 289
Elgerd, O. I., *49*
Elsgolts, L. E., *50*
encirclements, 103
energy:
  kinetic, 16
  potential, 16
energy approach, 13
engine governor, 37
equilibrium points, 386, 390, 403, 405
  of linear second order system, 396
equipotential contour lines, 409
error constants, 160
Euclidean space, 274
Evans, W. R., *52*
  basic equations of, 119
  root locus method of, 117
excitation function, 3, 4
exponential form, 144

F

factorization, 70
Faddeev's method, 323
Faraday's law, 5
feedback:
  definition of, 27
  system with, 119
flow graph, definition of, 356
flywheel, 37
Ficken, F. A., *53*
final value theorem, 57
focus, 391, 398, 403
Fomin, S. V., *58*
forced motion, 329
Foster's form, 302, 314, 440
Forsythe, G., *54*
FORTRAN programming language, 447
Fourier:
  integral, 179
  series, 170
    coefficients, 170, 197
  transform, 180
four-step technique for inverting a matrix,
  244
Frazer, R. A., *55*
freedom, degree of, 15
frequency method of Nyquist, 143
frequency response, definition of, 143
function transform table, 54
fundamentals of matrices, 329

G

gain margin, 155, 157
Gantmacher, F. R., *56*
Gamkrelidze, R. V., *112*
Gardner, M. F., *57*
gate function, 182
Gauss' reduction, 246, 265, 268
Geiss, G., *118*
Gelfand, I. M., *58*
general matrix block diagram, 307
generalization of the Heaviside expansion,
  322
generalized coordinates, 14
generalized conversion formula, 213
generalized feedback transformation, 304
geometrical interpretation:
  of controllability, 441
  of hyperplane, 276
  of postmultiplication, 258
Gibson, J. E., *59*
Gilbert, E. C., *60*
Goldman, S., *61*
Gould, L., *101*
governor, speed, 16
Graham, D., *62*
graphical methods, 379
graphical method for determining Fourier
  coefficients, 170
graphical representations of frequency
  response, 146
Grupta, S. C., *66*
Guillemin, E. A., *63, 64, 65*
  form, 301, 314

H

Haas, I. J., *25, 26*
Hadley, G., *67*
Hahn, W., *68*
Hall's chart, 204
Hamilton:
  equations, 291, 293, 317
  principle, 15
Hamiltonian, 291
Hayashi, C., *69*
Heaviside:
  decomposition, 185
  expansion, 64, 346
Henry, Joseph, 5
  law, 8
hierarchy of operations, 448
high-dimensional vectors, 271
high order equations, 280
Ho, Y. C., *56, 60*
homogeneity, 7
homogeneous differential equation, 46

homogeneous linear system, 278
Hooke's law, 5
Householder, A. S., *71*
Hsu, C., *27*
Hurwitz, 113
  criterion, 111
  polynomial, 106
hyperplane, 276
hypersphere, 275

I

identification based on Bode's diagram, 190
identification problem, 170
identity matrix, 242, 289
IF (sense switch), 453
Ingwerson, D. R., *72*
initial conditions, 94
initial slope, 215
initial state, 322
initial value theorem, 57
initial vector, 345
instability theorem, Chetayev's, 416
integration by parts, 422
integration in analog computation, 88
intersecting points, 124
invariance principle, 137
inverse Laplace transformation, 67, 343
inverse matrix, 243, 260
inverted pendulum, 405
iterative programming, 301

J

Jacobian matrix, 282, 394
James, H. M., *73*
Johnson, C. D., *117*
Jordan matrix, 342
Jordan's form, 305, 347

K

Kaiser, J., *101*
Kalaba, R., *8, 9, 10*
Kalman, R. E., *76*
Kalman-Bertram's Liapunov function, 432
Karman, V., *78*
Kidd, J. B., *79*
kinetic energy, 16, 289
Kirchhoff's laws, 8, 295
Koerner, J., *120*
Krasovskii, N. N., *80*
Kron, G., *81*
Kronecker's reduction, 265
Ku, Y. H., *82*
Kuo, B. C., *83*

L

Lagrange, 279
  equation, 3, 14, 16, 288
  equation for n degrees of freedom system, 20
  reduction, 246, 266
  variation of parameters, 49, 320
Lagrangian, 15, 291
Laguerre function, Laplace transform of, 201
Laguerre polynomial, 198, 199
Lanczos, C., *84*
Landau, L. D., *85*
Langhaar, H. L., *86*
Langill, A. W., *87*
Laplace, 279
  transform of a Laguerre function, 201
  transform table, 55
  transformation, definition of, 54
  transformation method, 320
LaSalle, J., *89, 90*
Lauer, H., *88*
leading principal minors, 267
Lee, Y. W., *90a*
Lefschetz, S., *90*
Legendre transformation, 291, 317
Lehnigk, S. H., *92*
LePage, W. R., *93*
Lesnick, R., *88*
Letov, A. M. *94*
Levy, E. C., *95*
  curve fitting, digital computer program of, 229
  technique, 223
Liapunov, A. M., *96*, 113
  asymptotic stability theorem, 410
  function, 411, 427, 434
  stability theorem, 414
Lifshitz, E. M., *85*
limit cycles, 392
linear systems, 6, 7
linear transformation, 239, 256, 336
linearization, 393
Lin's method, 70, 101, 119

M

M peak, 155, 157
magnetic amplifier, 7
magnitude response equation, 186
magnitude vs. frequency plot, 186
Mason, S. J., *98*, 356, 361
mass-dashpot-spring system, 118
matrices:
  addition of, 242
  fundamentals of, 239
  multiplication of, 240

matrices (*cont.*)
   subtraction of, 242
   transposition of, 241
matrix:
   definition of, 240
   rank of, 435
   for testing controllability, 437
   for testing observability, 442
matrix algebra, 239
matrix method, 364
   in flow graphs, 358
matrizant, 320, 330
Matson, L. E., *88*
Maxwell, 37
Mayer, R. W., *39*
McRuer, D., *62*
mechanical circuit diagram, 11
meter, 84, 307
Mikhailov, N. N., *99*
   criterion, 107
minor, 244
Minorsky, N., *100*
Mishchenko, E. F., *112*
modal matrix, 340
mode, 448
modified Vandermonde matrix, 342
Moler, C. B., *54*
motion space curve, 378
multi-input, 309
multiple root, 325
multiplication rule, 359
multiplicative property, 327
mutual inductance, 23

N

n degrees of freedom, 20
n-dimensional space, 322
negative definite, definition of, 269
negative semidefinite, definition of, 269
Newton, G., *101*
Newton, I., 5
Newton's method, 74, 101, 119
Nichols, N. B. *73*
   chart, 209
Nilsson, J. W., *17*
node, 357, 397, 403
   in phase plane, 389
node duplication, 359
node method, 10, 293
nonconservative system, 19
nonhomogeneous differential equation, 48
nonhomogeneous linear system, 278
nonlinear differential equation, 39
nonlinear systems, 6, 7, 310, 412, 415
nonstationary system, 278
*n*th-order differential equation, 45
null matrix, definition of, 242

number of degrees of freedom, 15
numerical integration, 360
numerical solution, 369
Nyquist, 143
   criterion, 149
   criterion, simplified, 156

O

observability, 434, 441
Ogar, G. W., *103*
Ogata, K., *102*
Ohm, George, 5
   law, 5, 8
Oldenburger, R., *104*
one degree of freedom, 15
open loop and closed loop frequency
      responses, 204
open loop transfer function, 203
open set, 275
operation of type I, 259
operation of type II, 259
operation of type III, 260
operation transform table, 56
Oppelt, W., *105*
Organick, E. I., *106*
oscillatory frequencies, 136
oscillatory mode, condition for, 135
over-all transfer function, 33
overshoot, 155, 162
outgoing branches, 369

P

Papoulis, A., *109*
parallel programming, 302, 440
parentheses, 448
Parker, R. R., *29*
Parks, P. C., *107*
partial fraction expansion, 64
particular integral, 48
Paskusz, G. F., *108*
pendulum, 405
   simple, 19, 279
performance, 155
performance equations, 296
periodic signal, 170
phase margin, 155, 157
phase plane, 378, 386
phase portrait, 388
phase response equation, 186
phase space, 381, 386
phase variable form, 301
phase vs. frequency plot, 186
Philip, B. L., *30, 31*
Phillips, R. S., *73*
physical interpretation of controllability,
      439

physical laws, 5
Pipes, L. A., *110*
planetary orbit, 279
Poincaré, 389
polar form, 144
pole zero form, 185
poles, 185
poles and zeros, definition of, 76
Pontryagin, L. S., *111, 112*
Popov, E. P., *113*
position error constant, 161
positive definite, 263
positive feedback, 215
positive semidefinite, 263
post multiplication, 256
potential energy, 16, 290
premultiplication, 246, 249
primer system, 336
Puri, N. N., *114*

Q

quadratic forms, 262
qualitative analysis, 138

R

Ralston, A., *115, 116*
  form, 433
rank of matrix, 435
Raue, D. S., *117*
  dissipation function, 18
  equation, 376
real part response, 196
real root branch, 123
real translation theorem, 58
rectangular form, 144
reduction:
  Kronecker's, 265
  Lagrange's, 264, 266
reduction method in flow graph, 358
Reiss, R., *118*
repeated roots, 47, 305, 342
representation problem, 3, 287
residues, 185
response function, 4
rise time, 155, 162
RLC circuit, 280
Robert, J., *119*
Robichaud, L. P. A., 119
root locus, auxiliaries for drawing, 122
root locus method:
  of Evans, 117
  of Routh, 134
Routh:
  algorithm, 134
  array, 112, 113

Routh (*cont.*)
  criterion, 111
  criterion, digital computer program of, 115
  criterion, double subscript notation for, 114
Routh-Hurwitz criterion, 427
  derivation of, 434
  symmetrical form of, 433
row matrix, 241
Roy, R. J., 44
rules of root-locus method, 125–131
Runge-Kutta method, 369
Russel, B., 108

S

saddle, 403
saddle-focus, 405
saddle point, 392, 397
Sanathanan, D. K., *120*
Savant, C. J., Jr., *121*
scalar product, 272
scaling theorem, 59
schematic diagram, 26
Schultz, D. G., *122*
Schwarz' form, 427
Sedlar, M., *124*
Seely, S., *125*
Seifert, W. W., *126*
self loop elimination, 359
self matrix loop, 367
sets, 274
settling time, 155, 162
servomechanism, 315, 380
Shannon, 356
Shen, D. W. C., *32, 127*
Shieh, L. S., *33, 34*
sign indefiniteness, 264
sign inversion in analog computation, 86
similar matrices, 338
simultaneous equations, 245, 357, 372
singularities, 389
sink node, 357
six degrees of freedom, 15
solution problem, 44, 319
source nodes, 357
speed governor, 16
spherical coordinates, 14
spherical elastic pendulum, 24
Spiegel, M. R., *128*
spirule, 134
square matrix, 240
stability, definition of, 99, 101
stability problem, 386
stability theorem, Liapunov's, 414
stable, 155

stable in the Liapunov sense, 405, 407, 409
Stanton, R. G., *129*
state, 15, 277, 298, 322
state variable equations, 280, 288
statement numbers, 449
stationary system, 278
steady-state error, 160
steady-state response, 143
steady-state values, 155
Steeg, C. W., *126*
subroutines, 451
subscripts, 451
subsidiary functions, 112
subtraction in analog computation, 88
subtraction of matrices, 242
summing symbol, 25
superposition theorem, 45
Sylvester:
   expansion, 323
   theorem, 266
symmetrical matrix, 242
system function, 3
system with feedback, 119

T

Takahashi, T., *130*
Takai, H., *131*
take-off symbol, 26
Taylor series expansion, 326
Teodorchik, K. F., 137, *11*
Thaler, G. J., *132*
Thevenin equivalent, 255
three degrees of freedom, 22
three-dimensional vectors, 270
time-frequency correlation, 181
topological approach, 293
Tou, J. T., *133*
trace, 324, 383
transfer, 449
transfer function, 144, 185, 203, 298
transform method, 59
transformation, linear, 239
transformation of quadratic forms, 264
transformation table, 53
transient part, 143
transition matrix, 320, 327
transpose, definition of, 241
trapezoid approximation, 182
triangular wave form, 172
triangularization, 247

trigonometric form, 144
Trimmer, J. D., *134*
trivial solutions, 405
Truxal, J. G., *135*
Tsang, N. F., *35*
Tsien, H. S., *136*
two degrees of freedom, 15
two-dimensional vectors, 270
two-*n* method, 296

U

undamped natural frequency, 162
underdamped case, 62
undetermined elements, procedure based on, 418
unit impulse, 100
unit impulse function, 160
unit ramp function, 160
unit step function, 160
unstable system, 409, 417

V

Van der Pol equation, 379, 393
Vandermonde matrix, 337, 342
Van Valkenburg, M. E., *137*
variation of parameters, 48, 334
vectors, 270
velocity error constant, 161
velocity regulator, 35

W

weighting function, 100
Weygandt, C. N., *114*
Wiener, N., *138*
Wiener-Lee decomposition, 185, 194
Wilf, H. S., *116*

Y

Yates, R. E., *36*
Yefimov, N. V., *139*

Z

Zadeh, L. A., *140*
zero, 185
zero input response, 320
zero state response, 329